# A Textbook of
# Environmental Studies

# About the Author

**Shashi Chawla** is currently Assistant Professor in the Department of Applied Sciences, Amity School of Engineering and Technology, New Delhi. He holds MSc and MTech degrees from IIT Delhi and a PhD from Jamia Millia Islamia, New Delhi, specialising in Chemistry, Polymer Science and Technology and Nanocomposites. With a total of 14 years of teaching and research experience, he has also worked in Lingaya's Institute of Management and Technology, Faridabad, as Lecturer, and in Jamia Millia Islamia as Visiting Faculty.

Dr Chawla has also worked on various R&D and consultancy projects for Inter University Accelerator Centre and IIT Delhi. Dr Chawla's research has included various modern topics in science and technology like radiation graft modification of polypropylene and development of PP based nanocomposites; swift-heavy-ion-induced structural, conformational and chemical changes in BOPP film; functionalisation of industrial polypropylene films via the swift-heavy-ion induced grafting of glycidyl methacrylate; swift-heavy-ion-induced grafting of glycidyl methacrylate onto latent ion tracks and surface of industrial polymeric films; and grafting of glycidyl methacrylate onto industrial polyproylene films irradiated with swift nickel ions.

He has presented many research and conceptual papers in various international and national seminars and Conferences. Recently, he spoke on the Role of Chemistry for Improvement in Quality of Water in the National Seminar on Role of Chemistry for Improvement in Quality of Human Life, organised on 17-18$^{th}$ March 2012 by Ujjain Engineering College, Ujjain (Madhya Pradesh).

He is a well-published author with books on Engineering Chemistry, Polymer Science and Engineering, Essentials of Experimental Engineering Chemistry and Materials Science and Engineering. Dr Chawla has been recognised as one of the top 100 educators of the world by the International Biographical Centre, Cambridge, England, in 2009. His biographical profile is published in *Marquis Who's Who in the World* (2009).

With a keen interest in sports, he is the Sports Advisor for all the students of ASET (1999–2012) and is a mentor in each academic session for the students of ASET (2000–2012).

# A Textbook of
# Environmental Studies

**Shashi Chawla**
*Assistant Professor*
*Department of Applied Sciences*
*Amity School of Engineering and Technology*
*New Delhi*

**Tata McGraw Hill Education Private Limited**
**NEW DELHI**

*McGraw-Hill Offices*

**New Delhi** New York St Louis San Francisco Auckland Bogotá Caracas
Kuala Lumpur Lisbon London Madrid Mexico City Milan Montreal
San Juan Santiago Singapore Sydney Tokyo Toronto

**Tata McGraw-Hill**

Published by Tata McGraw Hill Education Private Limited,
7 West Patel Nagar, New Delhi 110 008

**Environmental Studies**



ISBN (13 digits): 978-1-25-902953-0
ISBN (10 digits): 1-25-902953-0

Vice President and Managing Director—McGraw-Hill Education: *Ajay Shukla*
Head—Higher Education Publishing and Marketing: *Vibha Mahajan*
Publishing Manager—SEM & Tech. Ed.: *Shalini Jha*

Senior Editorial Researcher: *Harsha Singh*
Executive—Editorial Services: *Sohini Mukherjee*
Senior Production Manager: *Satinder S Baveja*
Production Executive: *Anuj K Shriwastava*

Marketing Manager—Higher Education: *Vijay Sarathi*
Product Specialist—SEM & Tech. Ed.: *Tina Jajoriya*
Graphic Designer—Cover: *Meenu Raghav*

General Manager—Production: *Rajender P Ghansela*
Production Manager: *Reji Kumar*



Typeset at Text-o-Graphics, B-1/56, Arawali Apartment, Sector 34, Noida 201 301 and printed at

Cover Printer:

*Dedicated to*

*My mother, Mrs Santosh Chawla,*
*for her support, enthusiasm, unfailing cooperation*
*and inspiration*

# Contents

# Preface

Environmental Science is an interdisciplinary academic field that integrates physical and biological sciences (including physics, chemistry, biology, soil, science, geology and geography) to the study of the environment, and the solution of environmental problems. Environmental science provides an integrated, quantitative and interdisciplinary approach to the study of environmental systems. Hence, preparing a textbook on the *Environmental Studies* course, which is based on the latest syllabus of UGC and also gives the reader sufficient information to understand the principles and importance of the subject, has been an eye-opening experience.

*A Textbook on Environmental Studies* is a replete and concise content endowing the essence of the theoretical concepts of the subject. It covers the course requirements as per the UGC curriculum for all the undergraduate students of all disciplines. I have made a sincere attempt to present the global subject of environmental studies in the Indian context with ample examples, flowcharts, illustrations and cases from India. I hope UG students of various Indian universities will find this book easy to read, understand and implement.

## Salient Features

This book is written in such a manner so as to

- Completely cover the entire syllabus of the 'Environment Studies' course of various universities as per UGC framework
- Develop your understanding of this crucial and logical subject through emphasis on fundamental principles, concepts and applications
- Promote visual and spontaneous learning through a simple and holistic approach, full of learning objectives, memory aids, solved examples, well-labelled illustrations, flowcharts and important definitions
- Enhance your knowledge with easy, concise and accurate *Case Studies* incorporating most recent developments
- Enable you to perform confidently and effectively in various examinations through its rich pedagogy for self-assessment, thought stimulation and skill evaluation:
    - 162 Theory Questions
    - 26 Solved Examples
    - 41 Case Studies

- 119 Fill in the Blanks
- 20 Match-ups
- 102 Multiple Choice Questions
- 85 True/False Questions

## Chapter Organisation

The content is divided into eight comprehensive chapters.

*Chapter 1* explains the multidisciplinary nature of environment studies. It outlines the nature of our environment, environmental studies, need for public awareness, environmental degradation, shelter security, economic security, social security, effects of housing on environment and effects of industry on environment.

*Chapter 2* discusses natural resources and includes topics like types of natural resources, for example forest, water, mineral, food-security, energy and land resources. It also discusses conservation of natural resources, sustainable lifestyles, Sustainable Water Management (SWM) and the biogeochemical cycle.

*Chapter 3* covers the ecosystem and explains ecological succession, food chain, ecological pyramids and the types of ecosystems like forest ecosystem, aquatic ecosystem, grassland ecosystem and desert ecosystem.

*Chapter 4* presents biodiversity and its conservation. It gives details on values or benefits of biodiversity, biogeographic zones of India, hot spots of biodiversity, endangered and endemic species, rare and threatened species, threats to biodiversity, human–wildlife conflicts and conservation of biodiversity.

*Chapter 5* handles environment pollution and its effects and elucidates on requirements of a nonpolluted environment, public health aspects, air pollution, water pollution, land pollution or soil pollution, marine pollution, noise pollution, thermal pollution, hazardous wastes, nuclear hazards (radiation pollution), solid waste and its management, role of individuals in pollution prevention and disaster management.

*Chapter 6* is on social issues and the environment. This chapter discusses sustainable development, urbanisation, water conservation, resettlement and rehabilitation of people; its problems and concerns, wasteland reclamation, acts for environmental protection, carbon credits, industrial symbiosis, initiatives and roles of Nongovernmental Organisations (NGOs) in environmental protection, issues involved in enforcement of environmental legislation and animal husbandry.

*Chapter 7* discusses human population and the environment. It explains population growth, family welfare programmes, environment and human health, fundamental rights, human rights, value education, HIV/AIDS, environmental education, women's education, and role of information technology in environment and human health.

Finally, *Chapter 8* presents some fieldwork activities like visit to a local area to document environmental assets: river/forest/grassland/mountain, visit to a local polluted site: urban/ rural/industrial/agricultural; study of common plants, insects, birds; and study of simple ecosystems: pond/river/hill slopes.

All the chapters contain learning objectives, memory aids, solved examples, well-labelled illustrations, flowcharts and important definitions.

## Acknowledgements

I feel greatly indebted to my parents, Smt. Santosh Chawla and Shri Thakur Das Chawla, who created in me an immense interest in this field. One of the main sources of inspiration for writing this book has been the many interested students in my classes. I am grateful to all the teachers, authors and environmentalists who helped me enrich my knowledge of this subject.

I express my sincere gratitude to Dr Ashok K Chauhan, founder President, Amity University; and Mr Atul Chauhan, President, RBEF and Amity University, for their constant inspiration and encouragement.

I am thankful to Prof. B P Singh, Senior Director and Prof. Rekha Agarwal, Director, Amity School of Engineering and Technology, for their continued guidance and encouragement.

Along with this, a special note of appreciation goes to reputed faculty members of various Indian universities who made suggestions, pointed out errors, responded to my questions and helped in numerous other ways.

I deeply acknowledge the contribution of my wife, Mrs Taruna Chawla, and her never-ending encouragement, moral support, patience and understanding. She has done a commendable job of editing this manuscript.

A number of experts took pains to provide valuable feedback about the book. My heartfelt gratitude goes out to those whose names are given below:

| | |
|---|---|
| **Sandeep Gupta** | *A K Garg Engineering College, Ghaziabad, Uttar Pradesh* |
| **Shashi Bala** | *Babu Banarasi Das National Institute of Technology and Management, Lucknow, Uttar Pradesh* |
| **Manisha Shukla** | *Babu Banarasi Das National Institute of Technology and Management, Lucknow, Uttar Pradesh* |
| **Shibu Krishnan** | *College of Engineering, Trivandrum, Kerala* |

Finally, I thank the publishing team of Tata McGraw-Hill for the enthusiasm and interest shown by them during all the stages of the preparation of this book.

## Feedback

I hope this book will serve its purpose and prove beneficial to its readers. Any suggestions and constructive criticism towards further improvement of the book are most welcome, all of which can be mailed to me at *shashichawla10@rediffmail.com*

**SHASHI CHAWLA**

# 1

# MULTIDISCIPLINARY NATURE OF ENVIRONMENTAL STUDIES

## *Learning Objectives*

After studying this chapter, you should be able to

- describe the term *environment* and its segments
- explain the multidisciplinary nature of environmental studies
- discuss the scope and importance of environmental studies
- explain the need of public awareness for the protection of environment
- write short notes on shelter security, economic security and social security
- discuss the effects of housing and industry on environment
- explain why is environmental education provided to engineers
- enumerate the objectives and guiding principles of environmental education

## 1.1 ENVIRONMENT

The term 'environment' originated from the French word ***environner*** or ***environ*** meaning 'to surround'. From this etymology, environment means the things or events that surround something else. In other words, environment means the area in which something exists or lives.

*Environment is defined as the social, cultural and physical conditions that surround, affect and influence the survival, growth and development of people, animals or plants.*

Environment includes everything around us. It encompasses both the living (biotic) and nonliving (abiotic) components of the earth.

The environment consists of four *segments*. These are briefly discussed below:

***Atmosphere*** It is the blanket of gases surrounding the earth.

***Hydrosphere*** It is composed of various water bodies on the earth. It includes the oceans, lakes, rivers, etc.

***Lithosphere*** It contains various types of soils and rocks on earth.

***Biosphere*** It is composed of all living organisms and their interactions with the environment, viz. atmosphere, lithosphere and hydrosphere. The biosphere is the earth's zone of air, soil, and water that is capable of supporting life.

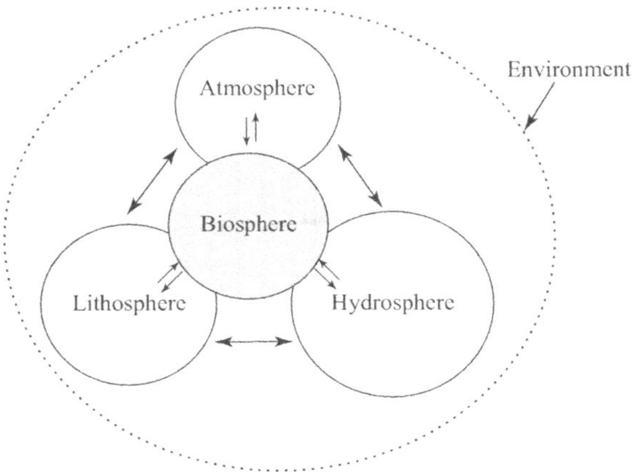


**Fig. 1.1** *Concept of environment as a functional system composed of organised, interacting and independent elements*

### 1.1.1 Composition of the Lithosphere

The lithosphere (from the Greek '*lithos*' for "rocky" and '*sphaira*' for "sphere") is the rigid outermost shell of a rocky planet. The lithosphere includes the crust and the uppermost mantle, which constitute the hard and rigid outer layer of the earth.

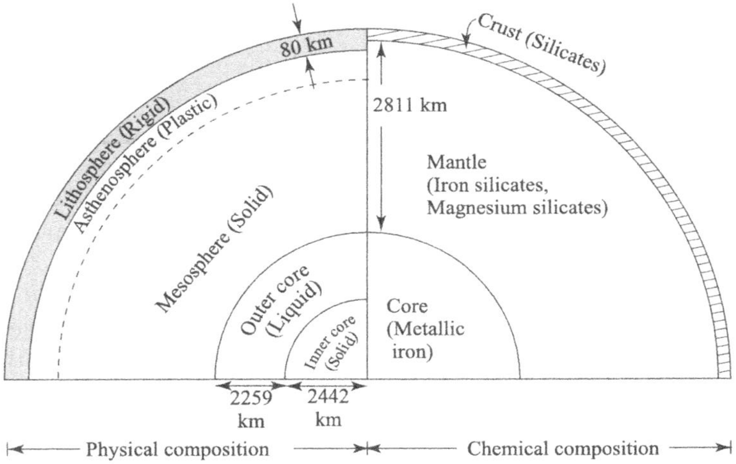


**Fig. 1.2** *Internal structure of the earth*

There are two types of lithosphere.

**(i) Oceanic Lithosphere** It mainly consists of mafic crust and ultramafic mantle (peridotite). It is associated with oceanic crust.

**(ii) Continental Lithosphere** It is associated with continental crust.

The oceanic lithosphere is more denser than the continental lithosphere.

## 1.1.2 The Biosphere

The biosphere is the earth's zone of air, soil, and water that is capable of supporting life. It is a zone which reaches about 10 km into the atmosphere and down to the deepest ocean floor. Processes in the biosphere include life and death, and evolution and extinction. Within the biosphere, the basic unit of analysis or study is the ecosystem. Thus, the biosphere is the global sum of all ecosystems.

The *atmosphere* forms a protective shell over the earth. The *troposphere* is the lowest layer and is only 12 km thick. It is the only part warm enough for us to survive in.

The *stratosphere* is 50 km thick and contains a layer of sulphates which is important for the formation of rain. It also contains a layer of ozone, which absorbs ultraviolet light known to cause cancer.

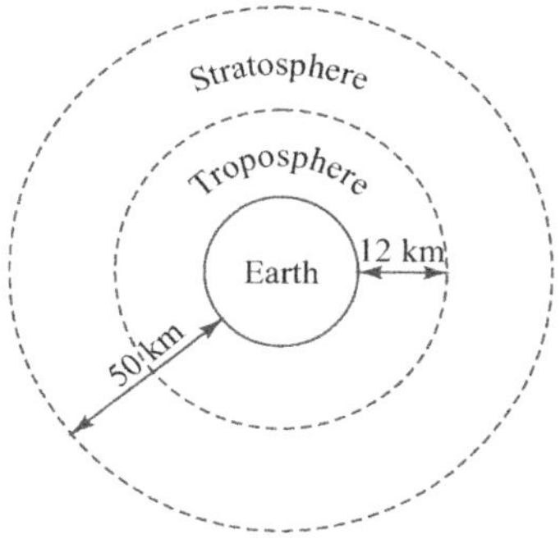


**Fig. 1.3** *Atmosphere around the earth*

## 1.1.3 Components of the Environment

Physical, biological and cultural environments are the three distinct dimensions of the environment. The study of *cultural environment* (i.e. social environment, economic environment and political environment) has been allocated to sociologists, economists and managers. Biologists and doctors are in charge of studying our *biological environment* [which comprises plants (flora), animals (fauna) and microorganisms]. Environmentalists take care of the *physical environment* (lithosphere, hydrosphere, atmosphere).

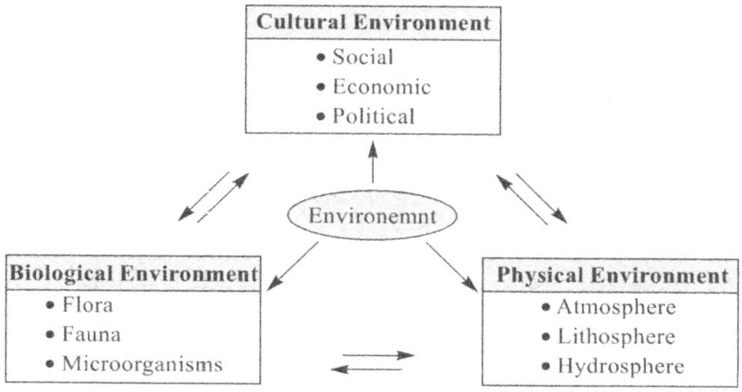


**Fig. 1.4** *Various types of environments and interactions between their various elements*

The components of the environment are classified in terms of biotic and abiotic components based upon living components and nonliving components respectively. It is from this component system that the study of the structure of ecosystems was evolved.

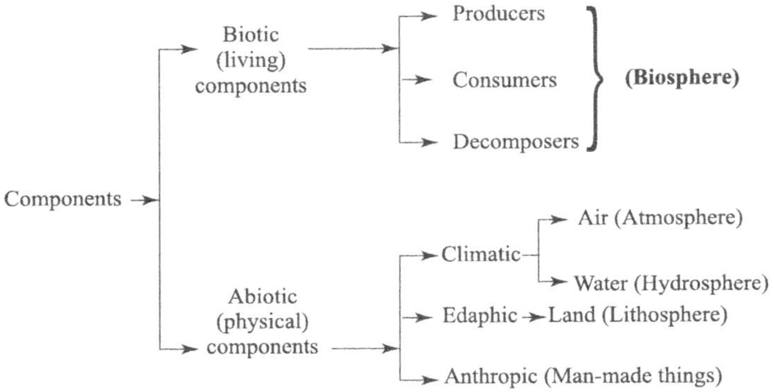


Fig. 1. 5 *Components of the environment*

## 1.1.4 Relationship between Humans and Environment

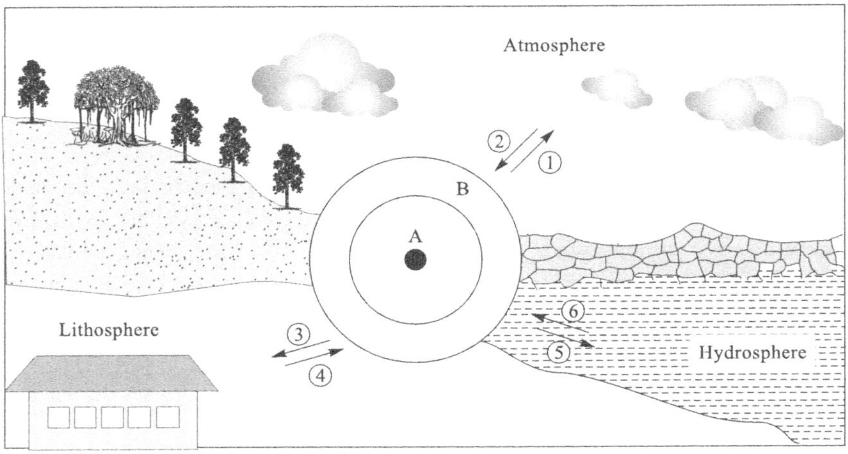


① Carbon dioxide for plants, air pollutants
② Oxygen for life, combustion of fossil fuels
③ Fertilizer/soil pollutants, industrial waste
④ Food/shelter/timber
⑤ Water pollutants/waste
⑥ Water, fishes, sea-food

**Fig. 1.6** *Human–environment relationship: (A) Urban growth (B) Industrial expansion*

With the development of human society, the ***human–environment relationship*** is changing. These relationships from prehistoric to modern periods is summarised below in four stages.

### *(A) Stage I: Period of Hunting and Food Gathering*

In this period, the basic requirements of primitive humans were limited to food and shelter. There was a very friendly relationship between humans and their environment because of limited requirements, very low population and disorganised society. However, the discovery of fire and invention of tools and weapons made humans capable of exploiting natural resources. Due to their carelessness while cooking, destruction of the environment started with inadvertent burning of forests.

Sometimes, humans intentionally used fire to clear the forest for creating habitats and to drive away dangerous animals from his surroundings.

### *(B) Stage II: Period of Animal Domestication and Pastoralism*

In this period, humans learnt to domesticate animals for milk, meat, etc. They also started living a community life. This ultimately led to the destruction of forests and exploitation of environmental resources. However, the changes brought about by human activities in the environment were well within the limits of selfregulatory mechanisms of the environment.

### *(C) Stage III: Period of Plant Domestication and Agriculture*

The emergence of socially organised human communities, human civilisations (specially, river-valley civilisations), farming practices (specially domestication of plants and primitive type of agriculture), gradual increase of human population and domestic animals led to the spread of human population and destruction of natural ecosystems. However, the human race continued to be guided by the physical environment and no serious damage was done to the natural environment.

### *(D) Stage IV: Period of Science, Technology and Industrialisation*

In this period, humans started exploiting natural resources in a reckless and indiscriminate manner for urban growth and industrial expansion. In this period, the hostile relationship between humans and their natural environmental was initiated. This has created most of the present ecological and environmental problems.

## 1.1.5 Impact of Technology and Development on Environment

Technology and development help us

(i) to provide better transport and delivery,
(ii) to speed up production and manufacture of goods,
(iii) to do efficient farming for making food, and
(iv) to make faster and reliable communication, etc.

Due to technology advancements, trade and commerce flourished at faster rates and brought about globalisation. As trade and commerce grew, it increased the greed of humans which resulted in direct and indirect impact on the environment. These are briefly discussed below:

**(i) Direct Impacts** These impacts are pre-planned. Both positive and negative consequences of any programme (launched to change or modify the natural environment for development of the concerned region) are known in advance. Suppose deforestation is done either for cropland or for commercial purposes. It leads to accelerated rate of soil erosion, resulting in loss of soil fertility and floods. However, these effects can be stopped by afforestation of the deforested area. Thus, direct impacts are reversible.

**(ii) Indirect Impacts** The indirect impacts of humans on the environment are mainly due to industrial development. These indirect impacts are not immediately noticeable but are experienced after long time. Sometimes, such effects are not reversible. They change the overall natural system and the resultant chain effects sometimes become suicidal for humans. Majority of the indirect impacts on the environment are related to pollution and environmental degradation.

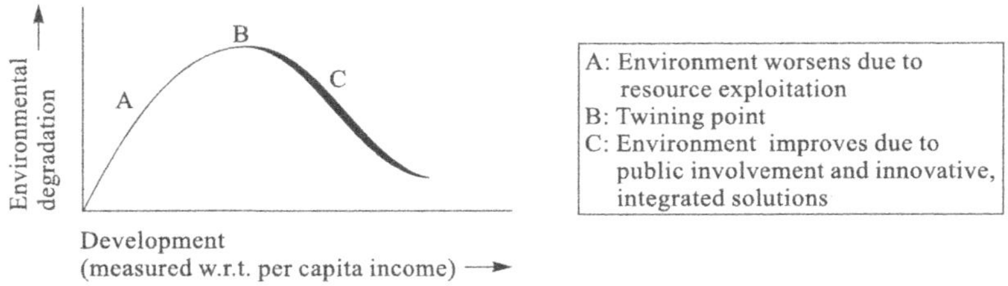


**Fig. 1.7** *Impact of technology and development on environment*

### *Green Technology for Lessening the Accumulated Negative Impacts*

Green technology is helping us in the following ways:

(i) It provides better solutions of generating heat and energy.
(ii) The sun's powerful UV rays are being harnessed through solar panels.
(iii) The kinetic powers of wind and water currents are being utilised to produce electricity.

These and some other green technological approaches help in lessening the demand for coal and fossil fuels, whose burning contributes to air pollution and consequent respiratory diseases and fatal illnesses.

## 1.2 ENVIRONMENTAL STUDIES

*Environmental education* refers to organised efforts to teach how natural environment functions and, particularly, how human beings can manage their behaviour and ecosystems in order to live sustainably.

*Environmental study* is the academic field which systematically studies every issue that affects an organism.

*Environmental science* is the systematic study of the scientific principles, economic influences and political action, and inter-relationship among living organisms (biotic components) and nonliving things (abiotic components) which affect the environment.

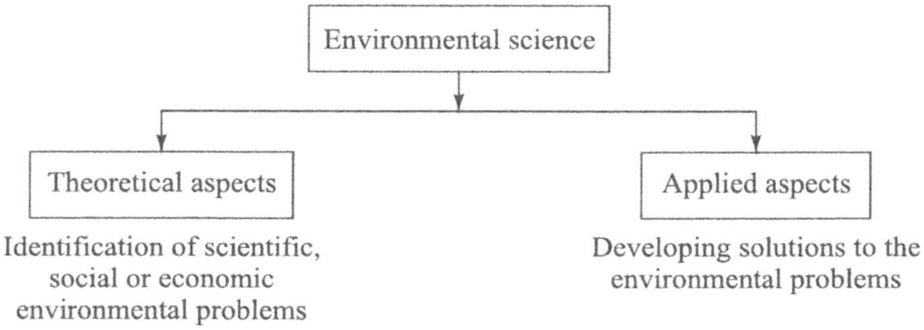


**Fig. 1.8** *Major aspects of environmental science*

Both theoretical and applied aspects of human impact on the world are studied in environmental science.

The *theoretical aspects* of environmental science identify threats to our survival and our future generations. For example, how ozone-layer depletion injures plant and marine life and even reduces our immunity to diseases is studied in theoretical aspects.

The *applied aspects* of environmental science suggest solutions to the identified environmental problems.

## 1.2.1 Multidisciplinary Nature of Environmental Studies

The environment is everybody's concern as we all live on the same planet. An understanding of the working of the environment requires the knowledge of various fields.

Air pollution is one of the important topics in environmental studies. Table 1.1 shows a list of topics commonly studied in air pollution and the related traditional fields of knowledge illustrating the interdisciplinary nature of the subject.

**Table 1.1** *Interdisciplinary nature of environmental studies—air pollution*

| *Environmental issue* | *Knowledge of subject required* |
|---|---|
| (i) History of air pollution and air pollution accidents | (i) History |
| (ii) Economic impacts of air pollution | (ii) Economics, demography |
| (iii) Nature and reactions of air pollutants | (iii) Chemistry and chemical engineering |
| (iv) Air-pollution control devices | (iv) Physics, chemistry and various branches of engineering |
| (v) Effects of air pollutants on human beings, animals, plants and materials | (v) Zoology, botany, physics, chemistry |
| (vi) Sociological impacts of air pollution | (vi) Sociology |
| (vii) Conservation of resources and pollution control | (vii) Various branches of physical and political sciences |
| (viii) Alternative fuels | (viii) Various branches of physical sciences |
| (ix) Ozone hole and global warming | (ix) Pure as well as social sciences |
| (x) Effect of climate on air pollution | (x) Mathematical modelling, meteorology, thermodynamics, geography, etc. |

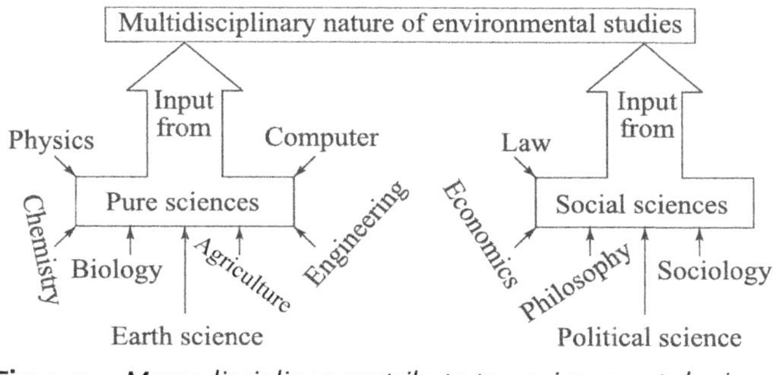


**Fig. 1.9** *Many disciplines contribute to environmental science*

The situation is similar in other topics of environmental studies.

To sum up, environmental science incorporates the principles of *pure sciences* (like physics, chemistry, biology, earth science, agriculture, engineering, computer science, etc.) with input from the *social sciences* (such as political science, law, philosophy, sociology and economics), thus creating a new interdisciplinary field.

It is essentially a multidisciplinary approach that brings about an appreciation of our natural world and human impacts on its integrity.

### 1.2.2 Scope of Environmental Studies

The scope of environmental studies is so wide that it is related to every science and scientific aspects in general and biology in particular.

*The scope of environmental studies in numerous fields* is given below:

(i) *Conservation and management* of *natural resources* (like forest resources, water resources, etc.)

(ii) *Conservation of biodiversities* (like conservation of genetic diversity, species diversity, ecosystem diversity, landscape diversity, etc.)

(iii) *Control of environmental pollutions* (like air pollution, water pollution, soil pollution, solid waste pollution, noise pollution, electronic waste pollution, e-pollution, etc.)

(iv) *Control of human population*

(v) *Replacement* of development (like green revolution, urbanisation,) economic growth, (industrialisation, etc.) with *sustainable development*

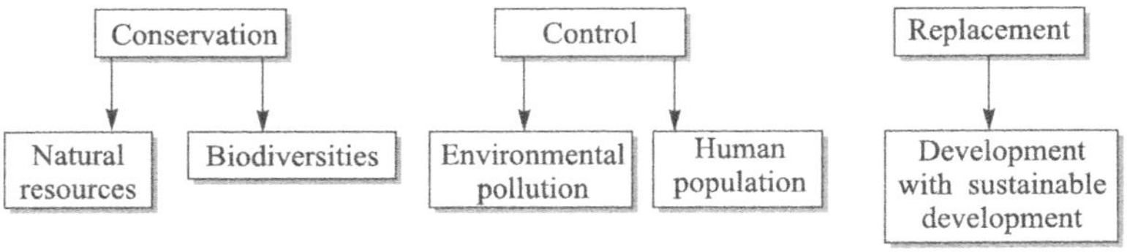


**Fig. 1.10** *Scope of environmental studies*

### 1.2.3 Objectives of Environmental Studies

The objective of environmental studies is to help social groups and individuals acquire an *awareness* of the environment as a whole and its related problems. They should

(i) acquire the *skills* for identifying and solving environmental problems

(ii) *participate* in improvement and protection of the environment

(iii) develop the ability to *evaluate* measures for the improvement and protection of environment

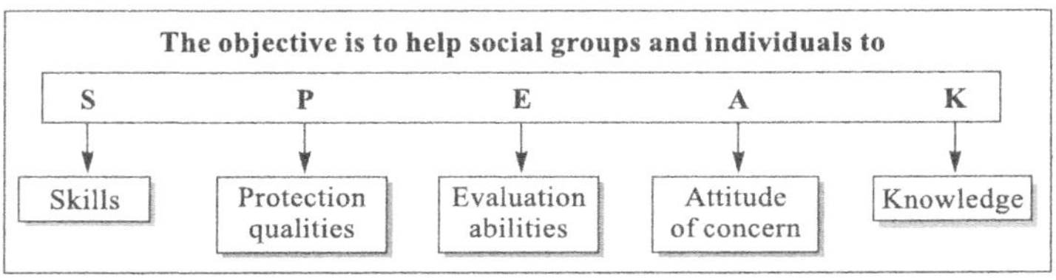

(iv) acquire an *attitude* of concern for the environment

(v) gain a variety of experiences and acquire a basic understanding and *knowledge* about the environment and its allied problems.

To sum up, the objective of environmental studies is to develop a world in which persons are aware of and concerned about the environment and the problems associated with it, and committed to work individually as well as collectively towards solutions of current problems and prevention of future problems.

### 1.2.4 Guiding Principles of Environmental Studies

The guiding principles of environmental education are as follows:

(i) Environmental education should help learners *discover* the symptoms and real causes of environmental problems.

(ii) Environmental education should have an interdisciplinary approach, and it should be continuous and compulsory, right from the pre-school to all formal as well as nonformal higher levels.

(iii) Environmental education should encourage *stewardship* to help reduce human impact on the planet.

(iv) Environmental education should emphasise the importance of economic development without degrading the environment (i.e. sustainable development).

(v) Environmental education should enable policymakers to include environmental-impact *analysis* in proposed developmental projects in order to minimise environmental damages.

(vi) Environmental education should emphasise the necessity of seeking international *cooperation* in environmental planning, and prevention and control of environmental problems.

(vii) Environmental education should lay more stress on practical *training* and practical *activities*.

(viii) Environmental education should promote the value and necessity to examine major environmental issues from the local, national, regional, and international point of view.

(ix) Environmental education should emphasise the complexity of environmental problems and the need to *develop* critical thinking and problem-solving skills.

### 1.2.5 Importance of Environmental Studies

For the survival of the present and future generations, environmental education is necessary.

*The important benefits of environmental studies are the following:*

(i) It directs attention towards the unlimited exploitation of environment (nature) by humans for greed or for the sake of development. Exploitation of nature has threatened the survival of all living organisms, including humans.

(ii) It generates concern for the changing environment, population explosion and throws light on the methods of solution.

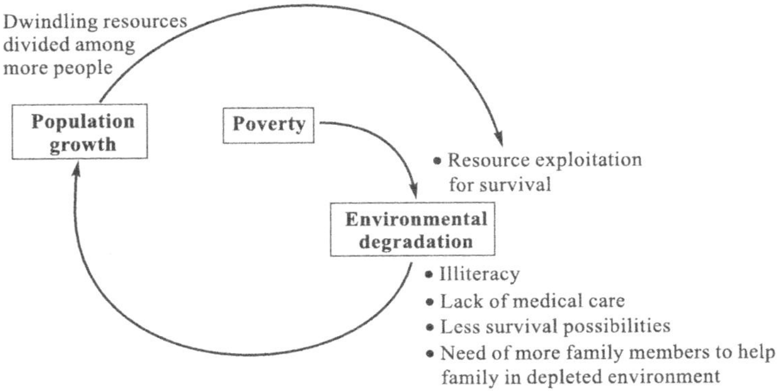


**Fig. 1.11** *A self-perpetuating vicious cycle of poverty, environmental degradation and population growth*

(iii) It helps to understand different food chains and to find ways and means to maintain ecological balance.

(iv) It helps in the maintenance of healthy life. Through improved health of people, economic productivity gets increased.

(v) It imparts knowledge about conservation of energy and reducing material dependence
   (a) by refusing to purchase things which are harming our environment
   (b) by reusing a product number of times
   (c) by motivating recycling of recyclable products

(vi) It helps in developing social responsibility towards protection of environment and control of environmental pollution.

(vii) It helps in appreciating and enjoying nature and working towards sustainable development.

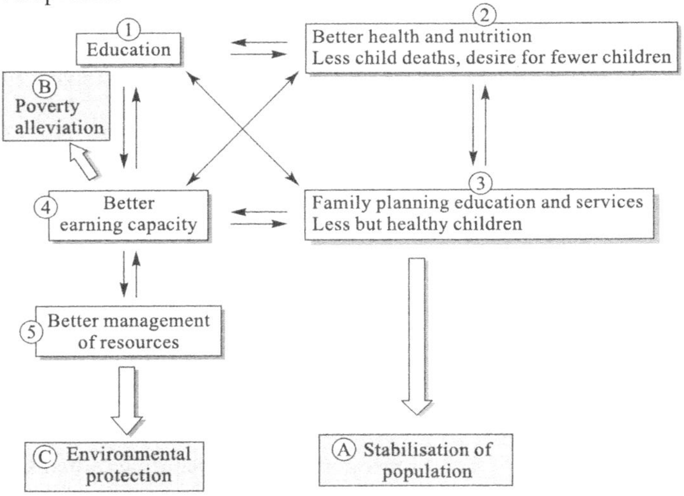


**Fig. 1.12** *Population stabilisation, poverty alleviation and environmental protection are mutually supportive of and dependent on one another*

### 1.2.6 Why is Environmental Education Provided to Engineers?

Engineering graduates entering industry get benefitted from environmental competence in their practice fields. This is because planning, extraction, design and manufacturing, all have some common and some unique regulations for their environmental effects.

Engineers are capable of finding creative solutions for climate change and other environmental problems. They will play a crucial role in helping the human race coexist with the rest of the world in decades to come. This is because, engineers use principles of chemistry and biology to prevent or solve environmental problems. They work in many areas, including industrial hygiene, land management, air-pollution control, toxic-materials control, etc. The duties of an environmental engineer range from planning and designing an effective waste-treatment plant, to study the effects of pollution on humans. A sales engineer in the environmental engineering field may be responsible for the sale of air pollution control products to factories. In addition to sale, they often assist with the design and modification of their products based on customer feedback.

## 1.3 NEED FOR PUBLIC AWARENESS

Humans are responsible for depletion of natural resources; degradation of the vital life-supporting systems (like air, water, soil, etc.); ecological imbalance; deteriorated environment, etc. Solid-waste disposal, oil pollution, water pollution, air pollution, Itai-Itai disease, Minamata disease, Chernobyl disaster, Bhopal gas tragedy, smog, etc., are some of the examples which are threatening human survival.

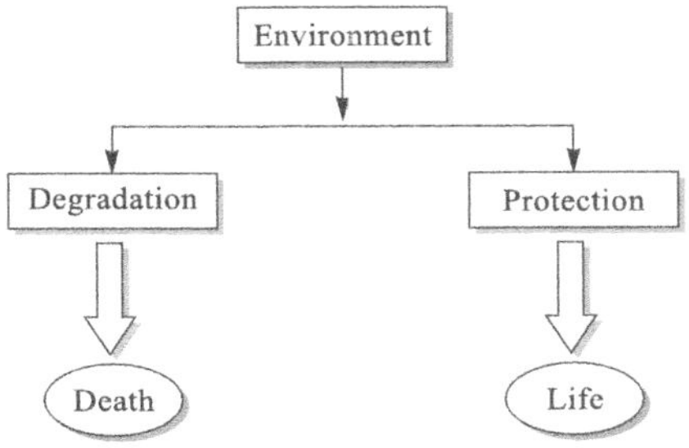


**Fig. 1.13** *Need for awareness of environment in public*

To get rid of these problems, environmental awareness is necessary. It ensures that everyone knows about the consequences of his/her activities on nature.

To protect global environment for sustainable development, we should do the following:

(i) ***Preserve Forest Cover*** Humans should minimise cutting of trees and using timber for aesthetic pleasure such as decoration of drawing rooms.

(ii) ***Preserve Natural Resources*** Humans should not unnecessarily and exhaustively extract natural resources such as mineral resources, water resources, etc.

(iii) ***Conserve Energy*** Humans should not harness too much energy from burning of fossil fuels.

(iv) ***Maintain Ecological Balance*** Humans should work to create synergy between green revolution and industrial evolution by compulsory growing green belts around industrial areas, maintaining wildlife sanctuaries and national parks.

(v) ***Practice Green Technology:*** Everyone from a farmer in the village to a policy planner in the government should use green technology that incorporates
   (a) treatment of air emissions,
   (b) treatment of waste waters,
   (c) waste reduction, waste or emission management, and
   (d) use of nonconventional, renewable energy resources like solar energy on priority, etc.

## 1.4 ENVIRONMENTAL DEGRADATION

The ability of an environment to sustain the resource demands of a species or a community without losing its ability to regenerate the resource is termed the ***carrying capacity.***

***Environmental degradation*** means that the carrying capacity is reduced by some natural or human phenomenon.

Environmental degradation is the deterioration of the environment through extinction of wildlife, depletion of natural resources and the destruction of ecosystems.

The main causes of environmental degradation are the following:

**(i) Population (P)** More population means less resource availability per person, which encourages couples to have more children to help gather resources leading to resource exploitation and environmental degradation.

**(ii) Affluence (A)** Rich people overuse the resources, and this lead to air, water, land and environmental pollution. Poor people exploit natural resources and so they also cause environmental degradation.

**(iii) Technology (T)** Heavy industrialisation, unplanned and heavy consumption of natural resources in which even renewable resources are not given time for renewal also lead to environmental degradation.

To sum up, environmental degradation (*ED*) is a function of *P*, *A* and *T*.

$$ED = f(P \times A \times T)$$

The causes and effects of environmental pollution are illustrated in Fig. 1.14.

## 1.5 SHELTER SECURITY

All of us need shelter security for food, water, sleep and warmth.

When humans feel safe and secure in their homes, and have the love and support of family and friends, they have the opportunity to grow in confidence, to gain the respect of others, and most importantly to achieve a high level of self-respect.

### 1.5.1 Problems Faced by Homeless People

If people do not have shelter security, they may have difficulty finding a school or a job, they may lose touch with family and friends, they may feel excluded, they may be unsafe, and they have reduced access to health care.

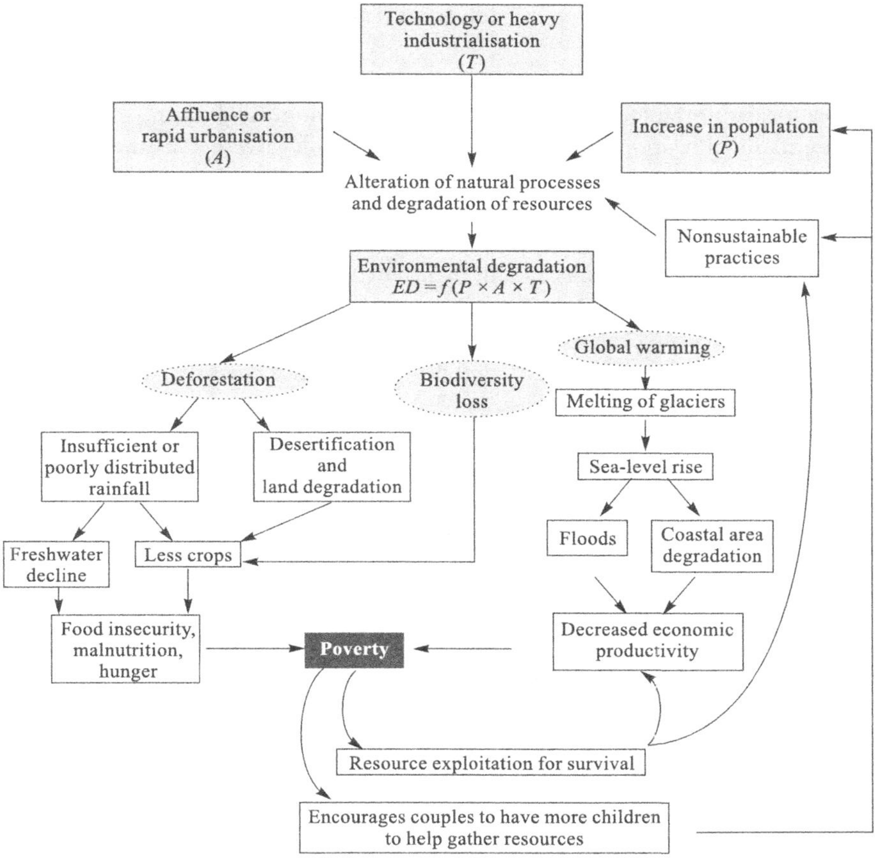


**Fig. 1.14** *Causes and effects of environmental degradation*

### 1.5.2 Causes of Homelessness

People become homeless for the following reasons:

(i) Unemployment
(ii) Lack of money to buy a property or to pay rent
(iii) Family breakdown due to domestic violence
(iv) Natural disaster or war
(v) Drug or alcohol addiction
(vi) Increased wealth disparity and income inequality leading to distortions in the prices of houses
(vii) Mortgage foreclosures on homes by banks

## 1.6 ECONOMIC SECURITY

Economic security refers to the condition of having a stable income (or other resources) to support a standard of living now and in the foreseeable future. It includes the society's production levels and monetary support for nonworking citizens.

When children, adults, and families are safe, healthy, educated and have sufficient money for comfortable living, it means they are economically secure.

International relations between countries are mainly governed by their economic securities.

If a foreign government gets unauthorised access to proprietary information or technology then the country's overall economic security is under threat.

Economic security of children and their parents is indicated by the income level and employment security of their families.

Economic security of retired people is based on social-security benefits, pensions, savings, earnings and employment, and health-insurance coverage.

*Economic security can be defined as the condition of an individual, household or community to be able to cover the essential needs (like food, shelter, access to health care, education, etc.) and unavoidable expenditures in a sustainable manner.*

If essential needs are not covered, a situation of crisis develops. The crisis can be of the following four types:

1. Pre-crisis
2. Acute-crisis
3. Chronic-crisis
4. Post-crisis

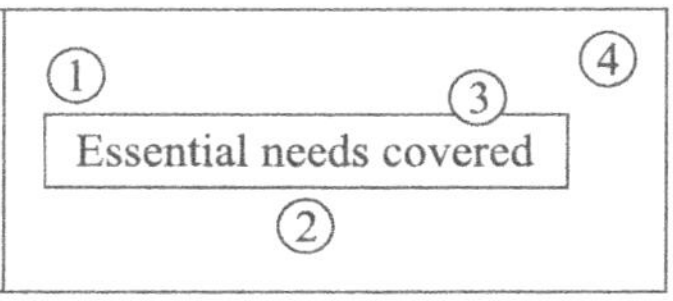


Essential needs are still covered, but are at risk of no longer being addressed in **pre-crisis.** Some essential needs are no longer covered in **acute-crisis.** Essential needs are insufficiently covered in **chronic-crisis.** Essential needs are covered by structure whose sustainability remains fragile in **post-crisis.**

The aim is to save lives in pre-crisis and acute-crisis. The aim is to support livelihoods in chronic-crisis and post-crisis.

## 1.7 SOCIAL SECURITY

*Social security means providing social welfare services to the poor, aged, disabled, widowed, retired, children or unemployed people by the society.*

***(A) Negative Effects of Social Security*** Social security may discourage people from working and saving to reduce international competitiveness and employment creation. It may also encourage people to take early retirement.

### *(B) Positive Economic Effects of Social Security*

(i) It may help maintain effective demand at the national level.
(ii) It may help create conditions in which a market economy can flourish, by encouraging workers to accept innovation and change.

## 1.8 EFFECTS OF HOUSING ON ENVIRONMENT

Traditionally, home is considered a haven, where humans are protected and nurtured. However, a house can also be a health hazard when factors such as poor design, environmental contamination and poverty combine to cause or exacerbate disease.

### 1.8.1 Housing and Health

Housing is an environmental health issue because of various socio-economic conditions as illustrated below:

(i) Lead-based paints were used for renovation and painting homes. They are major causes of childhood lead poisoning.
(ii) Poorly constructed houses lead to stress and children stay inside with increasing exposure to pollutants resulting in childhood asthma.
(iii) Substandard housing contributes to asthma incidences.

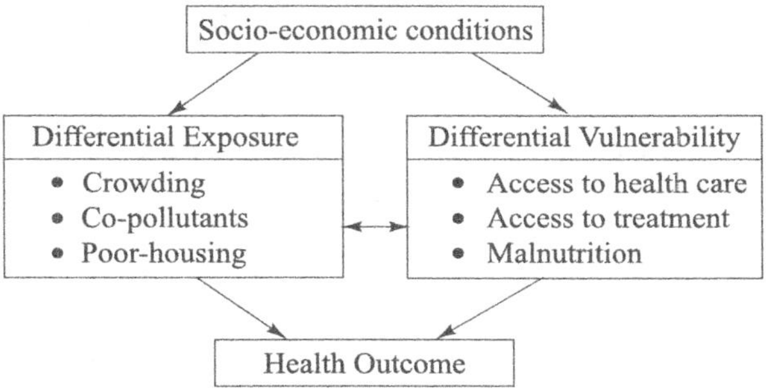


**Fig. 1.15** *Housing as an environmental health issue*

### 1.8.2 Housing and Environment

The quality of environment is determined by the following factors:

(i) Maintenance and arrangement of houses.
(ii) The presence, quality and accessibility of facilities
(iii) Security
(iv) Street cleanliness in the residential area, etc.

The material flows for houses during construction, maintenance and operation phases are illustrated below:

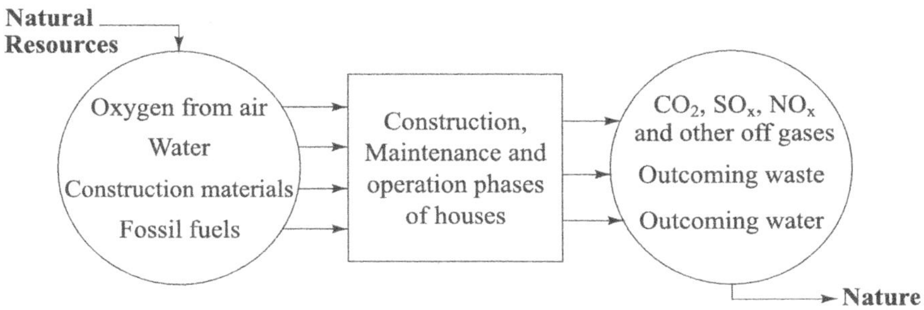


**Fig. 1.16** *Material flows for houses*

### 1.8.3 Strategies for Improvement of Environment

(i) Reduce the natural resource depletion regarding material use.
(ii) Reduce global warming via efficient energy and material use.
(iii) Reduce air, noise and water pollution by eco-designing of houses.
(iv) Reduce the environmental impact of housing through innovations in design, like green housing, zero emission housing, etc.

## 1.9 EFFECTS OF INDUSTRY ON ENVIRONMENT

For the development of a country and the prosperity of its people, industry is essential. Unfortunately, our technology choices have turned out to be wasteful because decisions are based on short-term and narrow goals of the enterprise rather than a holistic view of the full range of consequences from the use of a technology.

### 1.9.1 Adverse Environmental Effects of Industry

The total global flow of materials is around 500 billion tons a year. Only 6% actually ends up in consumer products whereas much of the virgin materials are being returned to the environment in the form of harmful solid, liquid, and gaseous wastes. Some of the adverse effects of industrial practices are summarised below:

(i) Industrial practices release enormous quantities of air and water pollutants.
(ii) They generate huge amounts of hazardous wastes.
(iii) Industrial effluents have polluted many lakes, rivers and coastal environments.
(iv) Industrial accidents, such as the Bhopal gas tragedy, often have tragic environmental consequences.

### 1.9.2 Approaches for Minimising Harmful Effects of Industry on Environment

The approaches for minimising adverse effects of industrial practices on environment are outlined below:

**(i) Pollution Prevention** Try to create less of the pollutant or waste or eliminate it.

**(ii) Recycling and Reuse** They not only reduce pollution, but they also conserve natural resources.

**(iii) Treatment** It is used to reduce the volume or toxicity of the waste.

**(iv) Disposal** Responsible disposal as per the law by both industries and municipalities should be done.

## Important Definitions

- *Environment* is defined as the social, cultural and physical conditions that surround, affect and influence the survival, growth and development of people, animals or plants.
- The biosphere is the earth's zone of air, soil, and water that is capable of supporting life.
- The objective of environmental studies is to develop a world in which persons are aware of and concerned about the environment and the problems associated with it, and committed to work individually as well as collectively towards solutions of current problems and prevention of future problems.
- The ability of an environment to sustain the resource demands of a species or a community without losing its ability to regenerate the resource is termed the *carrying capacity*.
- *Environmental degradation* means that the carrying capacity is reduced by some natural or human phenomenon.
- All of us need *shelter security* for food, water, sleep and warmth.
- *Economic security* can be defined as the condition of an individual, household or community to be able to cover the essential needs (like food, shelter, access to health care, education, etc.) and unavoidable expenditures in a sustainable manner.
- *Social security* means providing social welfare services to the poor, aged, disabled, widowed, retired, children or unemployed people by the society.

## EXERCISES

**1.** Define environment.

**2.** Briefly discuss four major global environmental problems.

**3.** What is the importance of environmental studies?

**4.** Discuss how formal environmental education can help in managing the environment.

**5.** **(a)** Write a note on the importance of education on environmental issues and concerns.

**(b)** Describe the multidisciplinary nature of environmental studies.

**6.** **(a)** Explain the impact of technology and development on environment.

**(b)** Give the composition of atmosphere and lithosphere.

**7.** **(a)** What are the components of environments? Explain by drawing a sketch.

**(b)** Discuss the physical components of the environment.

**8.** What are the objectives and guiding principles of environmental education?

**9.** Why is environmental education provided to engineers?

**10.** What is the meaning of environment? Discuss the relationship between different components/parts of environment.

**11.** Discuss the relationship between humans and environment. Why is environmental protection discussed recently?

**12.** What is meant by society and environment? How these are related to each other?

**13.** Explain the need for public awareness about the environment and its degradation.

# OBJECTIVE TYPE QUESTIONS

## I. Fill in the Blanks

**1.** The study of reciprocal relationship between organisms and environment is called ______ .

**2.** ______ teaches us how to respect nature.

**3.** The term *environment* has been derived from the French word '*environner*' which means ______ .

**4.** World Environment Day is celebrated on ______ .

**5.** The term ecology was introduced by ______ .

**6.** The United Nations Conference on Environment and Development (Earth Summit) was held at ______ in ______ .

**7.** The world's first Intergovernmental Conference on Environmental Education was held in ______ .

**8.** ______ is often identified as the founder of ecology.

**9.** The part of the earth and its atmosphere in which organisms live is known as ______ .

**10.** Troposphere is located ______ stratosphere.

**11.** World Food Day is recalled on ______ .

## II. Match the following terms.

Match the terms of column I with appropriate terms of column II.

**A.**

| Column I | Column II |
|---|---|
| 1. Living organism | (a) Stratosphere |
| 2. Air | (b) Biosphere |
| 3. Water | (c) Atmosphere |
| 4. Soil | (d) Hydrosphere |
| 5. Ozone | (e) Lithosphere |

**B.**

| Column I | Column II |
|---|---|
| 1. Industrial revolution (Father) | (a) $17^{th}$ and $18^{th}$ centuries |
| 2. Global warming | (b) James Watt |
| 3. Environmental movement | (c) $CO_2$ |
| 4. Industrial revolution took place in | (d) Rachael carron |
| 5. Industrial revolution took place during | (e) England |

## III. Multiple Choice Questions

**1.** The part of the earth and its atmosphere in which organisms live is the
- (a) biosphere (b) biomass
- (c) biota (d) biome

**2.** Which of the following conceptual sphere of the environment has the least storage capacity of matter?
- (a) Lithosphere (b) Atmosphere
- (c) Biosphere (d) Hydrosphere

**3.** Which one of the following is an abiotic component of the ecosystem?
- (a) Plants (b) Fungi
- (c) Humus (d) Bacteria

**4.** The most stable ecosystem is
- (a) mountain (b) desert
- (c) forest (d) ocean

**5.** Increase in fauna and decrease in flora would be harmful due to increase in
- (a) $CO_2$ (b) $O_2$
- (c) $N_2$ (d) $SO_2$

**6.** The largest reservoir of nitrogen in our planet is the
- (a) fossil fuel (b) atmosphere
- (c) biosphere (d) ocean

**IV. Indicate True or False for the following statements:**

**1.** Biosphere is made of atmosphere, hydrosphere and lithosphere. True/False

**2.** Environmental studies is multidisciplinary in nature. True/False

**3.** Public awareness is a must for the protection of environment. True/False

**4.** Temperature in thermosphere can be as high as 1200°C. True/False

**5.** The term ecology is not derived from the Greek word '*Oekologie*'. True/False

## Answers to Objective Type Questions

**I. Fill in the Blanks**

1. Ecology
2. Environmental studies
3. To encircle or surround
4. 5th June
5. Ernst Haeckel
6. Rio de Janerio, 1992
7. Tbilisi, Georgia
8. Eugen Warming
9. Biosphere
10. Below
11. October 16th

**II. Matching the terms.**

**A.** 1. (b) 2. (c) 3. (d) 4. (e) 5. (a)

**B.** 1. (b) 2. (c) 3. (d) 4. (e) 5. (a)

**III. Multiple Choice Questions**

1. (a) 2. (b) 3. (c) 4. (d)
5. (a) 6. (b)

**IV. True or False**

1. True 2. True 3. True 4.True
5. False

# 2

# NATURAL RESOURCES

## *Learning Objectives*

After studying this chapter, you should be able to

- explain the difference between renewable, nonrenewable and perpetual natural resources
- describe the impacts of over-utilisation of underground and surface water
- identify and explain the core causes of water crisis in the world
- explain the fluoride problem in drinking water
- discuss the measures to conserve water
- explain the various natural mechanisms involved in the self-purification of rivers
- describe the effects of mineral extraction on the environment
- explain the necessity of conserving mineral resources
- describe mineral resources of India
- define deforestation
- explain causes and ill effects of deforestation
- describe measures taken for conserving forest wealth
- name and explain various steps involved in the carbon cycle, nitrogen cycle, and sulphur cycle
- explain how elemental carbon is recycled in nature
- explain the various energy resources
- describe the energy scenario in the Indian context
- compare the alternative energy resources
- explain the nonrenewable energy resources
- describe the hydropower energy as renewable, clean and nonpolluting source of energy
- explain fossil fuels
- describe nuclear energy
- enumerate the application of solar energy in modern days
- explain the use of solar energy for space heating of buildings
- describe hydrogen as an alternative future source of energy
- explain the biomass energy as renewable energy
- describe a fuel cell
- explain the various ways in which energy from the ocean can be obtained
- discuss merits and limitations of geothermal energy

## 2.1 INTRODUCTION

*Resources* are objects, materials, creatures, or any form of energy found in nature that can be used to perform any useful function. They are or may become of potential economic interest due to their inherent properties.

*Reserves* are that part of a resource which has been fully evaluated and is found commercially viable to work on the consideration of mining, metallurgical, economic, marketing, legal, environmental, social and governmental factors.

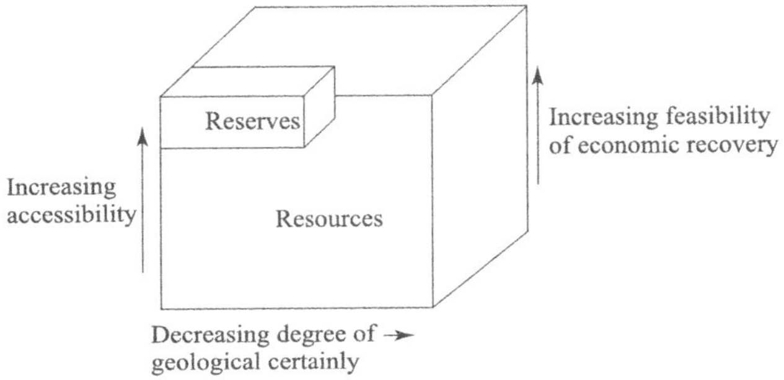


**Fig. 2.1** *Reserves and resources*

## 2.2 TYPES OF NATURAL RESOURCES

Based on their use, availability, origin and economic status, natural resources can be classified into the following types:

### 2.2.1 Perpetual, Renewable and Nonrenewable Natural Resources

Based on their availability or how human activities affect them, natural resources are of the following three types:

***(A) Perpetual Resources*** *Perpetual resources* are those natural resources that naturally perpetuate themselves and are not affected by human use.

***Examples*** Sunlight, wind, rainfall water and tides.

***(B) Renewable Resources*** *Renewable resources* are those natural resources that have the inherent ability to renew or replenish themselves if given a reasonable amount of time.

***Examples*** Soil, fresh water, forest, etc.

***(C) Nonrenewable (or Exhaustible) Resources*** *Nonrenewable resources* are those natural resources that cannot be regenerated or renewed or replaced within a time framework.

***Examples*** Fossil fuels (such as coal, petroleum, natural gas, etc.), nuclear power.

**Table 2.1** *Differences between perpetual, renewable and nonrenewable resources*

| *Perpetual Resources* | *Renewable Resources* | *Nonrenewable Resources* |
|---|---|---|
| (i) They are replenished naturally at a rate faster than their rate of consumption. | (i) The environment has the capacity to replenish them as long as they are properly conserved. | (i) They are being consumed or used up faster than they can be made by nature. |
| (ii) They last forever regardless of anything humans do to them. | (ii) Human activities can affect the supplies of renewable resources. | (ii) Once these resources are used up, they are gone forever. |
| (iii) Examples: Wind, sunlight | (iii) Examples: Soil, forest | (iii) Examples: Coal, petroleum |

***(D) Intangible Resources*** *Intangible resources* are those natural resources that are available in huge quantities, but at the same time can be destroyed easily.

The tourism industry is based on serenity, beauty, diversity, open space and satisfaction. However, a small piece of trash can easily destroy the beauty of any place. Thus, intangible resources are both exhaustible and inexhaustible.

### 2.2.2 Biotic and Abiotic Natural Resources

Based on their origin, natural resources are of the following two types:

***(A) Biotic Resources*** Biotic resources have originated from some living organism or have life.

***Examples***

- *Renewable*: Livestock, fisheries, flora, fauna and humans.
- *Nonrenewable*: Coal, petroleum, etc.

***(B) Abiotic Resources*** Abiotic resources are of nonliving origin.

***Examples*** Minerals, rocks, water, etc.

## 2.3 FOREST RESOURCES

### *(A) Types of Forest Resources*

Forests are broadly classified into there categories from the point of view of use as a resource:

**(i) Old-Growth or Ancient Forests** These are uncut forests that have not been seriously disturbed by natural disasters or human activities. As a result, they have attained great age, and thereby exhibit unique ecological features.

**(ii) Second-Growth Forests** They result from secondary ecological succession that takes place when forests are cleared and then left undisturbed for long periods of time.

**(iii) Plantations** These are (large, artificially established) forests of commercially valuable trees. These are created mostly by clearing old-growth or second-growth forests.

**Table 2.2** *Differentiation characteristics of forests*

| *Forest-Resource Type* | *Canopy Layer* | *Biodiversity* | *Prone to Disease* |
|---|---|---|---|
| (i) Old-growth | Several | More | Less |
| (ii) Second-growth | One | Less | More |
| (iii) Plantations | One | Least | Maximum |

**Example 1** *Automated analysis of forest satellite imagery is not reliable for estimating forest cover. Comment with respect to Indian perspective.*

**Solution** The Forest Survey of India (FSI) reported in 2009 that Indian forests had grown by almost 5% per year from the 1990s. They used automated analysis of forest satellite imagery.

The above method cannot differentiate between native forests and exotic tree plantations (such as rubber, teak, pine, eucalyptus trees). The plantation forest covers have very limited value for biodiversity.

Monoculture plantations are expanding by nearly 6,000 to 18,000 square kilometres per year in India.

If one subtracts plantations from total forest cover then India's native forests have actually declined at an alarming pace, from 0.8% to 3.5% per year from 2000–2005.

***(B) Functions of a Forest*** Forests help in production of timber, regulation of stream flow, control of erosion, recreation, provision of wildlife habitat, etc.

## 2.3.1 Use and Over-exploitation of Forest Resources

Beneficial functions of forests are the following:

(i) ***Influence on Climate*** The crowns of the trees hold the moisture in because the force of the wind is broken. It makes the forest cool in the summer and warm in the winter.

(ii) ***Control of Run-off*** Leaves and branches of trees break the impact of rain, causing it to drip rather than have a strong force. Rain is absorbed by the ground, reducing surface run-off.

(iii) ***Flood Control*** Forested watersheds help in avoiding extremes of water flow and so help in flood prevention.

(iv) ***Wildlife Habitat Provision*** Wildlife uses the products of trees and forests as food and shelter respectively.

(v) ***Prevention of Soil Erosion*** Water moves slowly through forested soils and stays free of sediments.

(vi) ***Reduction of Wind Erosion*** Trees are used as windbreaks and slow down the force of wind.

(vii) ***Removal of Pollutants*** The roots of trees absorb soil and water pollutants. Sulphur dioxide is used for metabolism of trees. Thus, forests aid in the cleansing of air, water and soil.

**(viii)** ***Noise Abatement*** Trees act as a sound barrier.

**(ix)** ***Recycling of Nutrients*** Forests help in nutrient recycling.

**(x)** ***Provisions for Healthy Survival of Local Communities and Mankind*** Forests provide employment and income, aesthetic pleasure and spiritual solace. They also provide food, fibre, honey, medicinal plants and minerals.

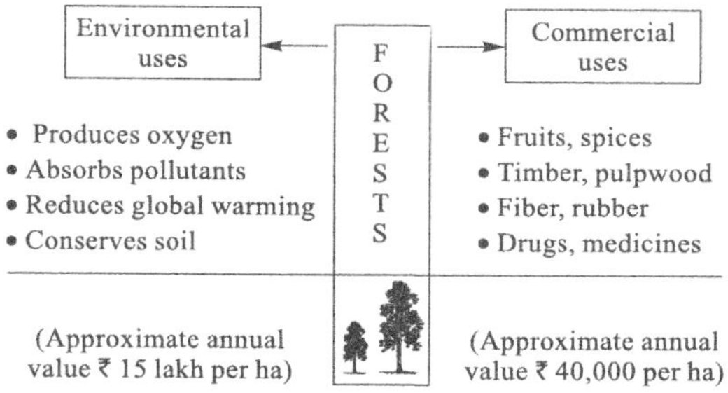


**Fig. 2.2** *Environmental and commercial value of forests*

### 2.3.2 Deforestation

*Deforestation* involves the cutting down, burning, and damaging of forests.

*Deforestation* can be defined as the change of forest with depletion of tree crown cover of more than 90%. However, depletion of forest-tree-crown cover less than 90% is considered as *forest degradation.*

***(A) Causes of Deforestation*** The main causes of deforestation are summarised below:

(i) Population explosion
(ii) *Agriculture*: shifting cultivation, overgrazing, cash-crop economy, etc.
(iii) *Commercial logging*: cutting trees for sale as timber or pulp
(iv) Poverty
(v) Mining
(vi) Dams
(vii) Infrastructure creation for logging
(viii) Forest fires
(ix) Acid rain
(x) Development projects and housing projects.

***(B) Ill Effects of Deforestation*** The ill effects of deforestation are summarised below:

**(i)** ***Soil Erosion*** Soil is exposed to wind, sunlight, evaporation due to deforestation. Soil fertility goes down due to soil erosion and rapid leaching of essential mineral nutrients.

**(ii)** ***Harm to Fisheries*** As the soil is eroded, it accelerates siltation in dams, rivers, and the coastal zone. The increased sedimentation harms downstream fisheries.

**(iii)** ***More Floods and Droughts*** Because of deforestation, there is no regulation of the flow into rivers. As a result, floods and droughts alternate in the affected areas.

(iv) ***Habitat Loss of Wildlife*** Butterflies, migratory birds and wild animals suffer due to the loss of their habitat.

(v) ***Extinction of Some Species*** Many species are affected and some become extinct.

(vi) ***Local and Global Climate Changes*** The rainfall pattern is affected as the forest is cut down. Local and global climate changes may result from deforestation.

(vii) ***Global Warming*** If the trees are burned, the carbon is released immediately as carbon dioxide which lead to global warming.

(viii) ***Danger for the Survival of Local Communities*** Communities lose their source of food, fuel, construction materials and areas for livestock grazing by deforestation.

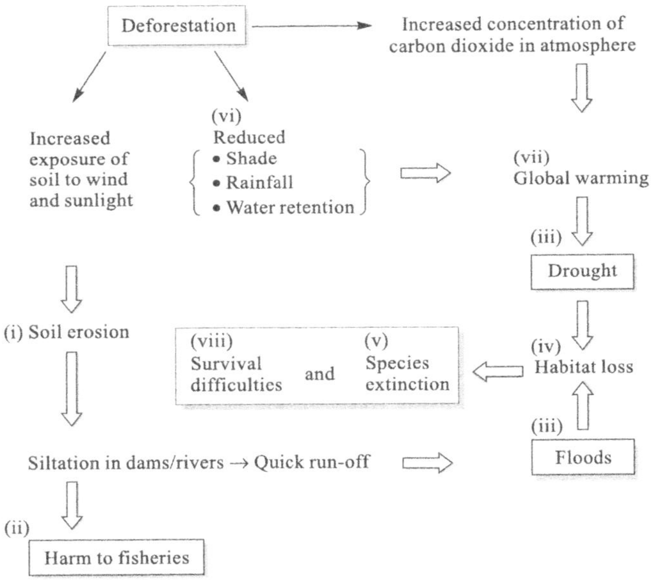


**Fig. 2.3** *Effects of deforestation*

***(C) Causes of Deforestation in India*** The deforestation in India is rooted in the commercially oriented forest use and ownership policies of the British government which continued even after India gained independence. (The other cause was the expansion of agriculture.) Immediately after independence, the other major causes of deforestation were

(i) State-sponsored agricultural expansion
(ii) Rapid industrialisation
(iii) Urbanisation
(iv) Growing consumerism
(v) Policies and programmes of unsustainable development like subsidies offered for making the paper and plywood industry a viable and profitable venture

(vi) Lack of education and awareness programmes regarding (a) real cost of the destruction of forests, and (b) legal provisions for the safeguarding of the forests
(vii) Absence of strict implementation of laws
(viii) Not including people at all levels in planning, decision making and implementation (i.e. absence of social engineering)
(ix) Not taking the correct decisions by decision makers on the basis of accurate knowledge and information
(x) The campaign to safeguard forests is not accompanied with social, economic and political reforms
(xi) Corruption of government institutions
(xii) Population growth and overpopulation
(xiii) The inequitable distribution of wealth and power
(xiv) Other causes: shifting cultivation, dams, weather, fires, etc.

***(D) Problems Created by Deforestation in India*** The following problems are created by deforestation in India:

(i) Decreasing levels of rainfall and rainy days
(ii) Increasing rate of soil erosion
(iii) Climate change
(iv) Loss of biodiversity
(v) Air pollution
(vi) Decline in watershed functions
(vii) Apparent loss of hardwood, fuel wood, and aesthetic stocks
(viii) Flooding
(ix) Desertification and sedimentation in rivers
(x) Long-term hydroelectric shortages

***Annual deforestation rate*** is calculated by using the formula (Dirzo, 1992; Vina, 1999; Ochoa-Gaona and Gonzales-Espinosa, 2000):

$$r = \left\{1 - \left(1 - \frac{A_B - A_E}{A_B}\right)^{1/t}\right\} \times 100$$

where $r$ = Annual deforestation rate (%)

$t$ = Number of years for the given period

$A_B$ = Area of forest at the beginning of the period

$A_E$ = Area of forest at the end of the period

- The deforestation trend is assessed by means of Indian topographical sheets and satellite images.
- *The socio-economic drivers of deforestation:* Population density, education, and infrastructure like road creation are socioeconomic drivers of deforestation. This can be confirmed by statistical analysis.

**Example 2** *Suppose for a particular area,*

$A_B = 29458.83$ ha; $A_E = 21397.96$ ha; $t = 10$ years

*Find annual deforestation rate.*

**Solution**

$$r = \left\{1 - \left(1 - \frac{29458.83 - 21397.96}{29458.83}\right)^{1/10}\right\} \times 100 = 0.315 \times 100$$

Thus, ***deforestation rate = 3.15 % per year.***

***(E) Economic Impacts of Deforestation*** As per BBC News, May 29, 2008; by 2050, damage to forests and other aspects of nature could (a) halve living standards for the world's poor, and (b) reduce global GDP by about 7%.

In developing countries, almost 3 billion people rely on forest wood for cooking and heating. People in developed countries continue to use timber for building houses and wood pulp for paper. In both developed and developing countries the forest products industry is a large part of the economy. By conversion of forest to agriculture, or over-exploitation of wood products, people get short-term economic gains. However, it simultaneously results in loss of long-term biological productivity, and hence a reduction in nature's services.

### *(F) Environmental Impacts of Deforestation*

**(i) Atmosphere** Less $CO_2$ taken in, burned trees add even more $CO_2$, which traps heat, causes more evaporation, and this leads to more precipitation. More sunlight reaches surface, less photosynthesis, increased risk of fire. This is responsible for global warming, which in turn causes deforestation (a).

**(ii) Hydrosphere** Run-off increases, turbidity increases. More sediment at mouth of rivers, more flash floods.

Increase of water temperature near river banks results in less availability of oxygen in water ways. This is responsible for degradation of aquatic habitat (b) & (d).

**(iii) Geosphere** Increased erosion from water and wind, top soil carried away, loss of minerals (C, N, etc.), soil depleted quicker, less wood for construction, fuel and other products (b).

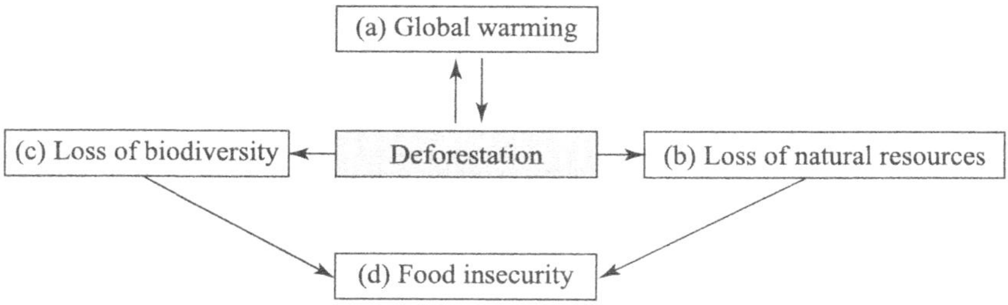


**Fig. 2.4** *Problems created by deforestation*

**(iv) Biosphere** Loss of vegetation, change of food supply, decreased habitat, decrease in number of species, decrease of diversity, decrease of pollinators and seed dispersers, loss of human cultural diversity (c) & (d).

***(G) Estimation of Socio-economic and Environmental Impacts of Deforestation*** The assessment procedure consist of

(i) Description of the environmental state before the deforestation, representing the baseline for assessment; refer (A).
(ii) Identification of the impacts of the deforestation on the environment; refer (B).
(iii) Qualitative environmental assessment and classification of the effects on the environment; refer (C).
(iv) Economic valuation of the environmental impacts, refer (D).
(v) Review for avoiding overlap with other sectors, refer (E).
(A) Goods and services provided by forest ecosystems must be clearly tabulated.
(B) Impact identification requires knowledge of *cause–effect relationship.* For example, More deforestation means more sunlight reaches the surface, less photosynthesis, more $CO_2$ in atmosphere, global warming and increased risk of fire (Fig. 2.4).
(C) To assess the effects of deforestation on natural capital (i.e. goods and services), one can begin by separating its components: physical medium (air, water, soil, climate), biotic medium (flora, fauna, human beings); perceptual medium (landscape, cultural and scientific resources) and interactions among the above-mentioned media.
(D) Environmental impact can be direct or indirect. Direct damage to the environment can be estimated as the value of the assets effected.
   - For permanent destruction, direct damage can be considered as the commercial value of the assets when a market exists for them. For example, if agricultural land is completely destroyed, the direct damage will be the value of the land.
   - When afforestation is possible, the direct damage can be approximated by estimating the cost of afforestation.
   - When a value cannot be assigned to assets for the estimation of direct damage, estimates must be made by indirect means.

(E) Review the damage estimation for avoiding counting many and varied cases of damage twice under the different social or economic sectors.

***Notes:***

(1) The above procedure is also applicable for estimating the socio-economic and environmental effects of disasters.
(2) The above methodology takes into account several major constraints, such as:
   (i) too little time available for carrying out the assessment,
   (ii) the lack of information on affected ecosystems, and
   (iii) the shortage of markets for most environmental services.

### (H) The Measures Taken for Conserving Forest Wealth

(i) ***Sustainable Forest Management (SFM)*** SFM is the use of the world's forests in such a way that they continue to provide resources in the present, without depriving future generations of their use.

(ii) ***Forest Certification*** Be responsible consumers. Buy wood only from companies that follow sustainable practices.

(iii) ***Involve Local Communities in Joint Forest Management (JFM)*** As local communities want to continue to get the benefits they previously enjoyed, they provide labour and help in conserving biodiversity. The Government should provide them *extractive reserves*. These are protected forests in which local communities are allowed to harvest fruits, nuts, medicines, fibres, rubber, etc., in ways that do not harm the forest.

(iv) ***Improve Governance and Accountability*** The Government must take bold political decisions and develop new civil society institutions to improve governance and accountability regarding forest use. Stop harmful subsidies to timber companies.

(v) ***Accelerate Education, Research and Training*** This is to ensure that SFM and JFM can quickly become a reality.

## 2.3.3 Timber Extraction

*Timber* is a term used to describe clusters of trees. It is also used to describe wood throughout its processing from the time it is cut down to the time it is used as a structural material. It is a durable wood of high quality used for making sports goods, doors, window frames, crates, plywood sheets, household utensils, coffins, furniture and other items.

*Timber extraction* is the removal of timber from forests. It requires various cutting, felling and hauling practices.

*Logging* is the work or business of felling and trimming trees and transporting the logs to a mill.

***(A) Classification of Timber Extraction Methods*** Usually, the following types of timber-extraction methods are used:

**(i) Clear Felling** It is a controversial logging practice in which most or all trees in a harvest are cut down. In clear felling, the aim is to create an even-aged group of trees with commercial species dominating by removal of noncommercial trees by cutting. Thus, clear felling means complete destruction of the native forest. All over the world, industrial timber logging is being done by clear felling.

**(ii) Handlogging** It involves timber felling by hand-held chain saws. The transport of logs from felling sites to log landing sites is also manual. It is practiced in those forests that are either seasonally flooded or permanently waterlogged (e.g. peat swamp forests). Many local people also use this method for clearing of forests for agriculture.

**(iii) Reduced–impact Logging** Reduced-impact logging means timber extraction is done selectively, carefully and in a planned manner. Reduced-impact logging practices are based on the Food and Agriculture Organization (FAO) guidelines. They help in minimising environmental damages through the selection of site-sensitive harvesting techniques.

The Forest Stewardship Council (FSC) is an international body that provides a framework for organisations certifying sustainable logging operations. The FSC criteria also help policymakers to define national standards of logging.

**(iv) Selective Logging** In *selective logging*, only few large individual trees of a few commercially marketable species are harvested. The other trees are left untouched till the next harvesting so that forests are designated to remain production forests. Per hectare, about 3 to 10 of the tallest trees are targeted in selective logging. However, the damage done to the total forest areas may be as high as 50% because of lack of planning, need to create access routes, dragging cut trees on the forest floor, etc.

**(v) Mechanized Logging** In it, heavy machinery is used for lifting, pulling and transporting cut-trees operations. Sometimes logs are transported by trucks from felling sites to log-landing sites. For this, roads are built to provide a network of access routes,

**Table 2.3** *Timber-extraction methods*

| C ↓ | H ↓ | RI ↓ | S ↓ | M ↓ |
|---|---|---|---|---|
| Clear Felling | Hand Logging | Reduced-Impact Logging | Selective Logging | Mechanised Logging |

## *(B) Impacts of Timber Extraction*

**(i) Deforestation** Logging roads are used by landless farmers to gain access to rainforest areas. They clear the forest by slashing and burning to grow enough food so as to keep them and their families alive. This '*subsistence farming*' and commercial logging is ranked as the biggest agent of tropical *deforestation.*

*(a)* ***Forest Degradation*** Poor logging practices results in a degraded forest.

*(b)* ***Forest Fires*** Timber extraction produces excess organic debris which makes a forest more vulnerable to destruction in the event of fire.

**(ii) Atmosphere** Trees act as sinks for carbon dioxide. In the global carbon cycle, forests provide the largest carbon-storage mechanisms. When the forests are destroyed by timber extraction, the carbon-storage capacity is lost. Moreover, through decay and burning, additional carbon is released into the ***atmosphere.***

| Impacts of Timber Extraction on | | | | |
|---|---|---|---|---|
| D ↓ | A ↓ | N ↓ | C ↓ | E ↓ |
| **D**eforestation | **A**tmosphere | **N**ature | **C**limate change | **E**rosion |

**(iii) Harm to Nature** Timber extraction results in forest fragmentation. It promotes loss of biodiversity because some species of animals and plants require large continuous areas of similar habitat to survive.

**(iv) Climate Change** Removal of forest cover leads to increased spring run-off coupled with summer drought, soil erosion and landslips. This also results in hotter summers and cooler winters.

**(v) Soil Erosion and Siltation** The tracks made by heavy machinery and the clearings left behind by loggers are sites of extreme soil disturbance. From these sites, heavy rain causes soil erosion. This causes siltation of the streams, rivers and the forests. As a result, the lives of indigenous people, life-support systems and the habitat of hundreds of birds and animals gets disrupted.

## Case Studies

### (i) Chipko Movement

'*Chipko*' in Hindi means hugging or embracing. Contractors used to make huge profits, from the feeling of trees in the hills. The Chipko movement was the hill communities' response to the unfair and destructive nature of this contract system. The Chipko movement spread through India during the 1970's.

As per the folk-poet Ghansyam Returi, Chipko movement is

"Embrace the trees in the forests and save them from being felled! Save the treasure of the mountains from being looted away from us!"

The slogan of the Chipko movement was

"What do the forests bear? Soil, water, and pure air!"

The Chipko movement ensured that the contract system was abolished and the indiscriminate felling of trees stopped. The Forest Development Corporation (FDC) department was formed which works for the welfare of hilly areas and the people living there. It enlightened the people about the necessity of ecological balance in the nature.

The Chipko movement took place under the leadership of Sunder Lal Bahuguna (an environmentalist and journalist) and Chandi Prasad Bhatt in Tehri Garhwal (Uttarakhand). Sunderlal along with his wife, Vimla, has given his time and talent freely to work for the good of India. He has been the catalyst of change, encouraging thousands of people to work without pay for the good of India's people and ecology, through non-violent resistance. As a Gandhian peace worker, they do not resort to violence to achieve the change.

Chandi Prasad Bhatt encouraged the development of local industries based on the conservation and sustainable use of forest weather for local benefit.

Forests in Uttarakhand district covered more than 81% of its geographical area in 1950. The Government initiated the process of development by allowing a pulp and paper mill, a plywood factory and a chain of hydroelectric dams on rivers. Over-exploitation of forest resources by these industries and submergence of huge forests and agricultural areas by dams resulted in shrinking of the forests to nearly 25% of the district's area by 1980. The poor, local population was forced to displace. The conversion of the natural mixed forests into eucalyptus and teak plantations dried up the water resources, directly affecting forest dwellers, and resulted in poverty instead of intended development.

The Chipko protests in Uttar Pradesh achieved a major victory in 1980 with a 15-year ban on green felling in the Himalayan forests of UP by the order of the then Prime Minister of India (Mrs Indira Gandhi). Since then, the movement has spread to many states in India. The movement has also helped in stopping deforestation in the Western Ghats and the Vindhyas.

### (ii) Appiko Movement

In Karnataka, the Chipko movement is known as Appiko movement, because in Kannada, the local term for 'hugging' is *appiko*. The main objectives of the Appiko movements are:

(a) *Ulisu* (to conserve),
(b) *Belesu* (to grow), and
(c) *Balasu* (rational use).

The importance of the Appiko movement can be understood from the fact that it is trying to evolve a sustainable development strategy for conservation and improvement of forest resources.

In September 1983, men, women and children of Salkani "hugged the trees" in Kalase forest and gave birth to a new awareness all over southern India through this Appiko movement. It uses various techniques to raise awareness like foot marches in the interior forests, street plays, folk dances, etc. As a result of the Appiko movement, the state government has banned felling of green trees in the same forest areas. Only dry, dying and dead trees are felled to meet local requirements.

The Appiko movement is also promoting afforestation on denuded lands. It is also active in promoting rational use of the ecosphere.

### (iii) The Bishnois

Jambhoji, a resident of a village near Jodhpur, had a vision in the fifteenth century that the people's interference with nature like felling of trees, killing of animals would result in drought. Thereafter, he become a sanyasi and initiated the Bishnoi sect. He came to be known as Swami Jambeshwar Maharaj. He laid down tenets (including a ban on killing animals, a ban to the felling of *khejri* and other trees) for his followers.

Once the Maharaja of Jodhpur sent his men to the area around the village of Jalnadi to fell the trees as he required wood for building a new palace. When Amrita Devi (a Bishnoi rural woman) saw this, she rushed out to prevent the men and hugged the first tree, but the axe fell on her. Before dying, she uttered the now famous couplet of the Bishnois, 'A chopped head is cheaper than a felled tree'. To prevent the Maharaja's men from felling the trees, people from 83 surrounding villages rushed to the spot and by the end of the day more than 350 had lost their lives.

When the Maharaja heard about this, he was filled with regret and came to the village to personally apologise to the people. He promised them that they would never again be asked to provide timber to the ruler, no *khejri* tree would ever be cut, and hunting of animals would be banned near the Bishnoi villages. The village of Jalnadi thus came to be called *khejari*. The Bishnois have proved that human lives are a small price to pay to protect the wildlife and the forests around them.

### (iv) The Green Belt Movement

Dr Wangari Maathai started the Green Belt Movement (GBM) in 1977 as a grassroots tree planting programme to address the challenges of deforestation, soil erosion and lack of water. Now, GBM is one of the most prominent women's civil society organisations, based in Kenya. The GBM advocates for human rights, supports good governance and protects the environment through peaceful democratic change.

In 2004, Dr Wangari Maathai became the first African woman and the first environmentalist to receive the Nobel Peace Prize.

Across Africa, GBM has helped in the plantation of more than 40 million trees. As a result, soil erosion has been reduced in critical watersheds, biodiversity-rich indigenous forests have been restored and protected in thousands of acres, and lakhs of women and their families are standing up for their rights and those of their communities to live healthier and more productive lives through training in forestry, food processing, beekeeping, etc.

In the next decade, the goal of GBM is to plant one billion trees worldwide. A healthy natural world is at the heart of an equitable and peaceful society.

### (v) Social Forestry

Thimmakka and Chikkanna were residents of Hulikal village in Karnataka, India, and were childless. Frustrated from the taunts of neighbours for being infertile, they decided to raise banyan (*Ficus religiosa*) trees as their children. Every year, for many years, they planted 15 to 20 new saplings of banyan trees along a hot, dusty 4 kilometre stretch between Huikal and Kudur villages in Karnataka. Every morning, the couple set out—Thimmakka (the wife) with a pot atop her head and another on her hip and Chikkanna (the husband) with two more pots hanging from a pole he held over his shoulder. They watered the tiny plants and placed thorn guards around their little wards to protect them from grazers. They watered the trees everyday till they flourished. They refilled the pots from wells and ponds along they way as their trees required about 50 pots of water a day. They visited their plants at least once a week until the trees were 10 years old.

More than 45 years later, the banyan trees (the adopted children of Thimmakka and Chikkanna) stretch all along the 4 kilometres between Hulikal and Kudur villages. The trees are in fact a proud and memorable mark of their 'parents' dedication. They provide shade for the villagers, who often had to work on the hot and dusty road. Chikkanna died a few years later and after five years, the couple started receiving recognition. Thimmakka has received the National Citizen's and the Prime Minister's Award for Social Forestry for the strong upbringing of her many offspring.

Thimmakka proved that environmentalism is not exclusively for the wealthy and privileged, but that it should be at the centre of everyone's daily lives without cultural restrictions or economic education.

## 2.3.4 Mining

Mining is the extraction (removal) of metals and minerals from the earth.

***(A) Sustainable Mining*** Extraction and beneficiation of raw materials has to be

| | |
|---|---|
| • Environmentally compliant<br>• Socially acceptable<br>• Economic | "Sustainable" |

*A mining operation is socially acceptable if*

(i) It obeys standards in occupational safety and health
(ii) It is accepted by the society
(iii) It obeys national and international guidelines and laws
(iv) It provides resettlement help
(v) It considers cultural and social constraints
(vi) It provides suitable working conditions for the workers

*A mining operation is economic if*

(i) It fulfills the needs of the society with material and immaterial goods
(ii) It is working for the long-term maximisation of revenues and profit

*A mining operation is environmentally compliant if*

It does not impose any harm to the environment.

***(B) Mining Process and the Environment*** Some of the major environmental impacts of mining are a result of associated mining operations as summarised in Fig. 2.5.

***(C) Environmental Impacts of Mining*** Some of the major environmental impacts of mining are the following:

### (i) Ecological Impacts

*(a)* ***Degradation of Land*** Due to leaching out of toxic elements, the growth of vegetation is adversely affected. Loss of fauna and flora is also observed.

### (ii) Socio-economic Impacts

*(a)* ***Pollution of Water Resources*** Even when drainage is controlled, some leaching and release of harmful elements (e.g. Pb, Cd, etc.) into the surface and groundwater occurs. It affects the ecosystem stability adversely due to alterations in water quality and availability.

*(b)* ***Pollution of Air*** Mining processes emit dust and gases which cause air pollution. These air pollutants have adverse impacts on historical monuments and religious places.

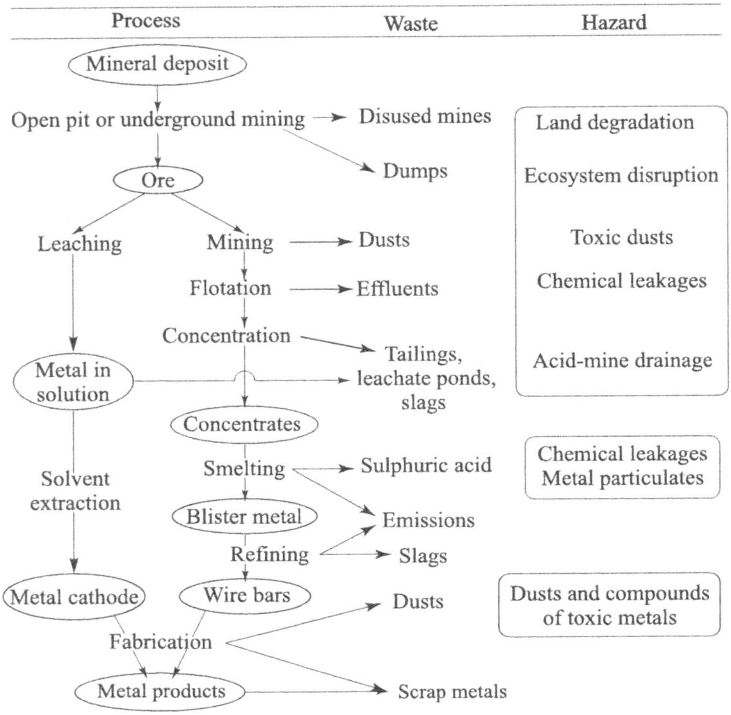


**Fig. 2.5** *Mining process and the environment*

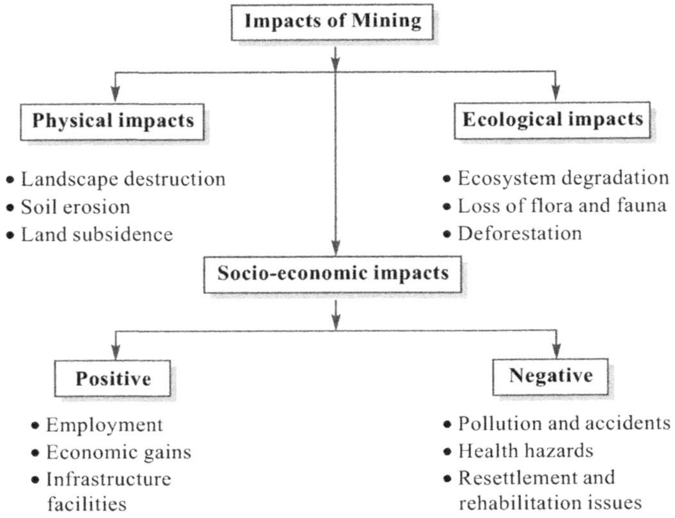


**Fig. 2.6** *Impacts of mining activities*

*(c)* ***Problems in Rehabilitation of Affected Population*** It is one of the biggest problems due to economic constraints.

To sum up,

> *Mines always cause environmental impacts*
> But ... it is usually cheaper to prevent the impacts than to deal with them after they have already happened. It is easier to prevent the impacts if one knows about them in advance.

# Case Studies

## (i) Impact of Coal Mining on Vegetation

Meghalaya, one of the seven states of North-East India, is honoured with rich natural vegetation as well as large reserves of mineral resources.

In the early 1970s, coal mining was initiated in the Jaintia Hills district of Meghalaya. Since then mining and the area affected by it is increasing day by day. The dense forest areas were converted into open forests and the considerable area of the forests was converted into a nonforest.

Extensive coal mining has led to landscape damage and damage to the biological communities in enormous ways. The number of trees and shrub species drastically decreased due to mining. The unfavourable habitat conditions prevailing in the coal-mining areas has reduced the chances of regeneration of species, thereby, reducing the number of plant species in the mined areas.

## (ii) Jhansi Open-Cast Mining Site: Uttar Pradesh, India

The Bundelkhand region occupying about 71818 $km^2$ in Uttar Pradesh is known for its rich deposits of graphite, saltpetre, sand, etc. Presently there are around 325 active mining sites in the Jhansi district alone (2010).

Mining and its allied activities significantly contributed towards infrastructure development and raising the living standards of the people.

Deforestation, dust generation, noise, air and water pollution as well as resource depletion are common hazards associated with open-cast mining widely prevalent in the Jhansi region.

## (iii) Marble Mining and Drying of Lakes in Rajasthan

The Aravalli Hills are the lifeline of Haryana, Rajasthan, and Gujarat. They control the climate and drainage systems of the region. The hills also act as a watershed for the region. The hills are also known for their rich deposits of teak, marble and granite. About 1,75,000 workers are employed in mining and related industries. About 9700 industrial units are connected with mining in Rajasthan alone.

Over the past 20 years, the forest cover has been depleted by 90% in Rajasthan due to large-scale mining. When the mines reach below the underground water level, a cone of depression is formed that sucks water from the surrounding areas, drying up wells and lakes and affecting agriculture.

The Rajasamand Lake in Rajasthan had not dried up for at least 300 years. However, in 2001, this finally did happen because of a decade of marble mining in the Rajnagar area. While mining has led to the depletion of water, mining waste has destroyed fertile land.

### (iv) Impact of Mining Activities on Labourers

Studies on mining activities in the Aravallis have shown that

(a) Labourers are not provided with any health care
(b) Tuberculosis, silicosis and other lung diseases are very common
(c) Diseases make labourers invalid and even kill them by the age of 40
(d) The condition of women workers (30–40%) and child labourers (10–15%) is the worst.

### (v) Impact of Mining on the Culture and Lifestyle of People

From the gold and copper project, 1000 $m^3$ of conc. cyanide were released into a river in 1984 on the OK Tedi Island in new Guinea. It caused extensive environmental damage, devastated the ecosystem and destroyed the culture and lifestyle of the native Wopkaimin people.

## 2.3.5 Dams and their Effects on Forest and Tribal People

A *dam* is a huge and giant barrier constructed across a river to obstruct its natural flow. Consequently, an enormously large artificial lake is created to store water. The water thus stored is utilised for multipurpose services such as power generation, irrigation, flood and drought control, etc.

Construction of dams in countries like India and China displace a large number of people because of the high population densities of these countries. In India, dams account for 75–80% of displacement of about 4–5 crore people. Out of the total people displaced, only 25% have been rehabilitated so far. Tribal people are economically, socially and politically the weakest and the most deprived community in India.

***Some of the problems associated with effects of dams on tribal people are listed below:***

*(i)* ***No Human Rights*** Human rights violations create unrest among tribals.

*(ii)* ***No Basic Amenities*** They are forced to migrate to urban slums in search of employment. They become landless labourers in rural areas. A majority of tribal people end up with less income than before, less resources of the common people, inferior houses, etc. They are forced to live without drinking water, sanitation, health care, and other basic amenities.

*(iii)* ***No Benefit Sharing*** They hardly get to share the benefits of development projects that cause their displacement.

*(iv)* ***No Home*** Tribal people have been forced to leave their ancestral homes and go elsewhere.

*(v)* ***No Cultural Identity*** Tribal communities get dispersed, traditional support systems get broken and cultural identity gets devaluated because of dams.

**Table 2.4** *Problems associated with effects of dams on tribal people: Loss of*

| R | A | S | H | I |
|---|---|---|---|---|
| ↓ | ↓ | ↓ | ↓ | ↓ |
| Rights | Amenities | Sharing | Home | Identity |

All major dams are constructed in mountainous regions, where there is plenty of rainfall. These places are covered with rich vegetation and forests.

***The major effects of dams on forests are summarised below:***

**(i)** The forests area which is supposed to get submerged is cleared off by the contractors.

**(ii)** The forest is also cleared for approach roads, offices, residences and for storage of construction material.

**(iii)** As more and more workers occupy the dam sites, forests are destroyed for getting fuel and timber.

**(iv)** ***Irrecoverable Loss to Ecosystems and Biodiversity*** Forest fragmentation causes serious irrecoverable loss of species and ecosystems. This is because some species of animals and plants require large continuous areas of similar habitat to survive.

## 2.4 WATER RESOURCES

*Water resources* are sources of water that are useful or potentially useful to humans.

Water is a prerequisite for the existence of life. Plants, animals, and human beings cannot survive without water. Water is used in agricultural, household, industrial, recreational and environmental activities. Water is essential for economic growth, environmental stability, biodiversity conservation, food security and health care.

***(A) The Water Cycle*** It describes the continuous movement of water above and below the surface of the earth. It is driven by the sun.

The sun heats water in seas and oceans. Water evaporates into the air as water vapour. Snow and ice can sublime directly into water vapour. Rising air currents take the water vapours into the atmosphere where cooler temperatures help them to condense into clouds. Air currents move clouds; they collide, grow, and fall out of the sky as precipitation. Some precipitation falls as snow, and can accumulate as ice caps and glaciers . Most water falls back into the oceans or onto land as rain where the water flows over the ground as surface run-off. Much of the run-off is soaked into the ground as infiltration. Some run-off is stored as fresh water in lakes. Some run-off enters rivers in valleys in the landscape. Some water infiltrates deep into the ground and replenishes aquifers. This helps in the long-term storage of freshwater. Some groundwater finds openings in the surface of land and freshwater springs come out. Some rainwater flows through rivers back into the ocean, where the water cycle begins again.

**Table 2.5** *Water cycle*

| *Significance of Water Cycle* | *Problems Arising from the Disturbances to the Water Cycle* |
|---|---|
| (i) The water cycle helps in the maintenance of life and ecosystems on the earth. | (i) Maintenance of life and ecosystems on earth get disturbed. |

(*Contd.*)

(*Contd.*)

| | |
|---|---|
| (ii) The water cycle helps in the transport of minerals from one part to different parts of the globe. | (ii) Mineral transport to different parts of the globe gets disturbed. |
| (iii) The water cycle purifies water by transferring water from one reservoir to another. | (iii) Water purification process gets disturbed. |
| (iv) The water cycle helps in the replenishing of the land with freshwater. | (iv) Replenishing of the land with fresh water gets disturbed. |
| (v) Processes such as erosion and sedimentation associated with the water cycle helps in reshaping the geological features of the earth. | (v) Processes for reshaping the geological features of the earth get disturbed. |
| (vi) Through the evaporation and condensation process, the water cycle helps in the cooling and warming of the environmental respectively. | (vi) Influence on climate gets disturbed. |

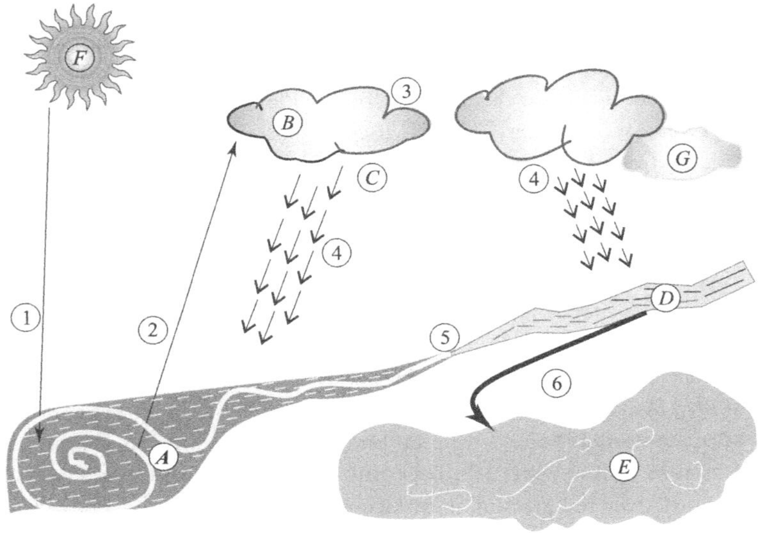


1. The sun heats the ocean.
2. Ocean water evaporates and rises into the air.
3. The water vapour cools and condenses to become droplets, which form clouds
4. Eventually the drops become heavy enough to fall to the ground as rain and snow
5. Some rainwater flows through rivers back into the ocean, rest rain collects in groundwells

A: Ocean
B: Cloud
C: Rain
D: River
E: Groundwater storage
F: Sun
G: Water storage in snow and ice

(2) **Evaporation:** The transformation of water from liquid to gaseous phase.

(2′) **Sublimation:** The state change directly from snow or ice to water vapour.

(3) **Condensation:** The transformation of water vapours to liquid water droplets creating fog and clouds.

(4) **Precipitation:** Condensed water vapour that falls to the earth's surface as rain.

(5) **Surface run-off:** The way by which water moves across the land.

(6) **Percolation:** Infiltration of surface water for groundwater storage.

**Fig. 2.7** *The water cycle*

**(B) Sources of Water** 97.5% of water on the earth is salt water in oceans. Only 2.5% is fresh water. Sources of fresh water are briefly described below:

*(i)* ***Surface Water*** Water in a lake, river or freshwater wetland is known as surface water.

*(ii)* ***Groundwater*** Fresh water located in the pore space of soil and rocks is called groundwater.

*(iii)* ***Ice Caps and Glaciers*** Fresh water from ice caps and glaciers is relatively inaccessible.

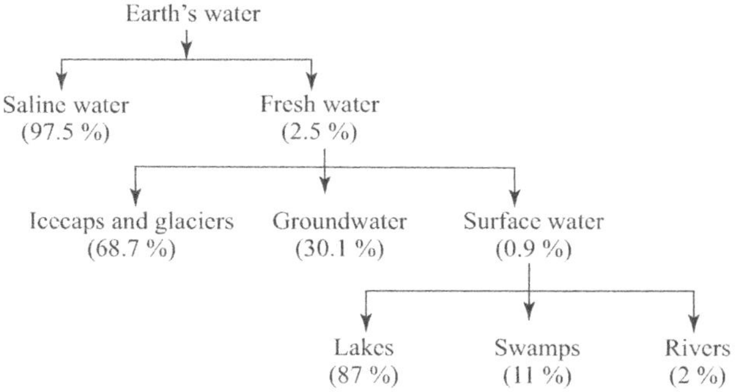


**Fig. 2.8** *Distribution of the earth's water*

**(C) Availability of Water** The total water in the world is estimated to be $1400 \times 10^6$ km$^3$. Unfortunately, 97.5% of this water is found in the oceans and is too salty to drink. Of the remaining 2.5% fresh water, 2% is locked up in relatively inaccessible ice caps and glaciers, and 0.5% is groundwater and most of it lies too far underground.

About $2 \times 10^5$ km$^3$ of freshwater is found in lakes and rivers and $14 \times 10^3$ km$^3$ of freshwater is found in the atmosphere. The available freshwater is distributed regionally as shown in Table 2.6.

The global overview of water availability versus the population (Table 2.6) stresses the continental disparities, and in particular the pressure put on the Asian continent. This is because Asia supports 60% of the world's population with only 36 per cent of the world's water resources.

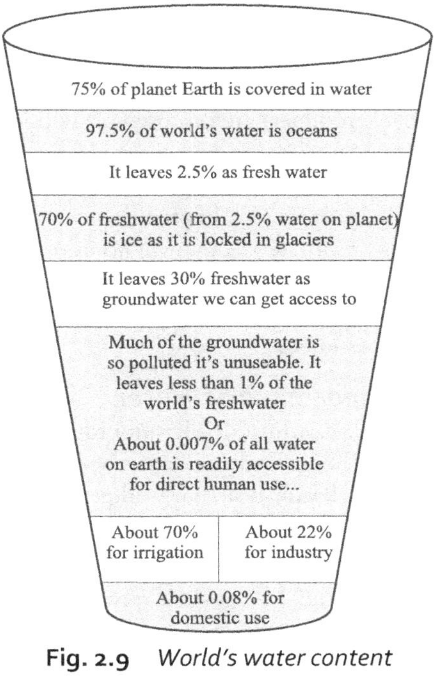


**Fig. 2.9** *World's water content*

**Table 2.6** *Water availability versus population*

| | *Continent* | *Population* | *Freshwater Availability* |
|---|---|---|---|
| (i) | North and Central America | 8% | 15% |
| (ii) | South America | 6% | 26% |
| (iii) | Europe | 13% | 8% |
| (iv) | Africa | 13% | 11% |
| (v) | Australia and Oceania | < 1% | 5% |
| (vi) | Asia | 60% | 36% |

**Availability of Water in India** India is the wettest country in the world, but rainfall is highly uneven with space and time. Rainfall is high in the North-East but extremely low in Rajasthan. Out of 4000 billion $cm^3$ rainfall received, about 600 billion $cm^3$ is put to use so far.

With 16% of the world's population, India has only 4% of global water resources.

The city of the Delhi gets 60 hours of rain a year, and only 11 hours of it are contained while the rest is wasted. Every monsoon we see flooded underpasses and buses floating by.

***(D) Causes of Water Crisis in the World*** The causes for shortage of water leading to water crisis are the following:

(i) *Growing population and with better lifestyles*, per capita use of fresh water is increasing, causing shortage of water.

(ii) *Spatial and temporal variations in available* water is also responsible for water crisis.

(iii) *Freshwater resources are reduced by pollution*. Industrial wastes, chemicals, human waste and agricultural wastes (fertilisers, pesticides and pesticide residues) are disposed off within water.

(iv) Increase in extreme weather conditions like *floods, droughts, typhoons, cyclones, etc.*, are also responsible for worsening of water quality and availability.

Recently, it is estimated that

- Climate change will account for about 20% of the increase in global water scarcity
- 50% of the population of developing countries are exposed to polluted water sources

***(E) Importance of Water*** Next to air, water is the most essential thing for our survival. We must drink water to avoid dehydration which means less or insufficient levels of water and important body salts of sodium and potassium in our body. The kidneys, brain, heart and other important body organs cannot function property without salt and water.

Water is also helpful in maintaining the relatively constant body temperature through the homeostasis process. It helps in avoiding upsetting of metabolic reactions by preventing sudden changes in temperature.

Water helps in the digestion process. Different types of food products, after being broken down to simple molecules (e.g. large starch molecules are broken down to simple sugars) are solubilised in the universal solvent 'water'. Different enzymes facilitate this digestion process.

Oxygen gas is also dissolved in water to some extent. This Dissolved Oxygen (DO) helps in the respiration process of many organisms who live in water and spend most of their time underwater.

*"Life is impossible without water. It is needed for health, ecosystem services, economic development, poverty reduction, protection of greenery, production of food and imparting of aesthetic beauty."*

### (F) Impacts of Over-utilisation of Underground and Surface Water

The over-utilisation of underground and surface water has the potential to alter, sometimes irreversibly, the integrity of freshwater ecosystems. Some of the major impacts are summarised below:

(i) ***Loss of Integrity of Freshwater Ecosystems*** Human activities for infrastructure development like creation of dams, land conversion, etc., are responsible for this loss of integrity of freshwater ecosystems. Water quality and quantity, fisheries, habitats, etc., are at risk due to this loss of integrity.

(ii) ***Risk to Ecosystem Functions*** Population and consumption growth increases water abstraction and acquisition of cultivated land. Virtually all ecosystem functions including habitat, production and regulation functions are at risk.

(iii) ***Depletion of Living Resources and Biodiversity*** Overharvesting and exploitation causes groundwater depletion, collapse of fisheries. Production of food, quality and quantity of water and supply of water gets badly affected by these depletions of living resources and biodiversity.

(iv) ***Pollution of Water Bodies*** Release of pollutants to land, air or water alters chemistry and ecology of water bodies. Greenhouse gas emissions produce significant changes in run-off and rainfall patterns. Because of water pollution, water supply, habitat, water quality, food production, climate change, etc., are at risk.

## 2.4.1 Uses and Overuses of Water Resources

***(A) Uses of Good-Quality Water*** Good-quality water is needed for all direct or indirect uses of water as illustrated in Fig. 2.10.

According to the Union Ministry of Water Resources (MoWR), 80 per cent of India's utilisable water is devoted to agriculture, mostly for irrigation. Demand from the domestic sector is about 5% of the annual freshwater withdrawals in India. The industrial sector in India is the second largest user of water.

As a result of growing sectoral demands and declining water supplies, competition is growing rapidly.

From public systems, allocation of water is not based on fundamental rights consideration or social, economic or environmental considerations, Thus, water allocation is inequitable. Growing inequity in access to and control over water leads to conflicts among different sectors.

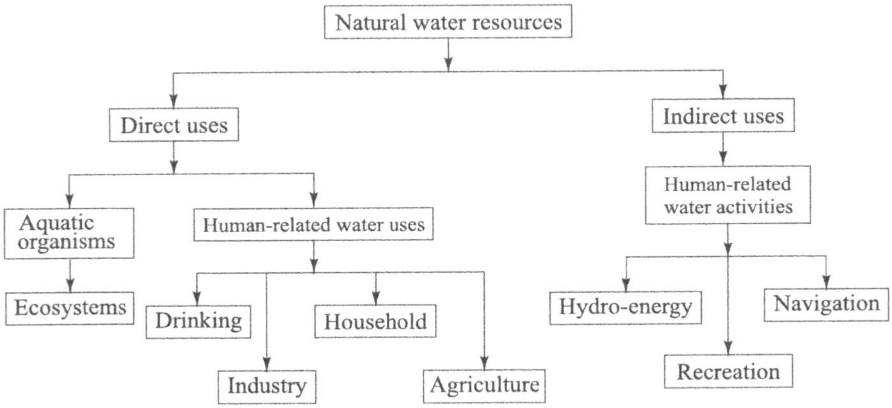


**Fig. 2.10** *Direct and indirect uses of water resources by humans and ecosystems*

The poor need water for both domestic as well as productive purposes which include growing food, fruit, vegetables, rear livestock, etc. The rich residents of cities consume around 200 litres per capita per day. It is believed that finding ways of providing similar quantity of water in support of the livelihoods of the rural poor is vital.

India's supply of water is rapidly decreasing mainly due to mismanagement of water resources, although over-extraction and pollution are also significant contributors.

As per Central Water Commission (CWC) assessment, 2011; water availability in India is 1869 billion cubic metres or 1869 $km^3$.

***(B) Sectoral Demand of Water*** As per the ministry of Water Resources (MoWR) assessment, 2000; water requirements for various sectors in India is tabulated here.

**Table 2.7** *Water requirement for various sectors*

| *Sector* | *Water demand in Billion Cubic Metres* | |
|---|---|---|
| | *Year 2010* | *Year 2025* |
| Irrigation | 688 | 910 |
| Drinking water | 56 | 73 |
| Industry | 12 | 23 |
| Energy | 5 | 15 |
| Others | 52 | 72 |
| Total | 813 | 1093 |

*Over-exploitations of water resources and the problem of availability of safe drinking water:*

**(i) Excessive Withdrawal from Surface Waters** Size of the sea is shrinking (e.g. the Aral sea in the former Soviet Union) primarily by the diversion of the inflowing rivers to irrigate water-intensive cotton and rice crops.

In 2007, about 60% of the Aral Sea's volume had been lost, its depth had declined by 14 metres. Moreover, its salt concentration had doubled.

**(ii) Inefficient Use of Freshwater** Excessive consumption by individuals, leakage in water delivery systems, inefficient use by industry and poor irrigation practices can all contribute to situations where there is not enough water for all uses.

**(iii) Excessive Withdrawal of Water from Underground Aquifers** Excessive freshwater abstraction along much of the west coast of India has allowed sea water to enter aquifers. It resulted in making the water saline and unfit for human use. The above problem has worsened due to leaching of excess irrigation water containing pesticides, fertilisers, etc., into these aquifers.

## *(C) Water Conservation*

"Water conservation is the most cost-effective and eco-friendly way to reduce our demand for water."

**(i) Need for Water Conservation** On an average, a citizen in most parts of the world is allocated 2.5 gallons of water per day for sustainability. However, the average American citizen uses 80–100 gallons of water per day. The poor do not have access to safe drinking water. More than 4000 children are dying every day as a result of diarrhoeal diseases caused from unsafe drinking water, lack of access to sanitation and inadequate availability of water. Thus, it is very essential to conserve water.

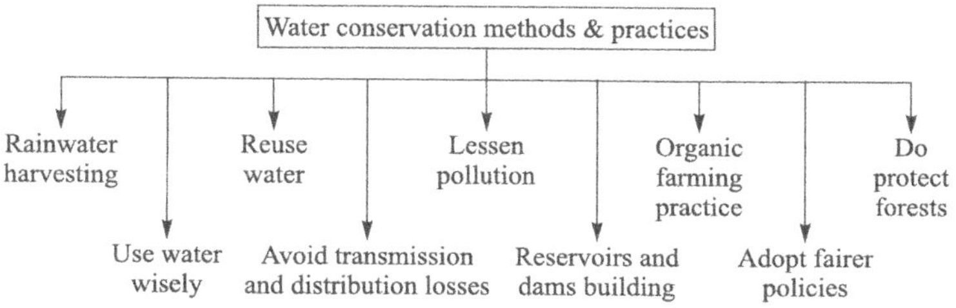


**Fig. 2.11** *Water conservation*

**(i) Measures to Conserve Water**

(a) Recharge groundwater by harvesting rainwater.
(b) Use water wisely for household, agricultural and domestic purposes.
(c) Reuse water whenever possible. For example, waste water after bath can be used for the toilet.
(d) Avoid transmission and distribution losses by checking leaks in pipes, hoses, etc.
(e) Prevent flow of untreated sewage to lakes and rivers. This will reduce the likelihood of water pollution and help in water conservation.
(f) Collect water by building dams and reservoirs, and digging wells.
(g) Use drip irrigation, precision sprinklers for agriculture. Practice organic farming.
(h) Adopt *fairer policies* for treatment, access and pricing of water.
(i) Prevent flow of industrial effluents to natural water resources to avoid water pollution.
(j) Do protect forests to protect rivers, lakes, wells and other sources of water.

## *(D) Quality Aspects*

**(i) Impacts of Poor Water Quality** Poor water qualities are responsible for

(a) public health hazards,

(b) damage to ecosystems, and
(c) adverse economic consequences.

A good status of water biology and healthy aquatic organisms are necessary for obtaining a good status of water quality.

### (ii) Characteristics of a Good-Quality Water

(a) It is transparent, colorless and odourless.
(b) It has sufficient oxygen concentration for marine life to survive.
(c) It is free from bacteriological contamination.
(d) It is free from any water pollution.
(e) It is free from excessive nutrients like N, P, etc., which are responsible for eutrophication.
(f) It is fit for the intended use.

***(E) Self-Purification of Rivers*** A variety of plant and animal species live in seas and rivers. If pollution does not attain a critical level, water can purify itself, i.e. progressively eliminate polluting agents. The phenomena of filtration and oxidation, combined with the action of organisms (insects, bacteria, plants, etc.) living in the water and on the banks helps water maintain its quality and preserve its ecosystem's balance.

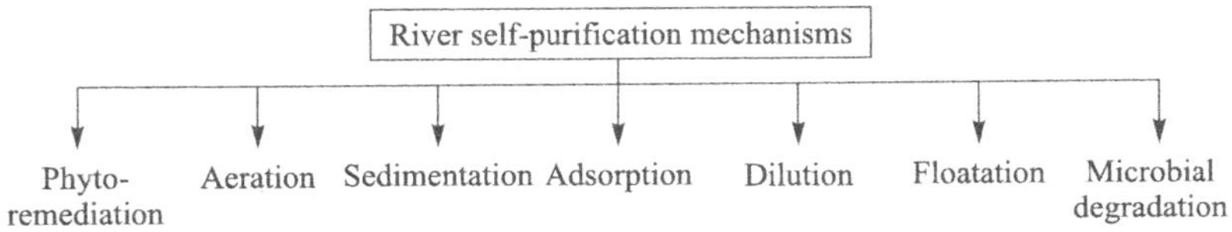


**Fig. 2.12** *River self-purification mechanisms*

Various river-self purification mechanisms are described below:

**(i)** ***Phyto-remediation*** Aquatic plants and vegetation on the river banks absorb (nitrate, phosphate and other nutrients) and remove pesticides and heavy metals from water. In this way quality of water in the river is largely improved.

**(ii)** ***Aeration*** When a river runs through hills, turbulence mixes air into water increasing the dissolved oxygen (DO). The increased DO concentration facilitates many chemical and microbiological processes in water to reduce the pollutant concentration.

**(iii)** ***Sedimentation*** In this mechanism, sand in the river bed acts as a sink for the pollutants. From the hills, when river reaches flat lands, it spreads, its velocity reduces and suspended pollutants settle on the sand bed.

**(iv)** ***Adsorption*** Pollutants are adsorbed onto sand particles, plant surfaces, rocks, etc., and thereby their concentrations get reduced in the river water.

**(v)** ***Dilution*** When a polluted river is joined by less polluted tributaries or during the rainy seasons, the volume of water in the river is increased. It reduces the concentrations of pollutants by dilution process.

**(vi)** ***Floatation*** After rapid mixing of water in falls, air bubbles act as vehicles to lift many pollutants to the water surface in the form of froth (or a layer of foam). This froth is exposed to the atmosphere, and it facilitates oxidation of pollutants to less harmful forms. The top layer is also directly exposed to sunlight; so either by increased temperature or due to various photochemical reactions, volatile organic compounds are removed from the top layer. At different sections of the river, various artificial traps help in the removal of this froth and thus rivers get self-purified.

**(vii)** ***Microbial Degradation*** The shallow and turbulent water results in high aeration of water. It helps in growth of bacteria and other microorganisms. They help in river purification by microbial degradation of pollutants.

***(F) Major Water-Quality Issues*** Wastes introduced by humans into rivers, lakes, groundwater aquifers and the oceans modify the environmental water quality and make huge quantities of water unsuitable for various uses.

### *(G) Major Factors Responsible for Water-Quality Degradation*

(i) Insufficient and incomplete treatment of domestic and industrial waste water
(ii) Eutrophication
(iii) Pathogens, and pesticide contamination
(iv) Stagnation of domestic sewage and contamination of groundwater

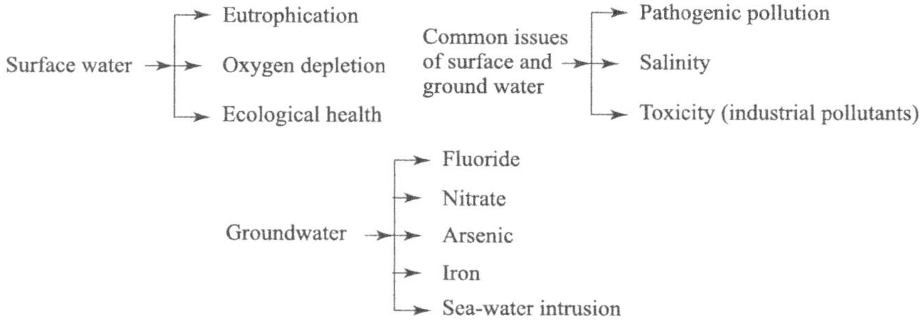


**Fig. 2.13** *Important quality issues of water*

***(H) Water-Borne Diseases*** *Water-borne diseases are illnesses caused by consuming water contaminated by pathogenic microorganisms.*

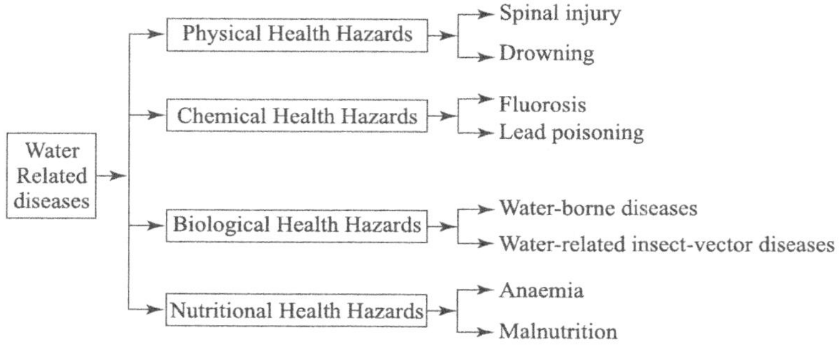


**Fig. 2.14** *Illnesses caused by consuming contaminated water*

Often lack of access to hygienic water, poor sanitation and rise in population of pathogenic microorganisms like protozoa, viruses, bacteria and intestinal parasites breeding in on water are considered the main causes of water-borne diseases.

According to the World Health Organization, diarrhoeal disease is responsible for the deaths of 1.8 million people every year and a majority of them are children in developing countries.

A few water-borne diseases are summarised in Table 2.8.

**Table 2.8** *Water-borne diseases*

| S.No. | Water-borne disease | Caused by | Symptoms |
|---|---|---|---|
| 1. | Giardiasis (Type of diarrhoea) | *Giardia intestinalis* | Severe abdominal cramps, diarrhea, nausea, greasy stool, gas, etc. |
| 2. | Amoebiasis (Type of diarrhoea) | *Entamoeba histolytica* (Protozoa) | Extreme abdominal discomfort, loose stools, bloating, weight loss, abdominal pain, etc. |
| 3. | Cryptosporidiosis (Type of diarrhoea) | *Cryptosporidium parvum* (tiny parasites) | Mild fever, weight loss, diarrhoea, vomiting, nausea |
| 4. | Cholera | *Vibro cholerae* (bacteria) | Sudden onset of acute diarrhoea, which may lead to excessive dehydration, kidney failure and finally, even death |
| 5. | Gastroenteritis or stomach flu | Noroviruses | Low grade fever, diarrhoea, frequent vomiting, dehydration, stomach or abdominal cramping |

The best ways to prevent water-borne diseases are

(i) avoid drinking untreated water,
(ii) avoid consuming undercooked food,
(iii) maintain good personal hygiene (e.g. wash hands before eating), and
(iv) educate for clean sanitation.

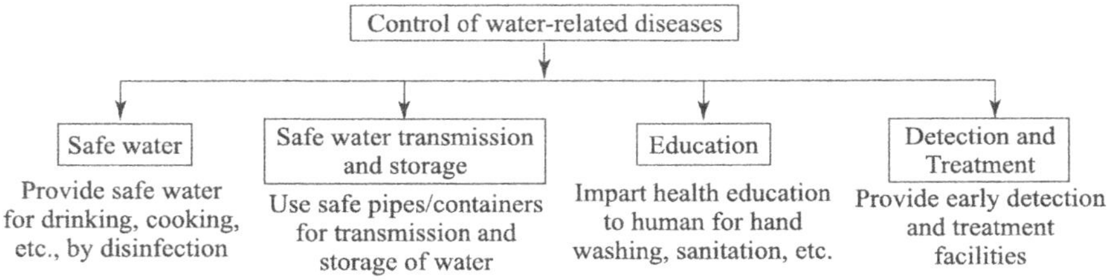


**Fig. 2.15** *Control of water-related diseases*

## *(I) Fluoride Problem in Drinking Water*

At low concentrations in drinking water, fluoride has beneficial effects on teeth. But excessive exposure to fluoride in drinking water can give rise to number of adverse effects. Although the *concentration* (mg/*litre*) of fluoride added to water can be controlled, but we cannot control the *dose* (mg/day). This is because one cannot control how much water people drink or how much fluoride they get from other sources.

### (i) Sources of Fluoride

(a) Fluoridated water supplies
(b) Food processed with fluoridated water
(c) Mouthwash enhanced with fluoride
(d) Toothpaste enhanced with fluoride
(e) Food supplements

### (ii) Fluoridation is not Necessary

(a) The level of fluoride in mother's milk is 0.004 ppm. It means a bottle-fed baby, where fluoridated tap water (with 1 ppm fluoride) is used to make up the formula milk, will get 250 times more fluoride than nature intended.
(b) Fluoride works from the outside of the tooth, not from inside the body, so it is not required to swallow fluoride or drink fluoridated water.

**(iii) Fluoride's Dangers** Fluoride damages teeth, bone, brain and endocrine system. It may cause osteosarcoma.

**Table 2.9** *Dangers of fluoride consumption*

| *Concentration of Fluoride in Drinking Water* | *Observed Effects* |
|---|---|
| (i) At 0.7 ppm | Early manifestation of skeletal fluorosis. |
| (ii) At 0.9 ppm | Lowered IQ in children with borderline iodine deficiency. |
| (iii) At 1.0 ppm | • Dental fluorosis impacts 30+% of children<br>• Increased cortical bone defects in children<br>• Osteosarcoma in young men. |
| (iv) At 1.5 ppm | Doubling of hip fracture rates. |
| (v) At 1.8 ppm | Lowering of IQ in children. |
| (vi) At 2.3 ppm | Lowering of thyroid activity |
| (vii) At 4.0 ppm | • Enamel damage with severe fluorosis (proven).<br>• Increased fractures in susceptible groups (probable).<br>• Skeletal fluorosis, stage II (possible) |

**(a)** ***Fluoride damages the teeth*** A permanent discoloration and mottling of the tooth enamel (*dental fluorosis*) is caused by a child's ingestion of fluoride (0.5–1.5 ppm) before its permanent teeth have erupted.
**(b)** ***Fluoride damages the bone*** In an area of high natural levels of fluoride (1.5–5.5 ppm), fluoride can weaken bone and increase the risk of fractures.
**(c)** ***Fluoride damages the brain*** Fluoride lowers the IQ of children, even when present at 1.8 ppm in water. It is apparent that fluorides have the ability to interfere with the functions of the brain.

**(iv) Defluoridation of Water** Defluoridation of water can be carried out by

(a) Reverse osmosis filtration
(b) Activated alumina defluoridation filter
(c) Nalgonda technique

## (J) Pesticide Removal Methods from Drinking Water

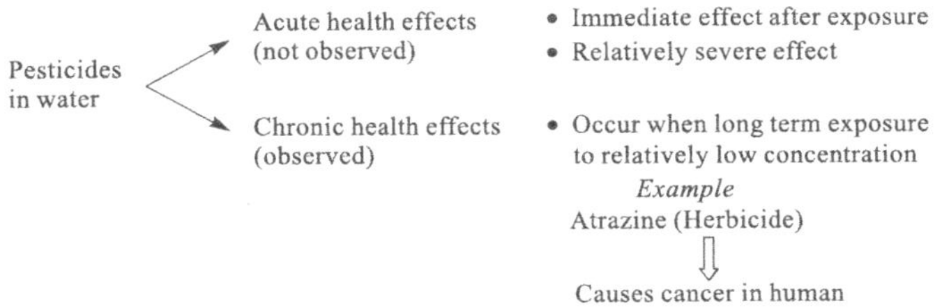


Water contaminated or suspected of being contaminated by pesticide and synthetic/volatile organic compounds can be purified using following methods shown in Fig. 2.16.

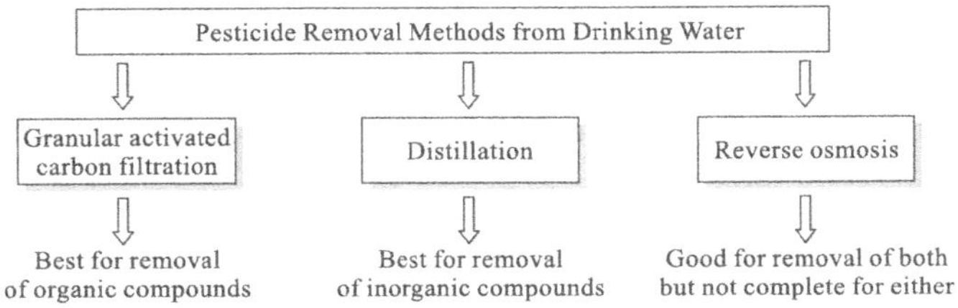


**Fig. 2.16** *Pesticide removal methods from drinking water*

## 2.4.2 Floods

Waterways are formed slowly over time, and their size is proportionate to the amount of water that normally accumulates in that area. Sometimes, due to excessive runoff from precipitation or snowmelt or by coastal storm surges or other tidal phenomenon, there is suddenly a much greater volume of water. As a result, the normal waterways overflow, and the water spreads out over the surrounding land. This anomalous accumulation of water in an area of land is called *flood.*

Flood can be defined as *a temporary rise of the water level, as in a river or lake or along a sea coast, resulting in its spilling over and out of its artificial or natural confines onto land that is normally dry* or *flood is a temporary covering by water of land not normally covered by water.*

### (A) Effects of Flood

The effects of flood are briefly described below:

**(i) Primary Effects** Flood can cause either physical damage or casualties.

*(a)* ***Physical Damage*** Flood can damage any type of structure resulting in physical damage to canals, bridges, sewerage systems, roadways, cars, buildings, etc.

*(b)* ***Casualties*** Humans and animals die due to drowning. Floods can also cause casualties through epidemics and water-borne diseases. The spawning grounds for fish and other wildlife can become polluted or completely destroyed.

**(ii) Secondary Effects** Secondary effects of floods are briefly summarised below:

*(a)* ***Contamination of Water*** Clear drinking water becomes scarce because of contamination of water due to floods.

(*b*) ***Spread of Water-borne Diseases*** Floods are responsible for unhygienic conditions leading to various diseases.

(*c*) ***Loss of Harvest*** The entire harvest can be lost due to floods leading to shortage of food crops and this badly affects the food supplies.

(*d*) ***Death of Some Nontolerant Tree Species*** Floods can lead to suffocation and death of some nontolerant tree species.

**(iii) Tertiary Effects / Long-term Effects** Floods are responsible for food shortage, leading to price increases. They are also responsible for temporary decline in tourism. Money is also, needed for rebuilding any type of structure damaged by flood. All the above effects (where economic hardship in concerned) is discussed in tertiary effects or long-term effects.

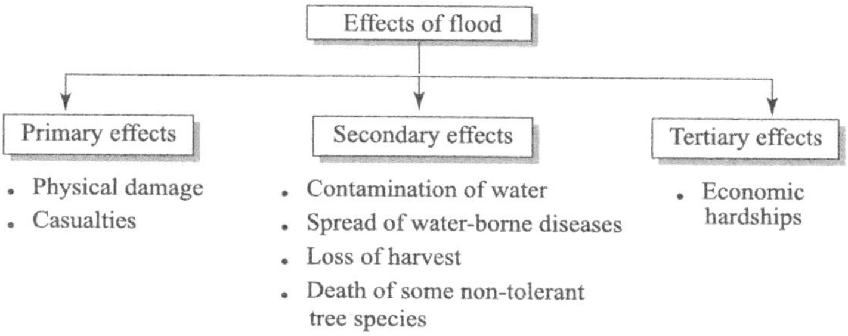


**Fig. 2.17** *Effects of flood*

***(B) Benefits of Floods*** The more frequent and smaller floods can bring following benefits:

**(i) Water Availability** Floods helps in recharging of groundwater. Flood waters provide much-needed water resources in arid and semi-arid regions where precipitation events are unevenly distributed throughout the year.

**(ii) Ecosystem Services** Specially freshwater floods play an important role in maintaining ecosystems in river corridors and in maintaining floodplain biodiversity.

**(iii) Increase in Soil fertility** Floods help in making the soil more fertile by providing nutrients to soil.

**(iv) Improved Fisheries** Flooding adds a lot of nutrients to lakes and rivers which help in improved fisheries for some years. Fish, like weather fish, make use of floods to reach new habitats.

**(v) Benefits to Birds** Birds profit from the boost in production caused by flooding.

**(vi) Higher Viability of Hydro-energy Projects** The viability for hydrological based renewable source of energy is higher in flood-prone regions.

## *(C) Flood Disaster Impact Minimisation*

(i) If it has been raining hard for several hours, or steadily raining for several days, individuals must listen carefully to the radio or TV flood forecasts issued by the Central Water Commission.

(ii) They should listen to and follow the instructions of the emergency services.

(iii) They should extend all possible help to the administrative and engineering agencies of the states/union territories to take appropriate measures.

## 2.4.3 Drought

*Drought may be defined as the deficiency of rainfall (relative to the statistical multi-year average for a region ) over an extended period of months or years.*

## *(A) Types of Drought*

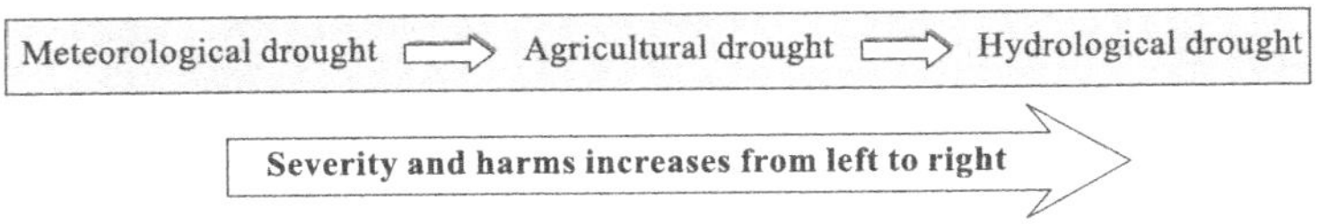


**Fig. 2.18** *Types of drought*

Drought can be classified into the following types:

**(i) Meteorological Drought** It is brought about when there is a prolonged period with less than average rainfall. As per the India Meteorological Department (IMD), meteorological drought occurs when the seasonal rainfall received over an area is less than 75% of its long-term average. The drought can be classified as 'moderate' or 'severe' depending on the rainfall deficit between 26% to 50% or exceeds 50% respectively.

Usually, meteorological drought precedes the other kinds of drought.

**(ii) Agricultural Drought** It is a drought that affects crop production or the ecology of the range. It is caused by extended period of below-average rainfall resulting in a shortfall in water for the crops. It is typically witnessed after a meteorological drought but before a hydrological draught.

**(iii) Hydrological Drought** It is brought about when the water reserves available in sources such as reservoirs, lakes or aquifers fall below the statistical average. It tends to show up more slowly because it involves stored water that is used but not replenished.

## *(B) Consequences of Drought*

The impacts of drought are briefly described below:

**(i) Impact on Agriculture** Droughts are responsible for diminished crop growth or diminished production yields due to lack of water for irrigation. Famine may also be caused.

**(ii) Impact on Environment** When drought hits an area suffering from desertification and erosion, **dust storms** and/or **dust bowls** result which further erode the landscape. Drought also affects both terrestrial and aquatic wildlife through habitat damage.

**(iii) Impact on Health** Drought is responsible for malnutrition, dehydration and related diseases. Drought can also reduce water quality, because lower water flows reduce dilution of pollutants and increase contamination of remaining water sources.

**(iv) Social Impacts** Subsistence farmers are forced to migrate during drought because they do not have alternative food sources. Mass migration results in internal displacement and international refugees. Droughts are also responsible for snake migration and increases in snakebites. In totality, a situation of social unrest arises because of drought. A war can also happen over natural resources including water and food.

**(v) Economic Impacts** Droughts lead to reduced electricity production due to reduced water flow through hydroelectric dams and insufficient available coolant for power stations. Droughts also lead to shortages of water for industrial users

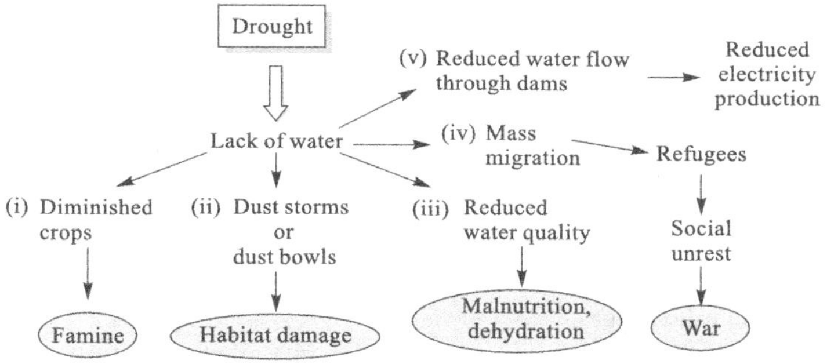


**Fig. 2.19** *Effects of drought*

***(C) Strategies for Mitigation of Drought Impacts*** Society's vulnerability to drought is minimised through the following actions:

(i) An artificial technique of *cloud seeding* helps in inducing rainfall.
(ii) For consumption or irrigation, *desalination of sea water* can be done in times of scarcity.
(iii) Carefully planned *crop rotation* can help to minimise soil erosion. This also allow farmers to plant less water-dependent crops in drier years.
(iv) Collection and storage of rainwater through *rainwater harvesting* is very useful.
(v) Regulating the use of water-intensive home maintenance tasks and the use of sprinklers, hoses, etc., on outdoor plants.
(vi) Redirecting rivers for irrigation in drought-prone areas.
(vii) Treatment and purification of sewage wastewater for *reuse*.

(viii) Continuous observation of rainfall levels and comparisons with current usage levels can help prevent man-made drought.

## 2.4.4 Conflicts Over Water

Water might be the source of the world's next big conflict. This is because fresh water availability is limited but its demand is rising day by day.

**(i) Infrastructure Failure** The cost and environmental risks due to failure of drinking and waste water infrastructure are very high, costing billions of rupees.

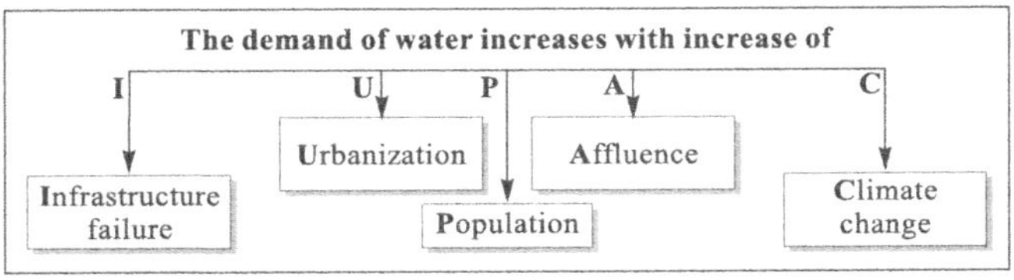


**Fig. 2.20** *Demand for water*

**(ii) Rapid Urbanisation** It requires significant investment in water infrastructure in order to deliver water to individuals and to process the wastewater so as to avoid unacceptable public health risks.

**(iii) Population Growth** The demand of water for residential and industrial purposes increases with increase in population.

**(iv) Increasing Affluence** Increasing affluence especially in India and in China, inevitably means more water consumption. Expansion of business activity requires more water supply.

**(v) Climate Change** Climate change poses a series of risks to water availability as a result of the following:

(a) Rising temperatures could increase the rate of evaporation from surface waters and reservoirs and lead to loss of freshwater held in glaciers.
(b) Increased rainfall might come in the form of storms that lead to floods.
(c) Climate change could increase annual precipitation and make more freshwater available in some places leading to difficulty in water management systems.

Where water crosses cultural, economic, political or legal boundaries, the stage is set for disputes between different users. The users try to safeguard access to a vital resource, while protecting the natural environment.

Management and transformation of water conflicts require implementation of right strategies for anticipation, addressing and mediation between competing users. If the strategies are not implemented, the conflicts over water are likely to become more frequent, more intense and more disruptive around the world.

**Water conflicts in India** can be classified as per the following themes:

Conflicts over (a) equity, access and allocations, (b) water quality, (c) dams and displacement, (d) privatisation, (e) contending water uses, (f) sand excavation and mining, (g) trans-boundary conflicts, and (h) micro-level conflicts are also present.

***Some specific examples of conflicts over water:***

**(i) Conflict between Poor and Rich** The rural peasants are poor and only require a pot full of drinking water. The urban elite is rich and uses large quantities of water for meeting the requirements of water-intensive sewage systems, space cooling, gardening, etc. This results in conflict.

**(ii) Conflicts between Agricultural Usage of Water** Conflict also exists between water-intensive cultivation of commercial crops for high cash returns and wise water use for protective irrigation of necessary food crops essential for survival.

**(iii) Interstate Conflicts** Sometimes water projects of upstream states influence the quantity and quality of water flow in the basin. This reduces the possibilities of water use by downstream states resulting in inter-state conflicts.

**(iv) Intrastate Conflicts** State-planned extraction of timber or minerals in the river catchment affect the river flow and generate conflicts downstream.

State-planned agricultural production based on large irrigation projects to generate marketable surpluses of cash crops conflicts with people's needs for local food production.

**(v) Across Borders Conflicts** India, Bangladesh and Nepal are disputing the best use of water of the Ganges–Brahmaputra basin. India and Nepal want to exploit the basin's huge hydroelectric power-generating potential, whereas Bangladesh wants the water management in such a way so as to minimise water shortages during dry months and flooding during monsoon months.

## 2.4.5 Dams—Benefits and Problems

### *(A) Positive Impacts of Dams*

#### (i) Ecological Impacts

***(a) Flood Control*** Dams help in controlling river flow and flooding.

***(b) Ecosystem Services*** Some dams help in creating new wetlands, new opportunities for fishing and recreation in the reservoirs.

#### (ii) Socio-economic Impacts

***(a) Hydroelectricity Generation*** Dams are useful in generation of electricity.

***(b) Help in Solving Problems of Hunger and Starvation*** About 16% of the world's food comes from land irrigated from dam reservoirs.

***(c) Water Supply*** Dams ensure a year-round water supply.

### *(B) Negative Impacts of Dams*

#### (i) Ecological Impacts

***(a) Seismic Tremors*** The hydraulic pressures generated by deep reservoirs is sufficient to change the seismicity of the region. Earthquakes of magnitude equal to or greater than of 5 on the Richter scale seem to occur if the rate of water loading in the reservoir exceeds 13 m per week.

***(b) Evaporation Losses*** The reservoir of the dam provides more surface area for evaporation. The loss of water due to evaporation is very high. As salt does not evaporate, the remaining water becomes more saline.

***(c) Salinisation of the Soil*** Use of saline water from dams for irrigation increases the rate of salinisation of the soil.

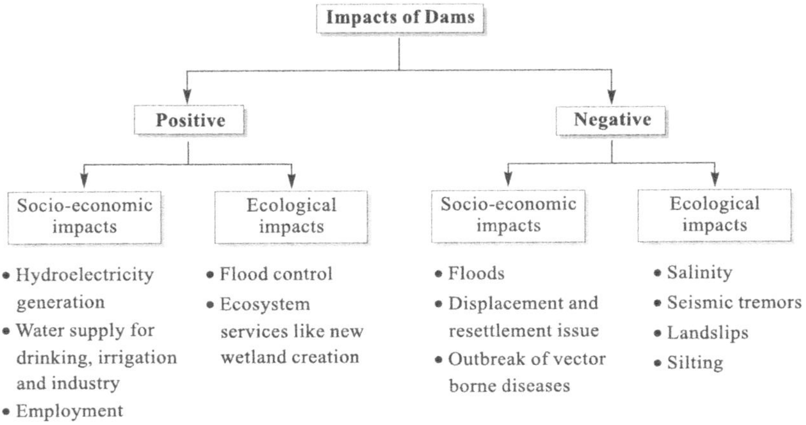


**Fig. 2.21** *Socio-economic and ecological impacts of dams*

***(d) Landslips*** The rise in water level can destabilise the geodynamic situation leading to substantial landslips.

***(e) Silting*** Generally, a turbulent stream feeds the reservoir of a dam. A rapid stream always carries some soil particles in suspension because of much up and down movement of water. However, water in the reservoir is calm and slow moving. As a result, most of the sediment that enters the reservoir from the run-off that feeds it, settles at the bottom at the rate of 10 cm per year. At such a rate, the lakes behind high dams can last up to hundreds of years, though not forever.

### (ii) Socio-economic Impacts

***(a) Increase in Water-related Diseases*** Water impoundments for dams may provide breeding sites for the vectors. This leads to transmission of malaria and schistosomiasis and spread of onchocerciasis in populations living near dam spillways.

***(b) Low Efficiency*** Most of the world's large dams have been unable to achieve the social, technical and economic objectives for which they were designed.

## 2.5 MINERAL RESOURCES

"Natural resources in the form of minerals are known as *mineral resources*." They include the ores of base metals such as copper, iron and lead as well as strategic and critical metals such as chromium, titanium, platinum, cobalt, manganese, palladium, etc.

## 2.5.1 Minerals and Their Classification

*Minerals* are naturally occurring, inorganic, solid, crystalline substances which contain a specific composition of elements.

A mineral which can be extracted and processed at a profit is known as an *ore.*

### *(A) Types of Minerals*

Minerals are broadly classified into two categories: metallic and nonmetallic. Metallic minerals are further sub-divided into ferrous and nonferrous materials. Nonmetallic minerals comprise of mineral fuels, precious stones, etc.

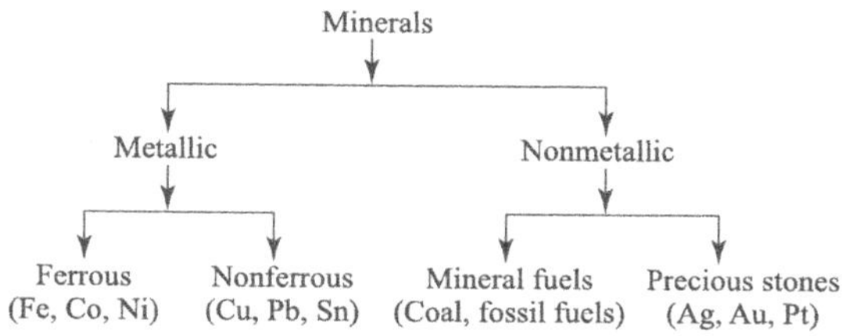


**Fig. 2.22** *Classification of minerals*

### *(B) Importance of Minerals*

(i) Almost all rocks are made of minerals.
(ii) They have high aesthetic value, e.g. gemstones.
(iii) They have natural resource value:
  (a) Minerals are sources of metals needed for electronic manufacture, airplanes, cars, etc.
  (b) Minerals are raw materials for making window glass, plaster, etc.

***(C) Mineral Resource of India*** India produces and works with roughly 100 minerals, which are an important source for earning foreign exchange as well as satisfying domestic needs.

We import graphite, mercury, cobalt, etc., and export iron ore, granite, bauxite, titanium, manganese, etc.

The distribution of minerals in the country is uneven and mineral density varies from region to region.

Coal, iron ore, manganese, mica, bauxite, copper, etc., are found in the North-Eastern peninsular belt located in Chhotanagpur Plateau and the Orissa Plateau covering the states of Jharkhand, West Bengal and Orissa. These regions are called the mineral heartland of India.

Gems, marble, coal, mica, graphite, manganese etc. exist in large quantities in Central Belt located in Chattisgarh, Andhra Pradesh, Madhya Pradesh and Maharashtra. The central Belt is the second largest belts of minerals in the country.

***According to the 2008 Ministry of Mines estimates:***

- India ranks 2nd in the production of chromite, barites
- India ranks 3rd in the production of coal and lignite,
- India ranks 4th in the production of iron ore,
- India ranks 5th in the production of bauxite and crude steel,
- India ranks 7th in the production of manganese, and
- India ranks 8th in the production of aluminium.

### 2.5.2 Environmental Effects of Extracting and Using Mineral Resources

The impacts on forest, land, occupation, water, ecological functions, rehabilitation of population, or impact on flowers due to pollution created during extraction and use of mineral resources are

(i) Deforestation including to loss of flora and fauna.
(ii) Degradation of land due to excavations.
(iii) Occupational health hazards.
(iv) Pollution of ground and surface water resources due to accidental or periodic discharge of pollutants.
(v) Damage to local ecological functions, nutrient cycling and biodiversity due to alterations in water availability or quality.
(vi) Problem in rehabilitation of affected population.
(vii) Pollution of air due to emission of dust and poisonous gases during mining and processing stages. Problems in providing living environment and clean water, air, etc., for the *survival* of large number of workers who have migrated nearby mine sites.
(viii) Problems in the safe disposal of tremendous amounts of solid waste generated during mining.

### 2.5.3 Conservation of Mineral Resources

The mineral resources are very essential for the growth and development of a country. The ever-increasing population in the world with improved lifestyles are responsible for the rapid consumption of mineral resources. The geological processes of mineral formation are so low that the rates of replenishment are very small in comparison to the present rates of consumption. Thus, mineral resources are valuable but they will be available for a limited time.

A sincere effort has to be made in order to use the mineral resources in a planned and sustainable manner. The following four steps are very useful for the conservation of mineral resources:

(i) Encourage use of improved technologies so as to *reduce* waste generation.
(ii) Encourage *recycling* of metals.
(iii) *Regulate* the use of mineral resources.
(iv) Reduce the purchase of unwanted products made from mineral resources.
(v) Encourage *research* for providing suitable ecofriendly alternatives for fossil fuels, metals, etc.

These are known as *4R's* for the sustainable use of mineral resources.

## 2.6 FOOD-SECURITY RESOURCES

*Food security is defined as the physical and economic access to sufficient, safe, nutritious food for maintaining a healthy and active life.*

### (A) Pillars of Food Security

There are three pillars on which food security is built:

**(i) Food Availability** It means sufficient quantities of food should be available to people to meet their dietary needs as well as their food preferences. Food can be made available from local, regional and international sources. Availability of food is governed by food production, food processing, food trade, food storage, etc.

**(ii) Food Access** It means people at all times should have physical and economic access to sufficient, safe and nutritious food. Food access is governed by marketing, transport, purchasing power, etc.

**(iii) Food Use** It means people must have knowledge of basic nutrition and care to maintain a healthy and active life. Food utilisation requires nutritious food choices, food safety and quality, clean water and proper sanitation.

### *(B) Food Security can be Achieved by Policies, Agreements, Imports and Distribution of Food*

(i) Implementing appropriate food, agriculture and rural development *policies* with conservation of natural resources that will facilitate food security in the long run.

(ii) Permitting appropriate international agricultural *agreements.*

(iii) Ensuring that the imported food products
- (a) are of acceptable quality and safe to eat,
- (b) does not reduce agricultural employment levels, and
- (c) permit protection of national food security.

(iv) Facilitating proper distribution of sufficient, safe and nutritious food.

***(C) Threats to Food Security*** Threats to food security are summarised below:

(i) Declining productivity, increasing poverty and declining income from traditional crops

(ii) Loss of preferences

(iii) Growing incidence of food-related diseases

(iv) High dependence on imported food

## 2.6.1 World Food Problems

According to Food and Agriculture Organization (FAO) estimates, on an overage, the minimum calorie requirement of a healthy man is 3000 kcal/day and of a healthy woman is 2200 kcal/day.

### *(A) Undernourishment*

If a person receives less than 90% of this minimum calorie requirement then the person is undernourished. Thus, *undernourishment* means to receive less calories than needed.

As per WHO, almost one-third of all children under five years of age are undernourished in developing countries.

#### Effects of Undernourishment

(i) Undernourished persons have less energy for doing any kind of work. They are susceptible to diseases, their body becomes weak and they frequently fall sick. They look old even at a young age.

(ii) Undernourished children suffer from the following problems: social inferiority complex, slow body growth, mental retardness, illness, delayed adulthood, etc.

## (B) Malnourishment

It means lack of essential nutrients (like proteins, vitamins, lipids, minerals, etc.) in the diet.

It may lead to the following problems:

Malnourished persons are more susceptible to diseases, they have less strength to function productively, and they face abnormal growth.

According to WHO, more than 3 billion people in the world are malnourished.

### Common Diseases of Malnutrition

***(i) Marasmus*** It is progressive deterioration caused by a diet low in total calories and protein.

- *Symptoms:* Pronounced slowing of growth, extreme wasting of muscles.
- *Remedy:* Adequate diet.

**BALANCED-DIET GUIDELINES FOR HEALTHY LIVING**

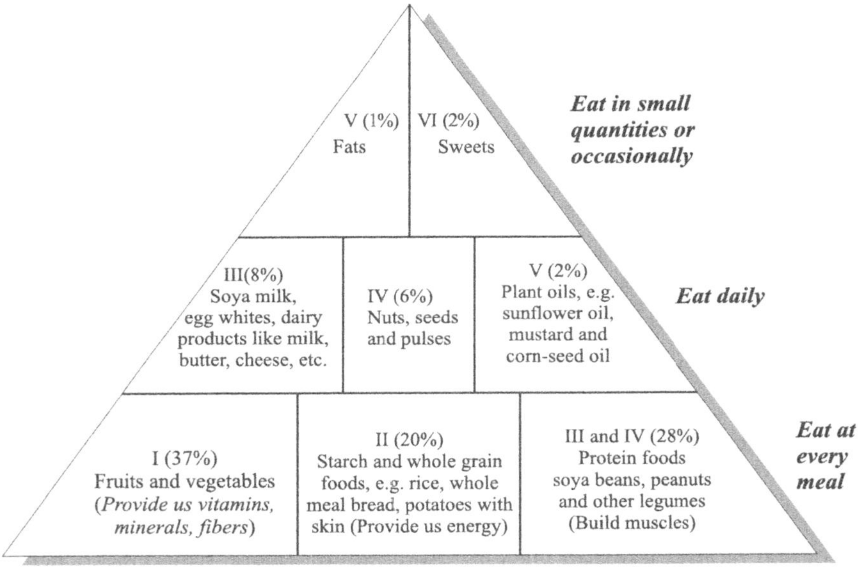


- To remove toxins, and to flush out waste products, drink 6-8 glasses of water per day.
- For easier absorption of iron from food, have some fruit or vegetable containing vitamin C(e.g. lemon) or some fruit juice with your meal.
- Small but regular exercise is must along with daily food and drink uptake for healthy life.
- Fibre in the food keep the guts healthy.
- Dairy products help in proper cell and body functioning.
- Vegetarian should eat 6% pulses, beans nuts, etc., and nonvegetarian Indians should eat 3% pulses plus 3% meat.

**Fig. 2.23** *Food triangle for balanced diet*

***(ii) Kwashiorkor*** It is caused by protein deficiency.

- *Symptoms:* Fluid retention, swelling of abdomen, dry, brittle hair, mental retardation and stunted growth.
- *Remedy:* Balanced diet.

***(C) Overnourishment*** It is due to overconsumption of saturated (animal) fats, sugar and salt in diet.

- *Symptoms:* Obesity, high blood pressure, diabetes, heart diseases, etc.
- *Remedy:* Balanced diet.

***Overgrazing*** Animal grazing is a natural process of forage utilisation, because herbivores produce in the environment where evolution formed them. When a plant is grazed on severely, it uses energy stored in its roots to support regrowth. As this energy is used, the roots die back. The dying of roots adds organic matter to the soil, which increases soil porosity, the infiltration rate of water and the soil's moisture-holding capacity. After enough leaves have re-grown, the roots will regrow as well. A plant is overgrazed when it is re-grazed before the roots recover. Overgrazing reduces root growth by 90%. Pastures are less productive because there is less root growth.

"*Overgrazing* can be defined as grazing plants (by livestock or wildlife) for extended periods of time or without sufficient recovery periods."

*Impacts of overgrazing* are listed below:

**(i)** Porosity of soil decreases.

**(ii)** The infiltration rate and moisture-holding capacity of soil decreases.

**(iii)** Soils have less organic matter and become less fertile.

**(iv)** Under the above adversed and undesirable conditions, desired plants become stressed but the growth of thorny shrubs/poisonous plants may increase.

**(v)** Overgrazing destroys the vegetation completely. The entire land area becomes prone to desertification and heavy soil erosion. This may cause siltation of any river leading to severe floods that may claim a large number of lives and property as well.

***Notes:***

1. Overgrazing is not a function of animal numbers. It is a function of time. Even one cow in a big pasture will overgraze plants if she is kept there for a longer duration or brought back too soon.
2. Worldwide, overgrazing is considered to be the major cause of soil degradation.
3. A grassland is overgrazed where a linked vegetation changes and loss of animal productivity arises from herbivores grazing of land.

***Control Measures for Overgrazing*** Economic sustainability cannot be achieved without environment sustainability. The latter requires that overgrazing must be stopped. Cattle owners need to make sure that they do not bring the animals before plants have recovered. When growth of plants is fast, recovery periods of (4–16) weeks may be adequate. Thus, overgrazing can be stopped with

(*Contd.*)

(*Contd.*)

8–10 paddocks. Overgrazing will not occur if livestock are kept in a paddock for less than 3 or 4 days. Awareness and some sort of legislation also help in stopping overgrazing. Negative impacts of overgrazing can be prevented and /or reversed by proper grassland-management practices.

## 2.6.2 Agriculture and Modern Agriculture

The word *agriculture* comes from the Latin words *ager* (meaning soil) and *cultura* (meaning cultivation).

*Agriculture* means the cultivation of soil and or production of crop plants or livestock products. It is synonymous with farming: the field-dependent production of food, fodder, and industrial organic material.

The term *modern agriculture* depicts the push for innovation, stewardship and advancements continually made by farmers to sustainably produce higher quality products with a reduced environmental impact. It is based on the following:

***(A) Agriculture Business and Marketing*** For solving the problem of hunger, systematic and efficient marketing of foodgrains play an important role. Poor cannot pay higher prices of foodgrains and they remain hungry. When the marketing system is not efficient, profits earned from the consumers are not adequately transferred to the producers. As a result, farmers do not get sufficient price incentive to increase the production of food items which are in short supply. To sum up, agricultural marketing plays a role in modern agriculture in fostering and sustaining the willingness critical to rural and economic development.

***(B) Organic Farming*** *Organic farming* is based on development of biological diversity and the maintenance and replenishment of soil productivity through crop rotation, use of animal and green manures, and some forms of biological control of pests.

Chemical fertilisers are extensively used in *intensive agriculture* for increasing the crop yields. But their use has spoiled the land, soil and water.

**Table 2.10** *Difference between intensive agriculture and organic farming*

| *Objective* | *Intensive Agriculture* | *Organic Farming* |
|---|---|---|
| (1) To increase crop yields | • Use of expensive inorganic fertilisers | • Use of economic manuers, farm organic resources and biofertilisers |
| (2) To manage disease and weed | • Use of pesticides | • Use of Biological pest. |
| (3) Effect on soil quality | • Soil quality gets reduced | • Soil quality improves. |

### Principles used in Organic Farming

**(a)** Nature does not use chemical fertilisers, pesticides and excessive water, but still it provides the best food. Thus, nature is the best role model for farming.

**(b)** Soil must not be considered as an inert bowl to dump chemicals. Soil is a living system.

(c) Living populations of microbes and other organisms present in the soil are main contributors to fertility on a sustained basis. Thus, at all costs, they must be protected and nurtured.

(d) From soil cover to soil structure, the total environment of the soil is more important than any fertiliser we may wish to pump into it.

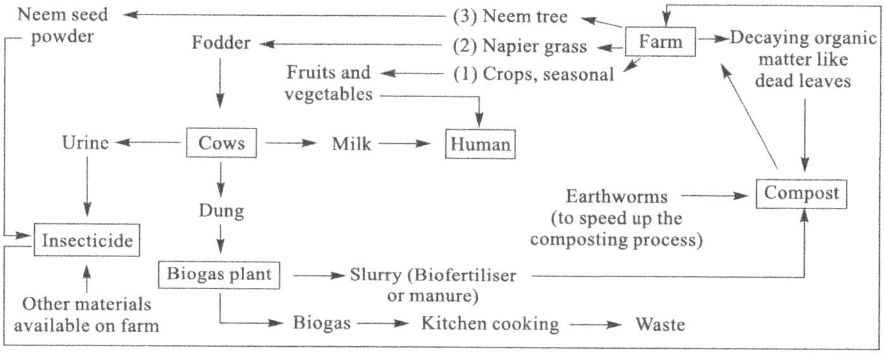


**Fig. 2.24** *Organic farming*

## *(C) Sustainable Agriculture*

"*Sustainable agriculture means meeting the food needs of the present generation without endangering the resource base for the future generation.*"

Sustainable agriculture requires focus on soil, plant, agriculture, water and nutrients through

(a) Conservation of soil and water to prevent degradation of soil productivity and lengthening of crop

(b) Integrated plant protection for reducing the effects of intensive agriculture

(c) Efficient use of irrigation water to prevent problem of soil salinity, alkalinity, etc.

(d) Integrated management of nutrients

**Strategies for Sustainable Agriculture** The following strategies are helpful for sustainable agriculture

(a) Management of soil, water, nutrients, etc., by farmers

(b) Regulations, formulations and implementations of appropriate policies by the government

(c) Use of right technologies for agriculture. It includes
- precision farming
- global positioning system, and
- animal feed-use efficiency.

***(D) Conservation Tillage*** *Conservation tillage* is a farming process which helps prevent land loss to erosion and water pollution. It also enhances carbon sequestration.

### (E) Integrated Crop Management

*Integrated crop management* is based on agro-ecological principles. It increases the yields of crop and reduces the environmental damage.

Farmers are supported by education and certification programmes. This ensures that they apply right agricultural practices, with care and only when required. They have the access to the technologies required to support modern agriculture practices.

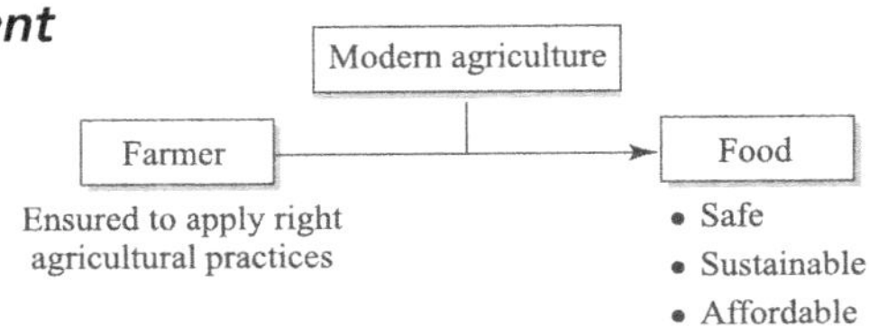


**Fig. 2.25** *Modern agriculture*

## 2.6.3 Effects of Agriculture or Environment

Humans get food from agricultural practices. However, the production, processing and distribution of food changes the environment.

Agriculture has both on-site and off-site environmental effects. These are summarised below:

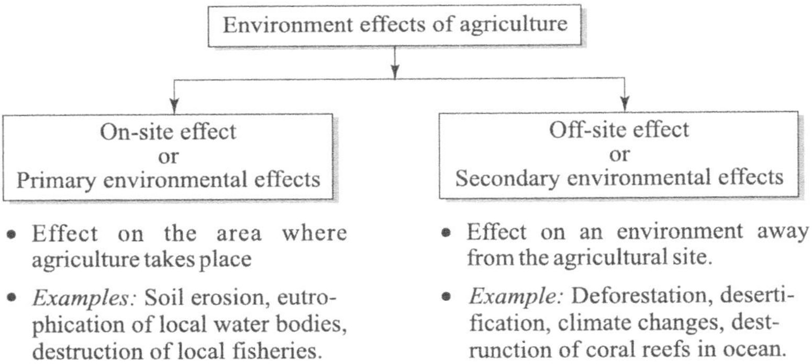


**Fig. 2.26** *On-site and off-site environmental effects*

Since the last few decades, the emphasis of most of chemical farms is exclusively on productivity through high input in exchange for high returns and high productivity. However, the following important considerations were overlooked:

### (i) What Happens to the Land?

**(a)** The continuous use of artificial fertilisers, together with a lack of crop rotation, reduces the soil's fertility and this ultimately leads to *land exhaustion*.

**(b)** Where repeated deep ploughing were used to turn over the ground, heavy rains carried away the top soil and left the ground useless for cultivation through soil erosion.

**(c)** In areas that were intensively farmed (in conventional tillage) using tractors, *soil compaction problem* was observed. Even a single tractor pass can compress the surface enough to reduce the porosity of the soil.

**(d)** Large and chemical farms tend to be monocultures growing the same crop and crop variety. This causes *loss of cultivated biodiversity* by 70%, increasing surface run-off and therefore, soil erosion.

### (ii) What Happens to the Environment?

**(a)** High crop-yield levels were produced by applying large quantities of artificial fertilisers. Natural fertility of the soil was not maintained. About half of the nitrate in the artificial fertilisers used in crops is dissolved by rain and irrigation water. The dissolved nitrate runs off the fields to *contaminate water resources.*

**(b)** Animals (to be used in modern farms) are generally crowded together indoors. Complex systems of machinery are needed to feed them. To prevent disease, constant medication is also given. The *cruelty to animals* involved in managing, breeding, growing and slaughtering is unimaginably horrifying. Furthermore, with so many animals forced to live in small areas (indoor), their waste accumulates at great speed. It is often poured into lagoons which leak into local *watercourses, contaminating* them with disease- causing pathogens and also contributing to algae-blooms.

**(c)** In places where stubble is burned, large amounts of potentially useful organic matter disappears into the sky in clouds of polluting smoke causing *air pollution.*

**(d)** The wild animals and plants which used to be around farms are deprived of natural habitat. This *habitat destruction* leads to their death.

### (iii) What Happens to Indigeneous Seeds and Plant Varieties?

Native animal breeds and cultivations lose out to hybrids and exotic species. Many native animal breeds are threatened with *extinction.* Moreover, within the space of one generation, many indigeneous seeds and plant varieties have *disappeared.*

### (iv) What Happens to the People?

**(a)** The supply and trading in agricultural inputs and food is in the hands of a few large corporations. This causes destruction of traditions and indigeneous knowledge. As a result, *food security is threatened* and the importance of the farmer and the consumer is reduced.

**(b)** Chemical agriculture is a threat to individual farmers and their livelihoods. They are forced to change their lifestyle, unfortunately not for the better.

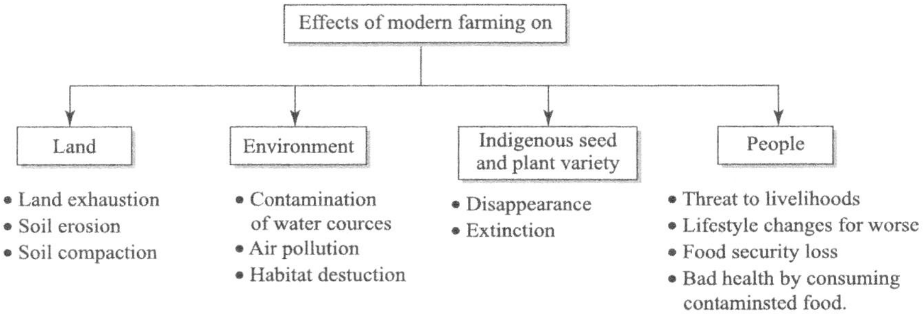


**Fig. 2.27** *How modern farming affects our world*

## *Effects of Modern Agriculture Practices on Environment*

**(i) Detrimental Effects** Increasing food demand, farming on environmentally sensitive land, technological changes, economical changes, and need to have more output per unit of land or labour (i.e. *intensification of agriculture*) has led to the following environmental harms:

(a) Water pollution
(b) Air pollution
(c) Loss of wildlife, habitats and landscape features
(d) Soil degradation
(e) Water depletion

**(ii) Environmental Benefits** In some circumstances, the following benefits are results of agricultural practices:

(a) Contribution to water accumulation and flood control
(b) Soil formation
(c) Nutrient recycling and fixation
(d) Carbon sequestration by trees and plants
(e) Wildlife protection
(f) Biodiversity protection
(g) Recreational provisions
(h) Aesthetic value provisions

## 2.6.4 Fertilisers and Associated Problems

*Fertilisers* are substances (natural or synthetic) that are added to the soil to restore and enhance the soil fertility to improve the quality and quantity of plant growth.

All fertilisers contain variety of micronutrients and macronutrients (nitrogen, phosphorus, and potassium). With proper application, these micro and macronutrients provide plants with the nutrients needed for healthy growth. However, improper application can lead to a variety of problems associated with fertilisers.

### *(A) Fertiliser Problems*

A brief description of *fertiliser problems* is given below:

**(i) Fertiliser Burn** An excess amount of fertiliser application causes fertiliser burn. It is identified by leaf-tip burn, premature yellowing of foliage, and patch discolouration,

- *Leaf burn* is due to overfertilisation, with symptoms appearing along the veins and leaf margins.
- *Discolouration* is due to the excess salts in the soil resulting from overfertilisation. Plants are not able to absorb enough water for heathy growth due to excess salts present in soil. This results in the yellow to brown appearance of tip burn.

By using the right amount of fertiliser (recommended for the specific plant), fertiliser burn can be prevented.

**(ii) Fertiliser Run-off** Due to overfertilisation, fertiliser run-off happens which can lead to a variety of environmental problems. For instance, phosphorus run-off into lakes and other water courses causes an elevated growth of weeds and algae. This unnatural growth of weeds and algae, saps oxygen from the water and threatens aquatic life.

***Eutrophication*** refers to a whole sequence of events, (illustrated in Fig. 2.28) starting with nutrient enrichment and proceeding to the growth and die-off of phytoplankton, the accumulation of detritus, the growth of bacteria, the depletion of dissolved oxygen and suffocation of fish, shellfish, etc.

The increased nutrient level in the water will promote a rapid and heavy growth of plants. Much of this plant life will be algae floating in the surface waters of the river, lake or pond. Their rapid increase in numbers cause a thick mat or ***algal*** bloom to spread across the surface of the water. This results in sunlight being cut off and death of vegetation beneath these mats. The increase in dead matter provides food for the bacteria which increase in numbers very rapidly, causing a rapid fall in oxygen levels. The reduction in oxygen then leads to the death of fish and other animals in the water.

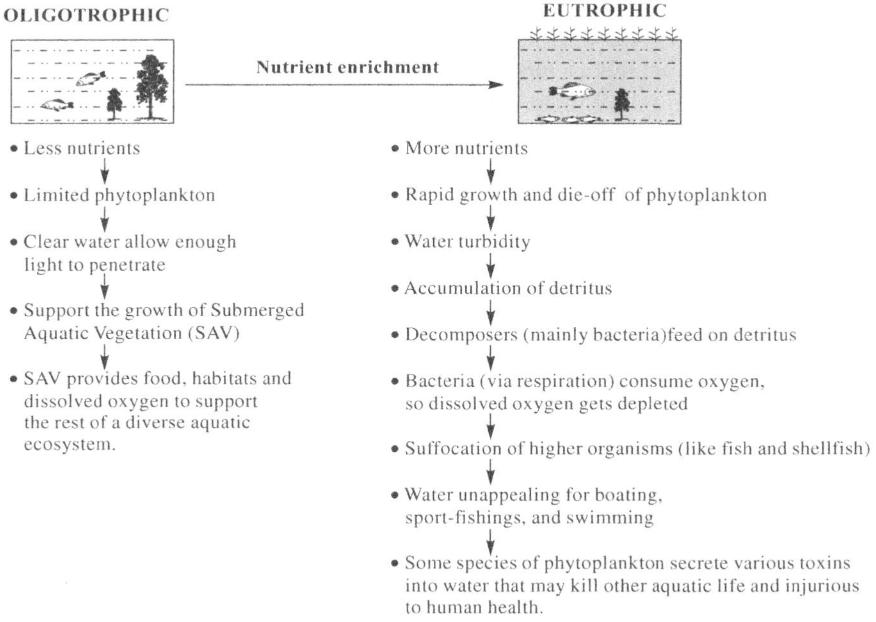


**Fig. 2.28** *Eutrophication*

To prevent run-off problems from overfertilisation, soil tests must be done. It helps in determining how much fertiliser needs to be added to soil.

**(iii) Increased Pests** In fertilisers, nitrogen is a primary macronutrient. It promotes cell division, helps plants gain energy, and causes leafy, green foliage growth. However, high nitrogen levels also results in increase in plant-damaging pests. Too much nitrogen causes an immense amount of weak foliage. The foliage growth may appear as healthy growth, but the thin cell walls make the plant vulnerable to insects and disease. Insect damage is a leading threat to weak but thick foliage as insects are attracted to foliage growth.

By using the right and recommended amount of nitrogen for the specific plant grown, it is possible to prevent increased pests.

**(iv) Soil Damage** The use of synthetic fertilisers in soils without a balanced addition of organic materials (which helps in formation of humus) leads to soil compaction. As a result, the soil loses its nutrient and water-retention capacity and is not able to make nitrogen readily available to the plants. As a consequence, requirement for synthetic fertilisers increases, which continues to make the problems worse. Gradually the soil becomes heavily mineralised and excessively susceptible to erosion.

**(v) Health Problems for People** Excessive use of inorganic fertilisers and resultant water pollution causes high consumption of nitrates through drinking water and leafy vegetables. In the intestines, these nitrates are reduced to nitrites by bacterial action. Nitrites, on reaching the bloodstream, become attached to haemoglobin forming methaemoglobin. This reduces the oxygen-carrying capacity of the blood and produces a condition known as *methaemoglobin anaemia* or the *blue baby disease*, characterised by bluish colouration of the skin.

In adult humans, if nitrates are converted into amines and nitrosoaminess, it causes gastric *cancer*.

***(B) Solutions for Fertiliser Problems*** For avoiding problems associated with fertilisers, the following guidelines are helpful:

(i) Use only a recommended amount of suitable fertiliser.
(ii) Do not apply fertilisers just before rain.
(iii) Use fertilisers at the minimum rate.
(iv) Poor aeration, wrong soil pH and diseases can all cause poor growth of plants. Thus, if a plant grows poorly, it may not necessarily need more fertiliser.
(v) Use compost and other organic fertilisers for economically improving soil texture and nutrient levels. Furthermore, organic fertilisers release nutrients more slowly and leach into water supplies less.
(vi) Select and use the right fertiliser for a given crop based on fertiliser number and apply it evenly.

**Table 2.11** *Fertiliser number*

| | *N* | *P* | *K* |
|---|---|---|---|
| Fertiliser code | 10 | 20 | 10 |
| Weight (kg) of constituent | (10/10) = 1 | (20/10) = 2 | (10/10 = 1) |

Suppose you have a 10 kg bag of 10-20-10 fertiliser. It means, the fertiliser has 1 kg of nitrogen, 2 kg of phosphorus, and 1 kg of potassium. This is because fertiliser number represents a ratio to inert matter. The first, second and third numbers tell us how much N, P and K is in the fertiliser bag.

*Nitrogen (N)* is the chemical which makes plants have dark luscious green colour. Nitrogen is required by chlorophyll which captures the sunlight in the photosynthesis process for making oxygen, carbohydrates and energy. Nitrogen is also used in reproduction. It also allows the plant to grow, produce more plant leaves, and create a darker green colour.

*Phosphorus (P)* is in the form of phosphate. It helps in early maturement of plants. As a result, flowering plants give flower early and of big size.

*Potassium (K)* helps in the development of healthy root system. The water dissolves the nutrients in the soil and is absorbed by the root system. Thus, potassium gives structure and strength to plant to sustain extreme weather conditions. Plants also become drought resistant with the right amount of potassium

## 2.6.5 Pesticides and Associated Problems

*Pest* is any organism that causes an economic loss or damage to the physical well being of human beings.

*Pesticides* are the chemical compounds that are used for killing of insects, rodents, etc., stopping vital physiological processes within the organisms. The use of pesticides help in increasing crop yields and in the eradication of diseases such as malaria.

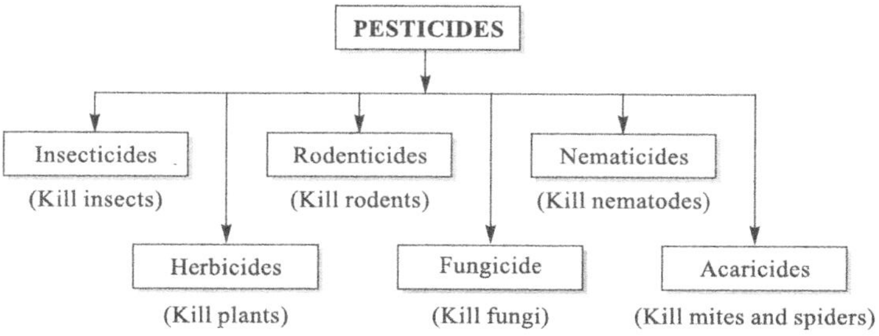


**Fig. 2.29** *Basic groups of pesticides*

***(A) The Pesticide Treadmill*** The term *pesticide treadmill* is used to describe attempts to eradicate pests with synthetic organic chemicals. However, these chemicals do not eradicate the pests. Instead, they increase resistance and secondary pest outbreaks. This leads to the use of new and larger quantities of chemicals, which in turn lead to more resistance and more secondary pest out-breaks, and so on. This nonsustainable process is an unending cycle constantly increasing contamination of foodstuff and ecosystems causing more risks to humans (Fig. 2.30).

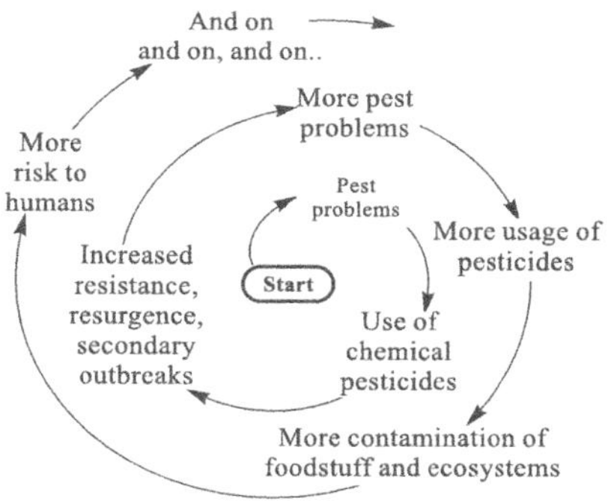


**Fig. 2.30** *Pesticide treadmill (as per late entomologist Robert Van den Bosch)*

### *(B) Effects of Pesticides*

The ecological fate of pesticides affects natural water supply, aquatic wildlife, insects and birds. This is briefly discussed below:

**(i) Water Supply** Various streams and groundwater supplies, drained watershed from agricultural, urban and mixed-use areas were found to contain pesticides. Even those pesticides were found whose use has been banned. This is because pesticides persist in the environment for longer times.

**(ii) Aquatic Wildlife** Pesticides are a potential cause of deformities and decline in amphibians. Pesticides are present in many streams at concentrations that may have effects on aquatic life or fish-eating wildlife.

**(iii) Insects** Pesticides are also responsible for dwindling populations of pollinator species (bees) and beneficial insects. *Colony Collapse Disorder (CCD)* means the sudden abandonment or evacuation of bee colonies through acute pesticide poisoning.

**(iv) Birds** Birds are *environmental sentinels* as they are much more sensitive to pesticides than mammals. They forewarn us about the potential hazards to our environment and our own health.

Pesticides affects birds indirectly by weakening them or reducing their food supply. Birds may prey on target pests or nontarget beneficials treated with pesticides. This leads to sudden exposure for a short period (acute) or prolonged exposure (chronic). As a result, population of birds gets reduced.

**(v) Humans** The human body is susceptible to the effects of pesticides. Over time, it can cause health problems.

***(a) Endocrine System Problems*** When inhaled or absorbed through the skin, the chemical structure of pesticides enable them to settle inside the fatty materials contained in the body. As a result, pesticide residue can accumulate inside the body gradually. The endocrine system of humans consists of chemical-hormone secreting glands. Pesticides change the body's normal hormone levels. As a consequence, reproductive processes, cell growth and repair as well as cell-metabolism processes are disrupted.

***(b) Neurotoxicity*** In the human body, pesticides act on cholinesterase (an essential neurotransmitter chemical in the brain and body tissue, it works in conjunction with the body's nerves and muscles). The body experiences a gradual decline in nervous-system function, nausea, muscle weakness, and chronic fatigue because the pesticide residues accumulate within body tissues.

Organophosphates and carbamate pesticides are responsible for chronic fatigue syndrome and Parkinson's disease.

***(c) Cancer*** When cells become unable to regulate growth functions, they start multiplying at abnormal rates resulting in *cancer*.

Accumulation of pesticides in the body causes abnormality in endocrine-system processes. This abnormality greatly contributes to a breakdown in cell function.

Pesticide 2, 4 D is associated with a 50% increase in lymphoma disorder. Other forms of cancer include lung, intestinal, prostate, brain and leukemia.

Risk = $f$(Hazard, Exposure)

- **Risk** (damage) by pesticide is determined by both toxicity (hazard) and the likelihood of exposure.
  A low level of exposure to a very toxic pesticide is less dangerous than a high level of exposure to a relatively low toxic pesticide.
- As per WHO estimates, about 1 million pesticide-poisoning cases and 20,000 deaths occur every year globally due to high pesticide residues in the food chain including fruits, vegetables, cereals, pulses, fishes, poultry, meat products, water, milk (including mother's milk) and milk products. The rate of utilisation of pesticides is such that it contaminates soil, water and food even in concentrations of parts per trillion to parts per million.

**Table 2.12** *Pesticides: Structure and uses*

| *Major Structural Class* | *Pesticides and its Structure* | *Uses* |
|---|---|---|
| (i) Chlorinated hydrocarbon | 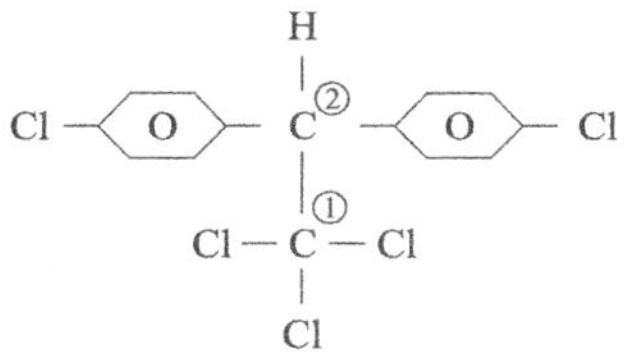 DDT (1,1,1-trichloro-2,2 bis (p-chlorophenyl) ethane | Broad spectrum—Peanut, soybean and cotton pests; mosquito control (use banned in some countries because it accumulates in food chain, cause adverse health effects on wildlife and persists in the environment) |
| (ii) Organophosphates | 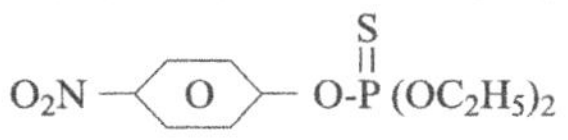 Parathion (O, O-diethyl O-p nitrophenyl phosphoro thionate) | Broad spectrum—Fruit and vegetable pests, larvicide for mosquitoes control |
| (iii) Carbamates | 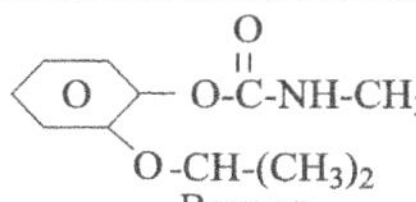 Baygon (2 - isopropyl N - methyl carbonate ) | Control of flies, mosquitoes, ants and cockroaches |
| (iv) Chlorophenoxy acids | 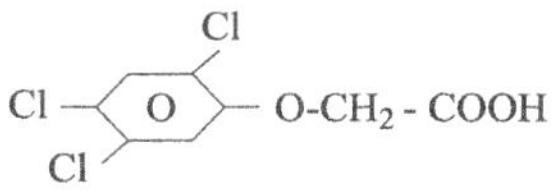 2, 4, 5 - T | Military defoliant, weed control |

## *(C) Solutions to Problems Created by Pesticides*

**(i)** Limit the use of pesticide products.

**(ii)** Initiate crop rotation in farms and other management approaches.

**(iii)** Clean and use nonchemical traps to avoid home pests and thus avoid use of pesticides. It helps in reducing your chances of getting cancer.

**(iv)** Support local initiatives to get a pesticide by-law enacted in your community as quickly as possible.

**(v)** Before applying liquid pesticide, remove food, toys, etc. Close doors to rooms that are not infested. Turn off central heating or cooling units. During application, apply at low pressure to avoid making small particles that can linger in the air. After application, open windows, turn on fans to speed up drying time. Keep children and pets out of the treated area until the pesticide has dried.

**(vi)** Use plant-based pesticides like neem-based insecticides which are eco-friendly. It has pesticidal, antibacterial and antiviral properties and has been developed in dust and spray forms for use.

**(vii)** Get complete toxicological information on the active ingredients of the pesticide manufactured and only use if found right.

**(viii)** Pesticide manufacturers must obey standard specifications formulated by competent authorities.

## *(D) Objectionable Properties of Chemical Pesticides*

**(i)** Chemical pesticides have high physiological and ecological specificity.

**(ii)** Chemical pesticides are resistant to biochemical degradation.

**(iii)** There is an increase in concentration of pesticides in successively higher trophic levels of a food chain or food web in a process known as *bio-magnification.*

In an aquatic ecosystem, zooplankton consume pesticides and other organic pollutants which may be consumed by the next trophic level (predators) such as small fishes. These predators consume many prey organisms and thus they accumulate and concentrate pesticides into a large extent. When these predators are consumed by the next trophic level, these pesticides are further concentrated and it goes on to successive trophic levels. Thus, there is a magnification of pesticides in higher-trophic-level organisms.

Biomagnification also occurs in terrestrial ecosystems.

**(iv)** Chemical pesticides have the capacity for delayed onset of intoxication.

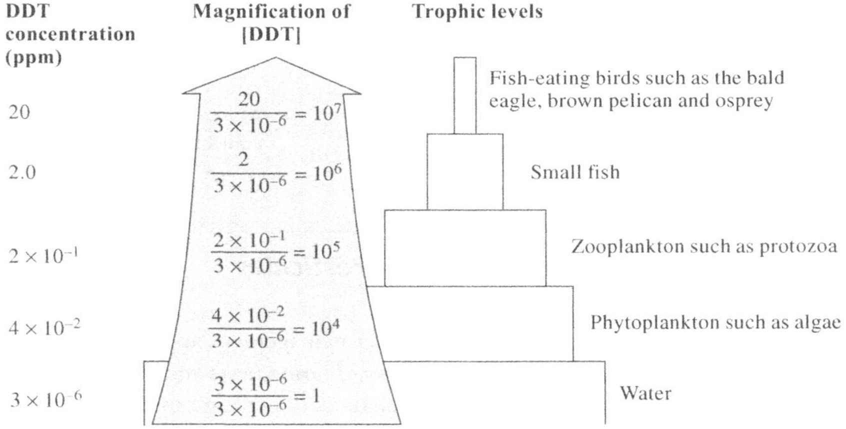


**Fig. 2.31** *Biomagnification*

### (E) How to Get Rid of Pesticides from Fruits and Vegetables

**Step (i)** Rinse your fruits and vegetables in tap water briefly. The majority of pesticide residue appear to reside on the surface of produce and is thus removed by the mechanical action of rinsing.

**Step (ii)** For fruits like apples or pears that are easy to handle, spray with (i) one per cent solution of dish liquid (i.e. detergent) and water (ii) 5% NaCl solution. Then rub the solution over the surface of the item with your fingers.

Leafy green vegetables and broccoli tend to hold dirt in their crevices and it is better to soak these often in a one per cent solution of dish liquid and water.

**Step (iii)** Rinse fruit or vegetable produce with tap water for about one minute. This will remove any traces of dish liquid from your fruits and vegetables.

**Step (iv)** Normal household procedures, viz. peeling and cooking, are also very effective.

## 2.6.6 Green Revolution

*Green Revolution means better food-production methods or better agricultural practices to secure the world food supply.* It consists of crop-variety improvements, expansion of irrigation and increased use of pesticides and fertilisers.

"*The Green Revolution refers to the rapid increase in food production mainly through the use of High-Yielding Varieties (HYVs) of seeds, chemical fertilisers, pesticides, agriculture machinery and water.*"

**Impacts of Green Revolution** Because of Green Revolution,

(i) Foodgrain prices have remain stable for the last 15 years
(ii) Food crisis has overcomed
(iii) Agricultural productivity has increased

## 2.6.7 Waterlogging

*It means too much water in the root zone of a plant so that roots cannot absorb enough oxygen to breathe.* The result is the plant stops growing and within a few days, it may die.

**Causes**

(i) Over-irrigation of soils
(ii) Farming on clayey soils; here water cannot move efficiently through the soil and cannot be adequately drained; eventually waterlogging occurs
(iii) Excessive rain or flood, etc.

## 2.6.8 Salinity

*It means accumulation of soluble salts in the soil due to over-irrigation.* When excessive quantities of soluble salts accumulate in the root zone, the crop has difficulty in extracting enough water from the salty solution. It results in adverse crop productivity. Fruit crops are the most sensitive to soil salinity followed by vegetables.

For the solution of this soil-salinity problem, root zone of the plants are flushed with excess water. However, it may contaminate the groundwater or irrigation canals.

**Example 3** *How has agriculture contributed to water pollution?*

**Solution**

(a) Pesticide run-off from agricultural land impairs drinking-water quality. It also harms water-based wildlife.

(b) Animal effluent from livestock and use of fertilisers in agriculture accounts for higher nitrogen and phosphate emissions in surface water. It results in eutrophication, which causes depletion of oxygen in water.

**Example 4** *How has agriculture contributed to air pollution?*

**Solution** In areas of intensive agricultural production, the following air-pollution problems are encountered:

(a) Emissions of methane, nitrous oxide, etc., from agriculture contribute to global warming and climate change.

(b) Offensive odours are produced by crop burning.

**Example 5** *Why do agriculture practices cause loss of biodiversity?*

**Solution**

(a) In countries where grasslands and wetlands are converted to cropland, several rare species of wildlife have declined.

(b) Intensification of farming practices lead to habitat degeneration.

(c) Reduced crop rotations, increased fertiliser and pesticide application are some of the main reasons for the loss of biodiversity.

**Example 6** *How do agriculture practices cause soil erosion?*

**Solution**

(a) The trees and plants hold the soil together. For agriculture, trees are removed to create fields. Once the cover is gone; wind, rain, etc., start soil erosion.

(b) Animals (like cows, sheep, etc.) compact the ground due to their hooves. It results in increased run-off of water, leading to soil erosion.

(c) Sheep also eat plants close to the ground, killing them and leaving the ground prone to more wind/water erosion.

## Case Studies

### (i) Effects of Modern Agriculture on Punjab

Dr Rayed Tirado, from the University of Exeter, UK, conducted the study in 50 villages in Ludhiana, Bhatinda and Muktasar districts under a Greenpeace Research Laboratories investigation in 2009. This study revealed radiation, biological and chemical toxicity rampant in Punjab. As per WHO the safety limit is 50 ppm w.r.t. nitrate levels in water. Twenty per cent of the sampled wells showed nitrate levels above this safety limit. The study connected it with high use of synthetic fertilisers. High nitrate levels in drinking water can be harmful to humans particularly for infants under 6 months where it is linked to methemoglobinemia or blue-baby syndrome. With increasing poisoning of the soil, the region once hailed as the home to the Green Revolution, now is being termed as the "Other Bhopal" due to excessive use of chemical fertilisers. A comprehensive study conducted by Post Graduate Institute of Medical Education and Research (PGIMER) has underlined the direct relationship between indiscriminate use of fertiliser and pesticides and increased incidence of cancer in several villages of Punjab including Khara, Bhimawali, Puckka, Koharwala, Jhariwala, etc.

**(ii) Waterlogging in India**

2,189.4 ha have been reported to suffer from waterlogging in irrigation canal commands in India. The increased level of the water table can lead to reduced agricultural production. Stagnant water tables at the soil surface are known to increase the incidence of waterborne diseases like dengue, malaria, etc.

**(iii) Salinity in India**

3,469.1 ha were reported to be seriously salt affected in India. Salinity induced by human activity is a desertification problem of increasing world concern as pressure increases on agricultural land for food production.

## 2.7 ENERGY RESOURCES

### 2.7.1 Growing Energy Needs (Energy Scenario in India)

Energy is essential for economic growth and sustenance of modern economy.

***(A) Energy Import*** India, though rich in coal and abundantly endowed with renewable energy in the form of hydro, wind, solar and bio-energy, has very small hydrocarbon reserves. India is a net importer of energy, more than 25% of primary energy needs being met through imports mainly in the form of natural gas and crude oil.

***(B) Energy Production Pattern*** Coal and oil account for 54% and 34% respectively, with natural gas, hydro, solar and nuclear contributing to the balance. Nearly 62% of power generation is from coal-fired thermal power plants and 70% of the coal produced every year in India has been used for thermal power generation.

***(C) Distribution of Primary Commercial Energy Resources*** The distribution is non-uniform. The northern and north-eastern regions account for 70% of the total hydro potential, whereas the eastern region accounts for about 70% of the total coal reserves in India. The southern region has most of the lignite deposits in the country.

***(D) Energy-Consumption Pattern*** The industrial sector consumes about 52% of commercial energy production. The energy consumption per unit of GDP (*energy intensity*) is one of the highest in comparison to other developed and developing countries. Thus, there is a huge scope for energy conservation in India.

***(E) Future*** The long-term availability of energy from sources that are affordable, accessible and environmentally friendly will govern the future economic growth in India.

***(F) Over-exploitation of Energy Resource in India*** Power demand in India touched an all-time high of 99,027 MW recently, a year-on-year growth of over 16%. In our country, economic growth is very fast which is responsible for the growing energy demands of the country.

India's demand for energy is set to increase to as much as 315,000 MW by 2017 at an average GDP growth rate of 8% per year as per the forecast from Mckinsey and company in 2008. To meet this energy requirement, the country has to over-exploit all its energy resources.

Presently the power needs of the country are primarily met by thermal power stations (about 60%), followed by hydro, wind and nuclear power stations.

India is experiencing shortages in the production of domestic coal. Many of the thermal power stations, including NTPC, faced critical coal stock levels and had to cut down their production levels. The obvious result was power cut. In 2008–09 due to non-availability of power, India Inc. has supposedly lost ₹43,205 crore as per a study. The coal reserves of India may get exhausted by 2040 at the current rate of coal consumption. India may have to depend entirely on coal imports.

***(G) Relationship Between Energy Resources and Development in India*** About 60 years back, at the time of independence, the installed power generation capacity was 1300 megawatt, (MW). In 2009, power demand in India touched 99,027 MW.

As on 31$^{st}$ march 2005, the installed power-generation capacity using various methods is tabulated below:

| | **The installed power generation capacity** |
|---|---|
| Thermal | 80,902.45 MW |
| Hydro | 30935.63 MW |
| Wind | 13065.37 MW (as on 31$^{st}$ Dec. 2010) |
| Nuclear | 2770 MW |

Energy is the key input to drive and improve lifestyles. Improvement in the living standards, industrialisation, education, health-care services, etc., all depend on availability of energy. Availability of energy, energy resources and consumption of energy all directly govern the progress of the country.

***(H) Problems due to Over-exploitation of Energy*** In the US, non-renewable fossil fuels provide 92% of the energy used. The US has 3% of the world's oil and petroleum resources, but consumes 25% of the crude oil extracted in the world. US citizens also waste tremendous amount of energy. Higher consumption of energy resources means larger emissions of greenhouse gases which lead to global warming and consequent problems.

Many developing countries are striving to reach the level of prosperity of the US. It has been estimated that if world starts consuming at the same rate as the US, the world could run out of fossil fuels in a few years.

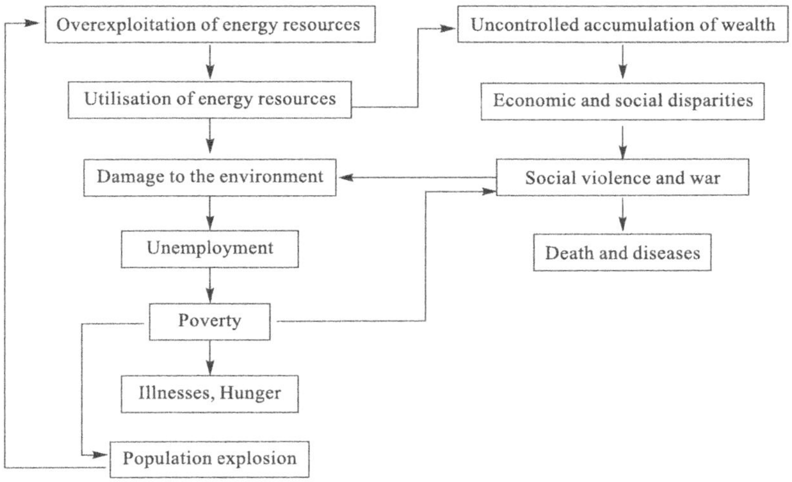


**Fig. 2.32** *Problems due to over-exploitation of energy resources*

## 2.7.2 Types of Energy

*Energy* is the ability to do work.

Broadly, there are six types of energies, viz. mechanical, radiant, chemical, heat, electrical and nuclear. *Mr CHEN* is a mnemonic for remembering the six forms of energy.

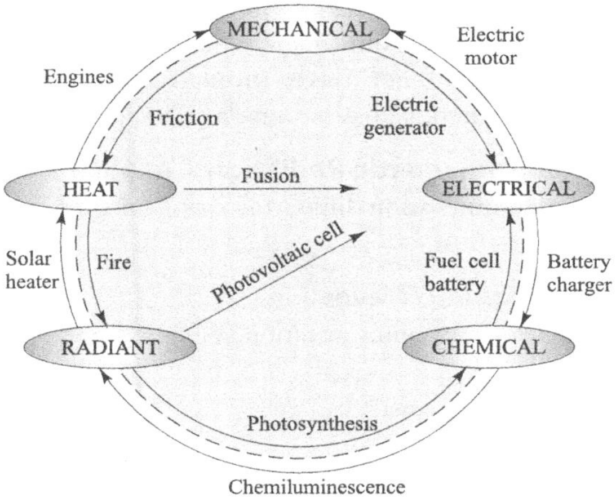


**Fig. 2.33** *Interconversion of different types of energies*

The different types of energies are briefly described below:

**(i) Mechanical Energy** Energy due to an object's motion (kinetic) or position (potential) is known as mechanical energy.
***Examples*** Windmill used to make electricity, water wheel used to grind grain.

**(ii) Radiant Energy** Radiant energy is electromagnetic energy. It includes energy from gamma rays, X-rays, ultraviolet rays, visible light, infrared rays, microwaves and radio waves.
***Examples*** Solar panel used to provide electricity and heat for a house; microwave oven used for cooking.

**(iii) Chemical Energy** Energy that is available for release from chemical reactions is termed chemical energy.
***Examples*** Gasoline is used to run a car.

**(iv) Heat Energy** The heat energy of an object determines how active its atoms are.
***Examples*** Coal was used to produce steam for steam engines of trains.

**(v) Electrical Energy** Energy caused by the movement of electrons is called electrical energy.
***Examples*** Electricity is used to run appliances in our homes.

**(vi) Nuclear Energy** Nuclear energy is the energy stored in the nucleus of an atom. It is the energy that holds the nucleus together. It is released when the nuclei are combined (nuclear fusion) or split apart (nuclear fission).
***Example*** A nuclear power plant uses the fission of atoms to create energy that is converted into electricity.

According to the ***first law of thermodynamics***, "Energy is never created nor destroyed, but it can be converted from one type into another."

## *Electromagnetic Radiation*

Electromagnetic waves are produced by the motion of electrically charged particles. These waves are also called "electromagnetic radiation" because they radiate from electrically charged particles. They travel through air, empty space and other substances. The electromagnetic radiation travels at $3 \times 10^8$ m/s in vacuum.

**(i) Examples of Electromagnetic Radiation** AM and FM radio waves, TV signals, cell phone communication links, microwaves, infrared radiation, light, X-rays, gamma rays.

**(ii) Uses of Electromagnetic Waves**

(a) One-way and two-way communication systems
(b) Radar
(c) Cooking (with microwaves)
(d) Medical imaging (X rays)
(e) Night vision (infrared)
(f) Astronomy (radio, gamma, UV, visible, IR, microwave)

All that we experience through our eyes is conveyed by electromagnetic radiation.

**Table 2.13** *Types of electromagnetic radiation*

| ***Non-ionising Radiations*** | ***Ionising Radiations*** |
|---|---|
| (i) They are low-energy radiations | (i) They are high-energy radiations |
| (ii) Overexposure causes skin burns | (ii) Overexposure causes genetic mutation or alteration in cellular structure |
| (iii) Examples: Visible and infra-red radiation, radio and microwaves | (iii) Examples: X-rays, Gamma rays |

## 2.7.3 Renewable and Nonrenewable Energy Sources

### *(A) Conventional and Nonconventional Sources of Energy*

*Conventional energy sources* are energy sources which are nonrenewable. However, *nonconventional energy sources* are energy sources which are renewable and ecologically safe (Fig. 2.34).

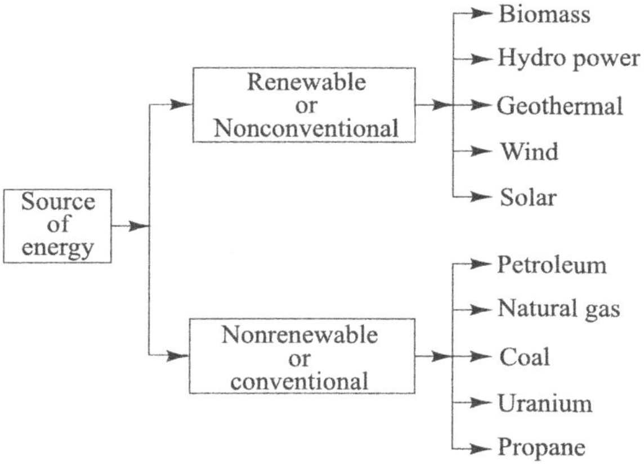


**Fig. 2.34** *Important energy sources*

The important differences between conventional and nonconventional sources of energies are summarised below:

**Table 2.14** *Differences between conventional and nonconventional sources of energies*

| ***Conventional Sources of Energy*** | ***Nonconventional Sources of Energy*** |
|---|---|
| (i) They are fully developed. | (i) They are still undergoing development. |
| (ii) They use nonrenewable resources. | (ii) They use renewable resources. |
| (iii) Inexpensive. | (iii) Expensive. |
| (iv) Require established technologies. | (iv) Require new technologies which are still under research and development. |
| (v) Ecologically not safe for usage. | (v) Ecologically safe to use. |
| (vi) Available in limited quantity. | (vi) Available in plenty. |

(*Contd.*)

(*Contd.*)

| | |
|---|---|
| (vii) Carbon and other greenhouse gas emissions from the combustion of coal, natural gas, etc., are known to have disastrous environmental and health consequences. These gases are also major culprit in climate change. | (vii) Free from such problems. |
| (viii) *Examples:* Petroleum, coal, etc. | (viii) *Examples:* Solar, wind and hydropower, etc. |

***(B) Renewable Energy Programmes in India*** Major thrust is given towards research, development, demonstration, commercialisation and deployment of

(i) New and renewable energy system/devices for transportation
(ii) Portable and stationary applications for rural, urban, industrial and commercial areas
(iii) Alternate fuels for surface transportation including electric/hydrogen/hybrid vehicles
(iv) Biofuel for motive power for stationary/portable applications
(v) Hydrogen energy
(vi) Cowdung based power plants (the new component in the biogas programme)

The above-mentioned new thrusts are being given to renewable energy programme in India to closely align them with the objectives of the common minimum programme regarding rural electrification, energy security and world-class infrastructure.

## 2.7.4 Hydro-Power Energy or Hydroelectric Energy

*Hydroelectricity or hydroelectric power is the electricity obtained by harnessing the power of water flowing down from a high level.*

It is a renewable, affordable and pollution-free source of energy.

### *Major Purposes of Dams*

Dams are used for storing water. This stored water is led down through large pipes or tunnels to lower levels. In the course of the fall of water, the water rotates turbines. The produced mechanical energy is converted to electricity by the generators connected to it. Transformers change the alternating current produced by the generators into currents of very high voltage for easy transmission through long distances (Fig. 2.35).

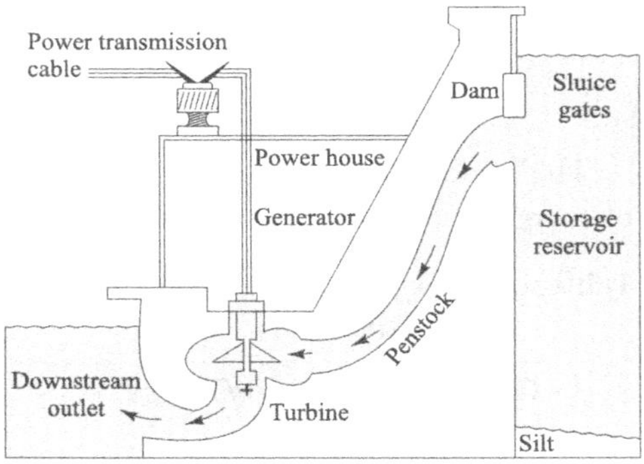


**Fig. 2.35** *Generation of hydro-electric energy using a dam*

### (i) Advantages

(a) ***Economical*** The cost of operating a hydroelectric plant is nearly immune to increases in the cost of fossil fuels. Operating labour cost is also low.

(b) ***Safe*** Hydroelectricity produces the least amount of greenhouse gases.

### (ii) Disadvantages

(a) ***Failure Hazard*** In the case of failure of dams, millions of people become homeless, sick and even die.

(b) ***Limited Service Life*** As rivers convey silt, higher the siltation, lower will be the service life of dam.

(c) ***Environmental Damage*** Dams are responsible for habitat fragmentation, aquatic ecosystem disruption, and greenhouse-gas emissions.

(d) ***Population Relocation Problem*** Millions of people need relocation and generally they do not get necessary compensation.

(e) ***Energy Production is Affected by Amount of River Flow*** Specially during drought period, hydroelectric power cannot be generated.

## 2.7.5 Fossil Fuels

Petroleum and coal are formed from the fossilised remains of animals and plants, hence they are known as *fossil fuels*. As they are used up much more rapidly than they are replenished by nature, it might ultimately result in fuel shortage.

### *(A) Coal*

*Coal* is defined as stratified rock, consisting of organic matter of fuel value derived from the partial decay and alteration of accumulated plant materials by the action of heat and pressure over millions of years.

Coal is of the following four types:

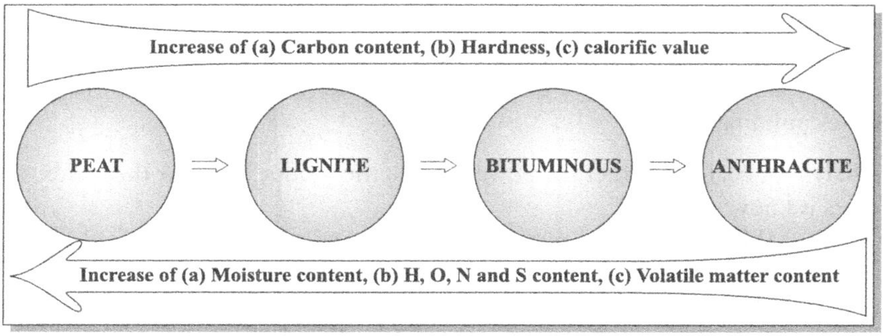


**Fig. 2.36** *Types of coal*

**(i) Advantages** The factors which are in fovour of usage of coals in huge tonnage quantities are availability, low cost, least risk of fire hazards and easy storage.

### (ii) Disadvantages

(a) Combustion of coal is a slow process.

(b) Combustion control is not easy.

(c) After combustion, ash is always produced and so its disposal is a problem. Smoke is also invariably produced.

(d) Use of coal in internal combustion engines is not possible.

(e) Calorific value and thermal efficiency is least.

## (B) Coal-based Thermal Power Plants (Environmental Problem)

About 80% of domestic reserves of coal in India are characterised by ash content of 35% to 45%. Burning of this inferior-quality coal results in lesser efficiency and higher emissions of pollutant gases including $CO_2$, $SO_x$ and $NO_x$. It also releases particulate matter, bottom ash and fly ash. Coal mining which is done to extract coal to feed into thermal plants is also hazardous. The environmental problems caused by mining and burning of coal are endless. A few of these are summarised below:

**(i) Carbon Dioxide ($CO_2$)** Coal-based thermal power plants are the major contributor of $CO_2$ (a greenhouse gas) responsible for *global warming*.

**(ii) Sulphur Dioxide ($SO_2$)** Depending on the type and rank of coal, sulphur in coal ranges from 0.1–3.5 %. During combustion of coal in thermal power plants, sulphur combines with oxygen to from $SO_2$. Thermal power plants are the largest emitters of $SO_2$. In atmospheric air, $SO_2$ is converted into sulphuric acid. It can precipitate down as *acid rain* leading to destruction of ecosystems.

**(iii) Oxides of Nitrogen ($NO_x$)** Coal-based thermal power plants are the second largest emitter of $NO_x$ which combine with water to form $HNO_3$. It is responsible for acid rain, smog and ground ozone. $NO_x$ is a hazardous pollutant and is responsible for visual and respiratory problems.

**(iv) Particulate Matter** Although there are arrangements in thermal power plants to collect the ash, still some ash goes out through the stack and is known as particulate matter emission. The particulate matter are dispersed over a very wide area from very tall stacks of thermal power plants. The particles of size less than 2.5 microns (called PM 2.5) are of great concern as they are responsible for respiratory illnesses in humans.

**(v) Ash** In coal-based thermal power plants, the residue left after the combustion process is know as *ash*.

A 500 MW coal-fired thermal power plant burning coal with around 20% ash results in the accumulation of 2 million tons of ash in 5 years. A small portion of this ash may be utilised by cement plants but disposing the bulk of it on a long-term basis can raise major environmental issues. The ash particles can be carried away to surrounding areas though wind. Toxic elements of ash can percolate into the water system causing harm to vegetation and humans.

## (C) Petroleum

The word petroleum is derived from Latin *petra* which means rock and *oleum* which means oil.

*Petroleum* is a complex mixture of paraffinic, olefinic and aromatic hydrocarbons with small quantities of organic compounds containing oxygen, nitrogen and sulphur. It is also called *mineral oil* because it occurs beneath the earth. Petroleum refining of crude oil or petroleum provides many liquid fuels that are in current use. A brief description of important liquid fuels, viz. gasoline, diesel and kerosene oil is given on next page.

**(i) Gasoline or Petrol** It is a mixture of hydrocarbons from pentane to octane. It is highly volatile and inflammable. It is used as a fuel for internal combustion engines. Its calorific value is about 11,250 cal/g.

**(ii) Diesel Oil** It is a mixture of higher hydrocarbons ($C_{15}$ to $C_{18}$). It is used as a fuel for diesel engines and its calorific value is about 11,000 cal/g.

**(iii) Kerosene Oil** It is a mixture of hydrocarbons ($C_{10}$ to $C_{16}$). Its calorific value is about 11,100 cal/g. It is used as domestic fuel in stoves.

### *(D) Gaseous Fuels*

**(i) Natural Gas** It is obtained from wells dug in the oil-bearing regions. It is mainly composed of methane, ethane and other hydrocarbons. It is also called *marsh gas* because it mainly consists of methane (about 88.5%). It is used as a domestic and industrial fuel, because of its high calorific value (8000–14000 kcal/m$^3$).

**(ii) Compressed Natural Gas (CNG)** The natural gas compressed at very high pressure of about 1000 atmosphere is called CNG. The calorific value of CNG is 31400–37700 kJ/m$^3$. The use of CNG as a fuel for automobiles has reduced pollution in urban cities. As it undergoes complete combustion in CNG engine so there is nil possibility of release of CO in the atmosphere. Further, CNG is much safer fuel with lower operating cost.

**(iii) Liquified Petroleum Gas (LPG)** The main constituents of LPG are *n*-butane, isobutane, butylene and propane. The calorific value of LPG is about 27,800 kcal/m$^3$. It is mainly used as domestic fuel. To help in the detection of gas leakage, a strong-smelling substance, viz. ethyl mercaptan, is added to the LPG gas cylinders. LPG is also used as motor fuel because it easily mixes with air and burns cleanly without residue and without knocking.

## 2.7.6 Nuclear Energy

The study of nuclear fuel is important because energy changes involved are many million times greater than in chemical fuels. The source of energy in nuclear fuel is nuclear fission or nuclear fusion reaction. These are discussed below.

***(A) Nuclear Fission*** When an unstable nucleus of a heavy atom (like uranium–235) is bombarded with neutrons, the former splits up into two medium weight nuclei with the liberation of an enormous amount of energy.

$$^{235}_{92}\mathrm{U} + {}^{1}_{0}n \xrightarrow{\text{Fission}} {}^{139}_{56}\mathrm{Ba} + {}^{94}_{36}\mathrm{Kr} + 3\,{}^{1}_{0}n + \text{Energy}$$

In all fission reactions, more neutrons are emitted than consumed. They, in turn, are capable of fission of more heavy atoms and a *chain reaction* is started. When this chain reaction is controlled, it can lead to *power generation* in a device called a *nuclear reactor.*

***(B) Nuclear Fusion*** When two lighter nuclei (like deuterium atoms) are heated to a very high temperature (~$10^6$°C), they fuse together to form a heavy, more stable nucleus and an enormous amount of energy is liberated.

$$^{2}_{1}H + ^{2}_{1}H \xrightarrow{\text{Fusion}} ^{4}_{2}He + \text{Energy}$$

High temperatures needed for fusion can be attained by using the heat evolved in a fission reaction. But, in this way, the fusion of hydrogen gets out of controlled and leads to *explosion* (*hydrogen bomb*).

## (C) Merits and Demerits of Nuclear Energy

### (i) Merits of Nuclear Energy

***(a) Availability*** Nuclear power plants could still produce electricity after coal and oil become scarce.

***(b) Less Fuel Requirement*** One ton of uranium produces more energy than is produce by several million tons of coal or several million barrels of oil.

***(c) Less Pollution*** Well-operated nuclear power plants do not release contaminants into the environment. As combustion is not done so no $CO_2$, $SO_x$, $NO_x$, are released. Thus, there is very little effect on environment.

***(d) Economical*** Cost of fuel is a much smaller percentage of the total cost and operating cost is about the same as coal-based thermal power plant. Generation of power is less susceptible to price fluctuations. One fuel pellet about the size of a pencil eraser produces the same energy as about 1 ton of coal, and after reprocessing $2/3^{rd}$ amount can be reclaimed.

***(e) Employment*** Energy generation from nuclear power plants creates high paying, skilled jobs.

***(f) Safe*** Safety record of nuclear power plants in the world is fantastic.

***(g) Reliable*** Nuclear power plants have very high capacity factors. Presently, 12% to 18% of the world's electricity is generated through nuclear energy.

### (ii) Demerits of Nuclear Energy

***(a) Large Initial Cost*** To develop a single nuclear power plant about 15 years to 20 years are required along with expenditure of huge amount of money. The nuclear reactors will work only as long as uranium is available.

***(b) Dangerous Waste*** The waste produced after fission reactions of uranium consist mainly of unstable, radioactive elements. It is very dangerous to the human health and environment for thousands of years. It needs professional handling and careful storage isolated from the living environments. In 1957, at a dump site in Russia's Ural mountains, dozens of people were killed by the mysterious explosion of buried nuclear waste.

***(c) Less Life of Nuclear Reactors*** They could only last for about 40 years to 50 years.

***(d) Chances of Worse Disasters*** *Meltdown* is one possible type of reactor disaster in which the fission reaction goes out of control, leading to nuclear explosion and the emission of great amount of radiation. In the *Chernobyl* nuclear power plant in Ukraine, a large amount of radiation escaped from the reactor in 1986. Hundreds of thousands of people were exposed to radiation, resulting in deaths of several dozens of peoples within a few days. Radiation-induced cancer resulted in deaths of thousands of more people. One cannot deny the possibility of repetition of such disasters in future.

*(e)* Domestic unavailability of safe storage and reprocessing facilities.

*(f)* *Fear security concerns,* terrorism and proliferation of nuclear weapons. They are a major threat to the world as they cause large-scale devastation.

### *(D) Major Concerns Regarding Use of Nuclear Power for Electricity Generation in India*

A brief description of the above is given below:

**(i) Deficiency of Uranium** The Department of Atomic Energy (DAE) has formulated an approach for the nuclear energy resource consisting of the following three stages;

(a) Setting up of natural uranium fuelled Pressurized Heavy Water Reactors (PHWRs): India has achieved maturity for this.
(b) Setting up of fast breeder reactor utilising a *uranium–plutonium* fuel cycle. The beginning was made in 2003 through commencement of construction.
(c) Setting up of breeder reactors utilising thorium fuel. India's thorium reserves, about 2.9 lakh metric tons, are the second largest in the world.

**(ii) Absence of Self-sufficiency** India's civil nuclear strategy has been directed towards complete independence in the nuclear fuel cycle. This self-sufficiency extends from uranium exploration and mining though fuel fabrication, heavy water production, reactor design and construction, to reprocessing and management.

**(iii) Small Contribution of Nuclear Energy to India's Energy Mix** The nuclear establishment in India has failed to deliver what the pro-nuclear lobby had promised nearly 60 years after its inception.

**(iv) Environmental Concerns Regarding Digging of Uranium Mines** India has only modest reserves of about 70,000 metric tons of uranium which may never be fully exploited because environmental consideration will not allow the miners to dig new uranium mines. Public protests against Uranium Corporation of India Ltd (UCIL) have prevented mining of uranium since 1985.

**(v) Safety Concerns** India still faces severe challenges regarding the operational safety of uranium mines, nuclear power stations, etc.

There are also serious problems to do with treating and disposing of the large volumes of highly radioactive wastes.

Terrorist attacks put additional tremendous pressure on nuclear reactor safety.

# Case Study

**Jaitpur Nuclear Power Project**

India has 20 nuclear reactors in operation generating 4780 MW nuclear power as of 2010. To generate additional nuclear power, seven other reactors are under construction. On the plains of Jaitpur, 420 km south of Mumbai, a six-reactor nuclear power complex is planned. If built, it would be one of the world's largest nuclear power complexes. In the wake of Japan's Fukushima nuclear accidents, environmentalists, fishermen and local farmers have been protesting over this project. This is because of the following reasons.

(i) To build the reactors, 931 hectares of farmland will be needed. This land is now home to 10,000 people, their main orchards, rice fields and cashew trees.
(ii) Livelihoods of fishermen in the region will be taken away.
(iii) Marine life will be adversely affected through the release of waste water from the plant into the sea.
(iv) The survival of the fishermen and local farmers will be very difficult.

(Hindustan Times 27/11/2000, Sydney morning Heralds 23/4/2011)

## 2.7.7 Solar Energy

*Solar energy* is the energy received by the earth from the sun that is converted into thermal or electrical energy. Solar energy influences the earth's climate and weather and sustains life. Although solar energy only provides 0.15% of the world's power, experts believe that sunlight has the potential to supply 5000 times as much energy as the world currently consumes.

Broadly speaking, solar energy is a term for describing a range of methods for obtaining energy from the sun. For instance, wind, biomass and hydropower are all forms of solar energy. Wind develops through lows and highs in temperature. Wind drives waves. Rainfall, created by sun-warmed evaporated water feeds the rivers that are sources of hydro power. Fossil fuels are also forms of stored solar energy. Coal, oil and gas formed hundreds of millions of years ago from decomposed plant matter, plant matter that grew by aid of the sun.

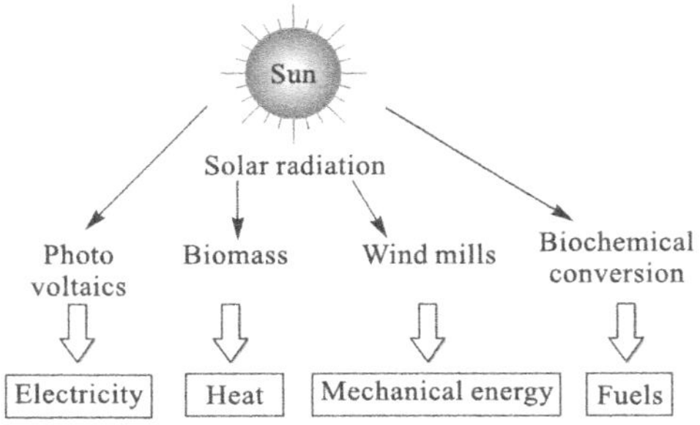


**Fig. 2.37** *Technologies and applications of solar energy*

***(A) Applications of Solar Energy in Modern Days*** Some of the important applications of solar energy are summarised below:

(i) Space cooling and heating through solar architecture
(ii) Potable water via distillation and disinfection
(iii) Solar cooking
(iv) Solar hot water

(v) Day lighting
(vi) High-temperature process heat for industrial purposes
(vii) Solar air-conditioning
(viii) Solar desalination
(ix) Solar electricity–photovoltaic
(x) Solar electricity–thermal
(xi) Solar vehicles
(xii) Solar chimney

## (B) Advantages and Disadvantages of Solar Energy

The *advantages* of solar energy are cost-effectivity, renewability, installation ease, low maintenance, pollution free. These are briefly discussed below:

**(i)** ***Cost-effectivity*** If you use a solar power system to reduce electric consumption from traditional power companies, the savings will be reflected in reduced electric bills.

**(ii)** ***Renewable Resource*** Solar power can be regenerated and replenished without fear of depletion because the sun is used as the energy source.

**(iii)** ***Easy Installation and Use of Solar Powered Products*** Solar panels, hot-water heaters, lighting, fountains, pumps, etc., are easy to install and/or use.

**(iv)** ***Long Life and Low Maintenance*** The life of solar panels is long (≥25 years) and maintenance free. Maintenance only requires that solar panels are kept dirt-free and snow-free for maximum contact with sunlight.

**(v)** ***Pollution Free*** Solar power energy is eco-friendly as solar energy does not pollute the air or produce greenhouse gases like traditional energy sources.

*The disadvantages* of solar energy are summarised below:

(i) As solar power uses the sun to produce electricity, generally, solar power cannot be created at night, during a cloudy day when sunlight is diminished or during a rainy seasons.

(ii) The initial cost of installation of solar panels is high.

(iii) Efficient collection of solar energy is a big challenge because solar radiations falls over a vast area in a scattered manner.

(iv) Solar cooking takes more time and all sorts of food cannot be cooked in a solar cooker, for example, the '*chapatti*'.

## (C) Use of Solar Energy for Space Heating of Buildings

The architectural design of buildings helps in their *passive space heating* using solar energy. Following strategies are useful for passive space heating:

(i) Provide large south-facing windows (Fig. 2.38).

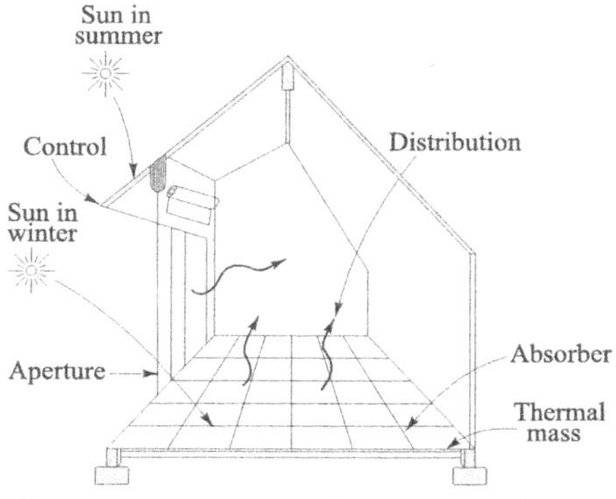


**Fig. 2.38** *Passive solar design technique*

(ii) Provide an entire wall of double-glazed windows.
(iii) Provide a heavy dark-coloured south-facing wall behind a layer of glass, with room air circulating by convection between the wall and the glass.
(iv) Provide a flat roof covered by a pond of water. Provision should be made for an insulating screen cover for cooling requirements in summer.

Note that no mechanical equipment is needed for passive solar heating.

An *active technology* of solar space heating needs a collector to absorb and collect solar radiation. Subsequently, fans or pumps are used to circulate the needed air or heat-absorbing fluid (generally water). Water systems are more common than air systems as they offer better heat-exchanger performance.

## (D) Use of Solar Energy for Generation of Electricity–Photovoltaics

Photovoltaics (PVs) are arrays of cells containing a solar photovoltaic material that converts solar radiation into direct current electricity. Solar cells produce direct current (dc) electricity from sunlight, which can be used to power bulb/equipment or to recharge a battery [Fig. 2.39]. However, for grid-connected power generation, an inverter is required to convert the dc to alternating current (ac):

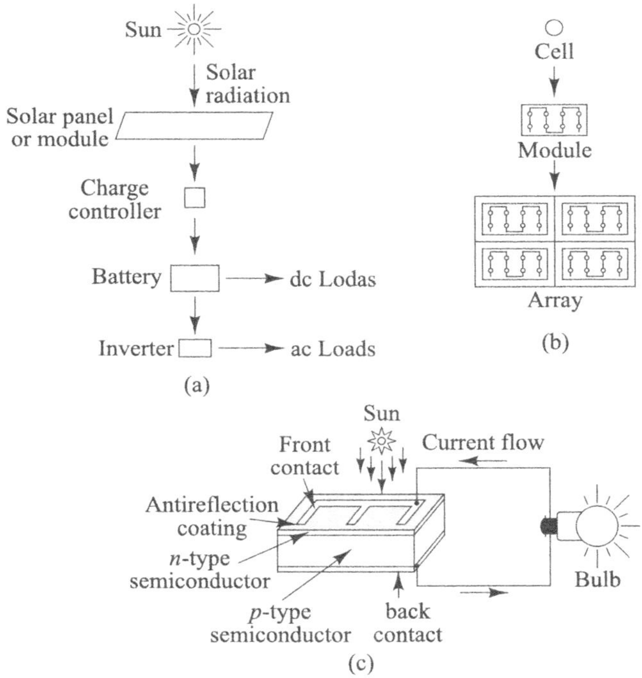


**Fig. 2.39** *Use of solar energy for the generation of electricity*

A number of solar cells, electrically connected to each other and mounted in a support structure or frame, is called a *photovoltaic module*. Multiple modules can be wired together to form an *array*. In general, the larger the area of a module or array, the more electricity that will be produced.

When $n$-type and $p$-type semiconductors are sandwiched together and irradiated with sunlight, the excess electrons in the $n$-type material flow to the $p$-type, and the holes thereby vacated during this process flow to the $n$-type. Through this hole and electron flow, the two semiconductors act as a cell, creating an electric field at the surface where they meet (known as $p$-$n$ junction). It is this field that causes the electrons to jump from the semiconductor out toward the surface and make them available for the electrical circuit [Fig. 2.39].

### (i) Advantages of Solar Photovoltaics

(a) Easy installation and maintenance
(b) Pollution free
(c) Long life
(d) Viable for remote and isolated areas, forest, hilly, desert regions

**(ii) Disadvantages of Solar Photovoltaics** The high initial cost, specially of the silicon wafer is the major constraint in the widespread use of solar cells.

**(iii) Applications of Photovoltaic Cells and Solar Panels** Photovoltaic cells are used in watches, pocket calculators and toys. Solar panels are useful to light up a house, run an irrigation pump, operate traffic lights, etc.

***(E) Solar Water Heater*** Solar water heater consists of a flat-plate collector, with a black bottom, a glass top, and water tubes in between. The collector is placed at a suitable angle to catch the sun's energy. The black bottom of the collector gets hot by absorbing solar radiation. The heat warms up the water in the tubes. The insulated storage tank is placed above the collector, the cool water moves down into the tubes and the water moves into the tank by natural convection.

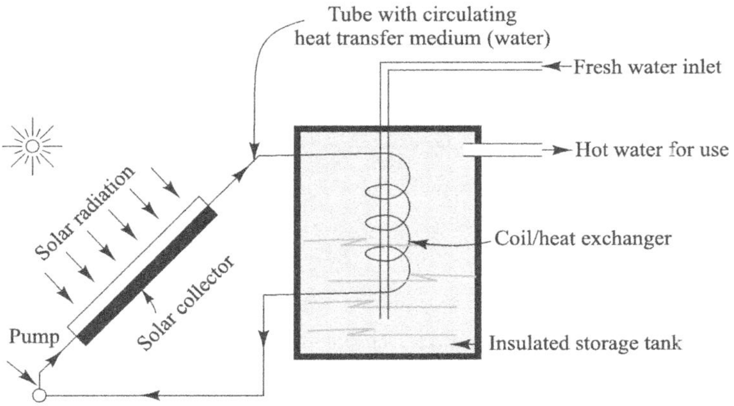


**Fig. 2.40** *Active solar water heater system*

As the energy is coming from the sun, utility bills are much lower and within a few years the installation cost is recovered.

## 2.7.8 Biomass

The term *biomass* is used for the dead plants and trees (e.g. wood, crop residue, etc.) and the waste material of living organisms (e.g. cattle dung, sewage, etc.). *Biomass energy* or *bioconversion* means the direct burning of waste paper, wood, cattle dung or converting them to a fuel.

The various ways of using biomass as a fuel:

(i) Biomass can be directly used as a fuel.
***Example*** Burning of biomass like cattle dung in *chulhas*.

(ii) The biomass is first converted into a fuel and then these fuels are used for heating purposes, more effectively.
***Example*** Conversion of cattle dung into biogas.

## 2.7.9 Biogas and Biofuel (or Green Fuel)

***(A) Biogas*** Biogas consists mainly of methane. It burns with a blue flame and its average calorific value is about 5300 kcal/m$^3$.

*Gobar gas* is the cheapest and most easily available biogas. It is made in a gobar-gas plant. Cattle dung is mixed with water (in equal parts) to form a slurry. It is then poured in a 'well' constructed of a masonry work (i.e. 'digester'). There anaerobic fermentation occurs at 34–48°C. The gas generated due to continuous decay is mostly methane (Fig. 2.41) and is collected in a gas holder. One kg of dung generates 160 litres of biogas and 164.6 kcal of extra heat is generated by this indirect burning.

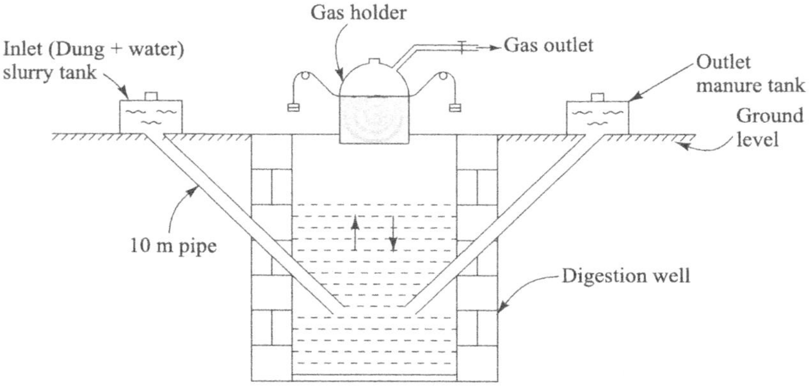


**Fig. 2.41** *Gobar-gas plant*

### (i) Advantages

(a) More heat is generated by burning gobar gas instead of cattle dung.

(b) It is free from smoke, dust, dirt, etc. Thus, the environment and utensils remain comparatively clean.

(c) In addition to cattle dung, a gobar-gas plant can also digest human refuge, poultry, etc. Thus, production of gobar gas means optimum utilisation of waste.

**(ii) Limitation** Gobar gas should be used within 10 metres of the gobar-gas plant.

**(iii) Applications**

(a) Biogas is used as domestic fuel in many villages.
(b) Biogas is also used for lighting purposes.
(c) A biogas plant also simultaneously gives good quantity of excellent manure. This manure has 2% nitrogen content as against 0.75% in farmyard manure.

***(B) Green Fuel*** "*Green fuel* (or biofuel) is a type of fuel obtained from nonfood sources like green algae which is more environmental friendly than the widely used and quickly disappearing fossil fuels."

In recent years, the processing of sugar and starch plants into ethanol has come under heavy criticism because

(i) This results in food shortages
(ii) The fermentation process causes air pollution
(iii) The obtained fuel on combustion may emit formaldehyde, ozone, and other carcinogenic substances
(iv) One acre of corn produces 200 times less oil as can be obtained from one acre of algae

## Case Study

The Central Salt and Marine Chemicals Research Institute (CSMCRI) has become the first Indian body to successfully produce ethanol from a seaweed, *Kappaphycus alvarezii*, in the year 2010.

Scientists of CSMCRI drove an ambassador car with a fuel blend of 10 % ethanol and 90 % petrol in a major ecofriendly step towards renewable energy generation and use in India.

CSMCRI plans for mass-scale production of ethanol in three years through offshore cultivation on a five-hectare area on the Tamil Nadu coast with the following major advantages.

**(i)** No pressure on agricultural land.
**(ii)** No requirement of fresh water, fertiliser and pesticides for irrigation.
**(iii)** Seaweed processing also produces **sap** (a biofertiliser especially used for sugarcane crop)

### 2.7.10 Hydrogen as an Alternative Future Source of Energy

*Hydrogen* is a very light gas, and its density is eight times less than that of natural gas. There are no significant problems with regard to storage, transportation, dispensation as well as end use of hydrogen.

Hydrogen is either produced through the steam reformation of natural gas or through electrolysis of water with renewable energies such as solar, wind and geothermal.

By burning fossil fuels, more than 90% of the world's energy requirements are met. This leads to emissions of carbon dioxide, which is responsible for the greenhouse effect and therefore responsible for the warming of our planet. Global warming at accelerated rates is inevitable if we do not reduce the emission of greenhouse gases. A hydrogen-based transport system has the potential to play an important role in reducing greenhouse gas emissions. To propel vehicles, the hydrogen can be burnt directly in either Internal Combustion Engines (ICEs) or used as a fuel for producing electricity in fuel cells. The electricity is produced by an electrochemical reaction. The electric power is then used to power an electric motor in the vehicle. Fuel-cell driven vehicles have great potential to be more efficient and ecofriendly than conventional fuel-driven vehicles. The vehicles will emit only steam but no greenhouse gas.

Hydrogen is considered a secondary source of energy (or energy carrier). Electricity is also an energy carrier. They are used to move, store and deliver energy in a form that can be easily used.

### (A) Advantages of Hydrogen as an Important Energy Carrier in the Future

(i) ***Pollution free*** Hydrogen is a pollution free fuel.

(ii) ***Economical*** It is less costly to ship hydrogen by pipeline than sending electricity over long distances by wire in some instances.

(iii) ***Various Applications*** It can be used for transportation, heating and power generation in places where it is difficult to use electricity.

(iv) ***Easy Storage*** A large volume of hydrogen can be easily stored in a number of different ways.

(v) ***High Efficiency*** Hydrogen is considered a highly efficient fuel.

### (B) Future Applications of Hydrogen

(i) ***Energy Carrier*** Hydrogen will join electricity as an important energy carrier in the future. This is because it can be made safely from renewable energy sources and is virtually nonpolluting.

(ii) ***Electricity Production*** Hydrogen will be used to produce electricity in fuel cells.

(iii) ***As a Fuel*** Hydrogen will be used as a fuel for 'zero-emission' vehicles, and for aircraft.

(iv) ***In Heating*** Hydrogen will also be used to heat offices and homes.

***(C) Infrastructure*** Many new facilities and systems must be built before hydrogen can play a bigger role and become a widely used alternative to gasoline.

***(D) Limitations of Hydrogen Energy*** The use of hydrogen as an alternative future source of energy is limited by

(i) Its low availability in pure $H_2$ form in the environment

(ii) Difficulty in handling, storing and transportation of $H_2$

(iii) Requirement of energy for the production of $H_2$

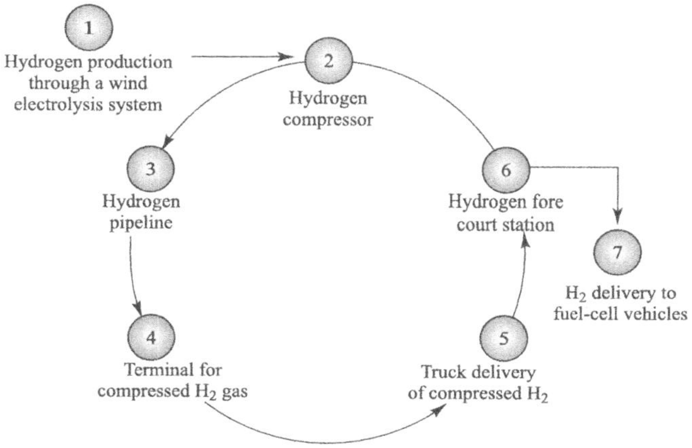


1. Use a wind electrolysis system to produce hydrogen.
2. Use compressor to compress hydrogen up to pipeline pressure.
3. Fed compressed hydrogen to a transmission pipeline.
4. The pipeline transports the hydrogen to a compressed gas terminal for its compression and loading in compressed gas tube trailors.
5. Delivery of compressed gas tube trailors using trucks to a forecourt station.
6. Further compression of hydrogen in forecourt station.
7. Storage and dispension of hydrogen to fuel cell vehicles.

**Fig. 2.42** *Infrastructure requirement for hydrogen as a source of energy for propelling vehicles*

## 2.7.11 Ocean Energy or Marine Energy

*Oceans* cover more than 70% of the earth's surface making them the world's largest solar collectors. Just a small portion of the heat trapped in the ocean could power the whole population across the globe.

### *(A) Various Ways in which Energy from the Ocean can be Obtained*

From the oceans, electrical power can be derived by the following methods:

(i) Wave power
(ii) Ocean Thermal Energy Conversion (OTEC)
(iii) Tidal power
(iv) Ocean currents
(v) Ocean winds
(vi) Salinity gradients

The first three technologies, viz. wave power, OTEC and tidal power, are the most developed technologies. These are briefly described below.

***(B) Wave Power*** The energy in waves comes from the movement of the ocean and the changing heights and speed of the swells. It is estimated that the total power of waves breaking on the world's coastlines is 2 to 3 million megawatts. An average 4 foot, 10-second wave striking a coast puts out more than 35, 000 horse power per mile of coast. This is equivalent to about 65 megawatts of wave energy density per mile of coast.

One of the approaches to capture wave energy is the use of *oscillating water columns (OWC)*. Its simple design is illustrated in Fig. 2.43. The OWC generates electricity from the wave-driven rise and fall of water in a cylindrical shaft. The rising and falling water column drives air into and out of the top of the shaft, powering an air-driven turbine.

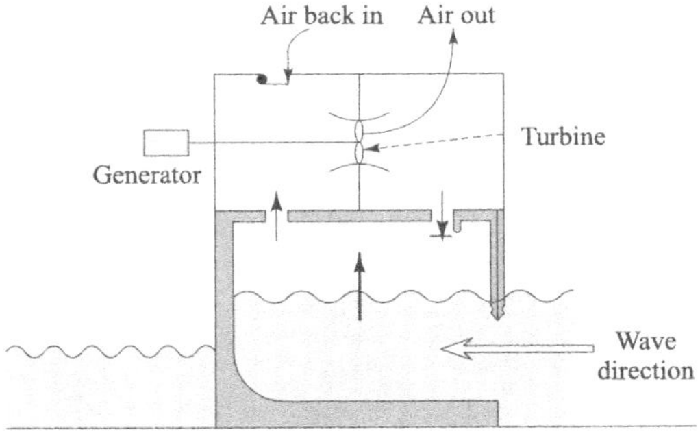


**Fig. 2.43** *Oscillating water columns for harnessing wave energy of oceans*

The theoretical potential of renewable energy from wave energy is estimated to be of the order of 8000–80000 TWh/year.

***(C) Ocean Thermal Energy Conversion (OTEC)*** The oceans absorb enough heat from the sun every day to equal the thermal energy contained in 250 billion barrels of oil. Thus, a great amount of heat is stored in the world's oceans. This thermal energy is converted into electricity by using ocean thermal energy conversion systems. *Closed-cycle plants* are one type of OTEC systems. Its simple design is illustrated in Fig. 2.44. They circulate a working fluid (which has a low boiling point, such as ammonia) in a closed system, heating it with warm surface sea water, flashing it to vapour, routing the vapour through a turbine, and then condensing it with cold sea water. The rotating turbine then activates a generator to produce electricity.

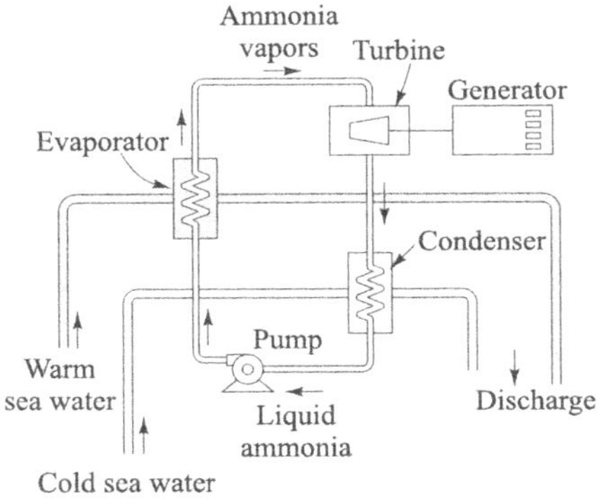


**Fig. 2.44** *Ocean Thermal Energy Conversion (OTEC) power plant*

The theoretical potential of renewable energy from OTEC is estimated to be of the order of 1000 TWh/year.

## 2.7.12 Geothermal Energy

*Geothermal energy is a clean, renewable and environment-friendly energy source based on the heat inside the earth.*

The word *geothermal* comes from the Greek words:

*Geo* = Earth ; *Thermal* = Heat

Thus, the energy that can be extracted from the heat inside the earth is geothermal energy.

In the earth's core, about 4000 miles below the surface of the earth, geothermal energy is continuously produced by the slow decay of radioactive particles. The regions where the earth's tectonic plates collide and one slides beneath another create the conditions which are most favourable for the geothermal activity.

The geothermal heat gets up to the earth's surface by the following mechanisms:

(i) Heat from the earth's interior flows outward. It is transferred by *conduction* to the outer layer of rock on the crust.

(ii) In some regions, the mantle beneath the crust may be hot enough to partly melt and create magma. Magma rising upward out of the mantle (*convection*) can bring intense shallow heat into the crust.

(iii) Through pores and crevices in the crust, rain water seeps down to depths of a mile or more and gets heated. The heated water may be stored at depths in geothermal reservoirs, or the hot water may flow upwards out of the reservoir to the surface as hot springs, or boil near the surface to create geysers, mudpots, and fumeroles.

In the past, people have used geothermal energy for the following purposes:

(i) *Bathing* Hot springs were used by ancient civilisations for bathing,

(ii) *Heating* Geothermal energy was used by early Romans to heat their homes.

(iii) *Cooking* Geothermal water was used for cooking by native Americans.

(iv) *Medical Therapy* Geothermal water was used by early Romans to treat eye and skin diseases.

***(A) Present Uses of Geothermal Energy*** Electricity generation, space heating/cooling, greenhouses, aquaculture, drying of fruits/vegetables, and industrial uses like manufacture of paper, washing wool, drying of cloth, etc.

In fact geothermal resources are available in the following temperature ranges (Fig. 2.45).

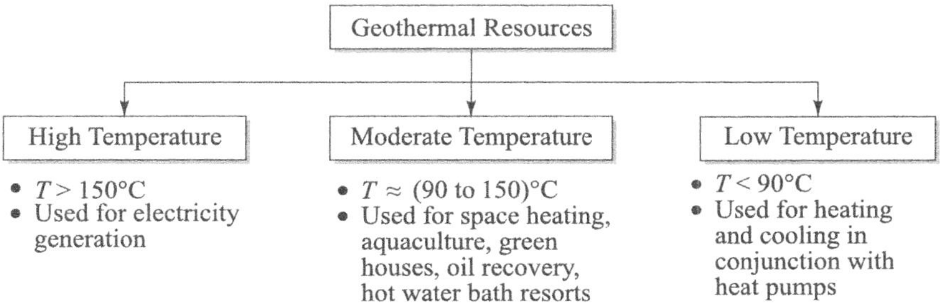


**Fig. 2.45** *Availability of geothermal resources in different temperature ranges*

***(B) Merits of Geothermal Energy*** Some of the advantages of using geothermal energy to generate electricity are given below.

**(i)** ***Flexible*** With growing demand for energy, additional units with modular designs can be installed easily, because there is no major land requirement.

**(ii)** ***Affordable*** The cost of electricity production is almost competitive with conventional energy sources.

**(iii)** ***Sustainable*** It is believed that enough heat will be radiated from the centre of the earth to fulfill human energy demand for all the times to come.

**(iv)** ***Clean Technology*** No emission and safe to use.

**(v)** ***Extraction*** Useful minerals (like zinc and silica) can be extracted from underground water.

**(vi)** ***Reliable*** Energy can be harnessed continuously, Only water is used that can also be recycled.

## (C) Limitations of Geothermal Energy

(i) If the cooled water is not injected back into the reserve after the heat is extracted, the following harmful effects are observed:

(a) Brine can salinate soil.

(b) Land subsidence can occur leading to an increase in seismic activity.

(c) Large quantities of $H_2S$ "the rotten-eggs" gas can be released and inhaling it in too much quantities is fatal.

(ii) Geothermal hot spots are scattered and are at faraway regions than the areas that need energy.

(iii) The overall production efficiency is lower.

(iv) At geothermal sites, drilling operations cause noise pollution.

## (D) Electricity Generation

**(i) Dry Steam Plants** They use underground steam to directly turn the turbines (Fig. 2.46).

**(ii) FlashSteamPlants** These plants pull deep, high-pressured hot water ($T$ = 360°F) to the surface. This hot water is transported to low-pressure chambers and the resulting steam drives the turbines. The remaining water and steam are then injected back into the source from which they were taken (Fig. 2.47).

***Note:*** A temperature gradient of 40°C/km exists is hot dry rock when dug underground. If one digs 20,000 feet underground, the large temperature difference is sufficient to produce electricity.

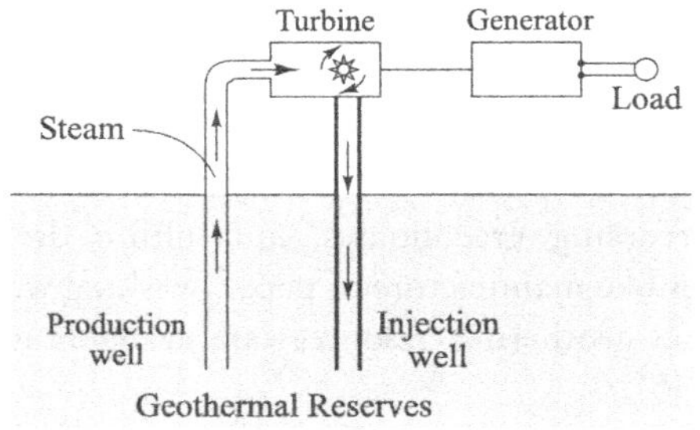


Fig. 2.46 *Dry steam plant*

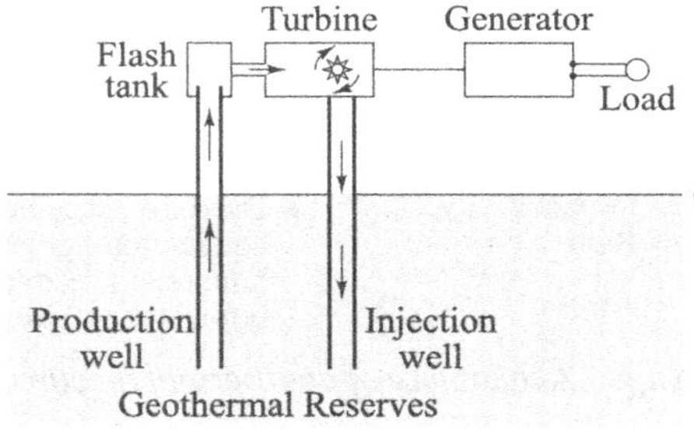


Fig. 2.47 *Flash steam plant*

***(E) Role of Geothermal Energy in India*** Global warming and climate change is and will continue to be one of the key challenges the world has to face along with India in the coming years. Geothermal energy could play a very important role in mitigating these challenges by reducing dependence on fossil fuels and provide clean energy.

India's first geothermal power plant with an initial capacity of 25 megawatts will be coming up in Andhra Pradesh's Khammam district by 2012.

Geosyndicate Power Pvt. Ltd. is an incubated company of the Indian Institute of Technology Bombay. APNPDC is Northern Power Distribution Company of Andhra Pradesh.

India's first geothermal Power Purchase Agreement (PPA) was signed in August 2010 between GeoSyndicate and APNPDC, under the aegis of the Nonconventional Energy Development Corporation of Andhra Pradesh Limited (NEDCAP).

GeoSyndicate is India's one and only Indian geothermal company who will play a significant role in reducing the country's carbon footprint by setting up geothermal plants.

Geothermal power plants emit 0.893 kg $CO_2$ per MWh while coal power plants emit 953 kg $CO_2$ per MWh. The combined geothermal potential of India amounts to $18348 \times 10^{14}$ MWh.

$245 \times 10^6$ MWh (33%) of **electricity** generated from coal based thermal power plants is utilised by **domestic and commercial building sector.** A major amount of this energy is spent for hot-water supply, refrigeration and space cooling. This amounts to emission of $234 \times 10^9$ kg of $CO_2$. India can save about $234 \times 10^7$ euros under CER by utilising geothermal sources. This resource can provide a stable supply of energy, in contrast to many alternative domestic renewable energy resources like wind, hydro and solar photovoltaic.

As per IEA 2007, the Indian food sector uses about 13% of the electricity amounting to $63 \times 10^6$ MWh. This energy is obtained from coal-fired thermal power plants. By using geothermal sources instead of this, $600 \times 10^6$ euros can be raised through Clean Development Mechanism (CDM) and ploughed into this industry. Thus, implementing CDM in **food processing sector** will help in reducing $CO_2$ emission and earning carbon credits.

If emission of $CO_2$, $SO_2$ and $NO_x$ is not controlled, it may lead to severe droughts in India and other developing countries and reduce supplies of clean, fresh water to the point where there are major threats to public health. With global water resources already under severe strain from rapid population growth and expanding economic activity, the danger is clear.

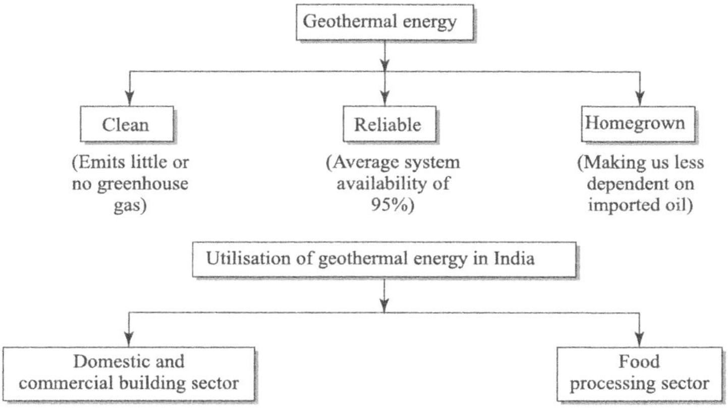


**Fig. 2.48** *Utilisation of geothermal energy in India*

In India, there are nearly 400 low to medium enthalpy thermal springs. These are distributed in seven geothermal provinces. These provinces include the Himalayas, West Coast, Godavari, Mahanadi, Sohana, Cambay, Son–Narmada–Tapi rift zone (SONATA).

The surface temperatures of these thermal springs vary from 47°C to 98°C. These provinces are the sites for commissioning small-scale power projects using binary plant technology. They are also the best sites for direct application technologies in which heat energy is directly used by a variety of small-scale industries. Dehydration of agricultural produce, food processing and food production can utilise this energy with maximum profits (about 5 % to 8% total operating costs).

## 2.7.13 Wind Energy

Air in motion is know as *wind*. It is caused by the uneven heating of the surface of earth by the sun. During the daytime, the air above the land heats up more quickly than the air over water. The warm air over the land expands and rises, and the cooler, heavier air rushes in to take its place, creating wind.

At night, the air cools more rapidly over land than over water; so the winds are reversed.

*The land near the earth's equator is heated more by the sun than the land near the North and South poles. This creats the atmospheric winds that circle the earth.*

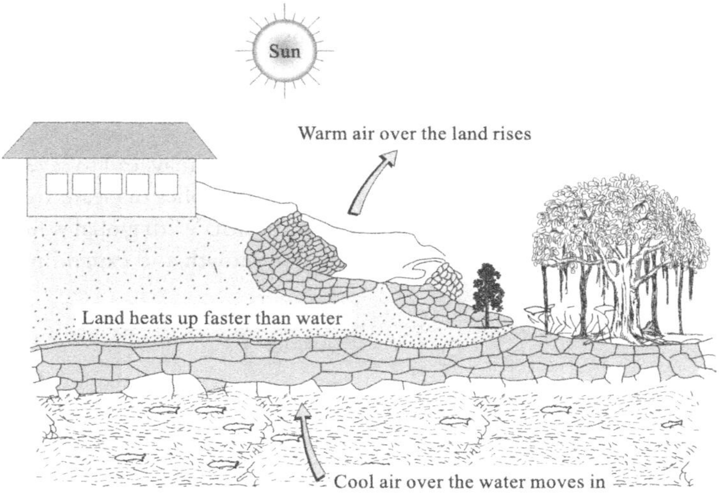


**Fig. 2.49** *Generation of wind*

Wind is a *renewable* energy source because the wind will blow as long as the sun shines.

***(A) Good Sites for Wind Plants*** Over open areas that have no windbreaks and with altitude, wind speed increases. Thus, the tops of smooth, rounded hills, mountain gaps that produce wind funneling, open plains or shorelines are good sites for wind plants.

### (B) Advantages of Wind Energy (a Clean Source of Energy)

(i) Windmills do not release emissions that pollute the water or air.

(ii) They do not require water for cooling.

(iii) Windmills help in reducing the amount of electricity generation from fossil-fuel based thermal power plants. Therefore, they help in reducing air pollution and carbon dioxide emissions.

(iv) Many windmills are located on farm, grazing and forest land. The extra income from windmills help farmers live a better life.

(v) Wind power projects are considered to be best alternatives to mountain-top removal coal-mining projects for thermal power generation.

(vi) Offshore wind turbines on the ocean are considered even more environment friendly than windmills on land.

(vii) Wind is an important renewable and sustainable source of energy, available free of cost.

(viii) Power generation using windmills is very economical. This is because running cost for windmills is almost negligible.

(ix) In large coastal, hilly and desert areas, laying of power transmission lines is extremely difficult and costly leading to unavailability of electric power. In such circumstances, wind energy is very useful.

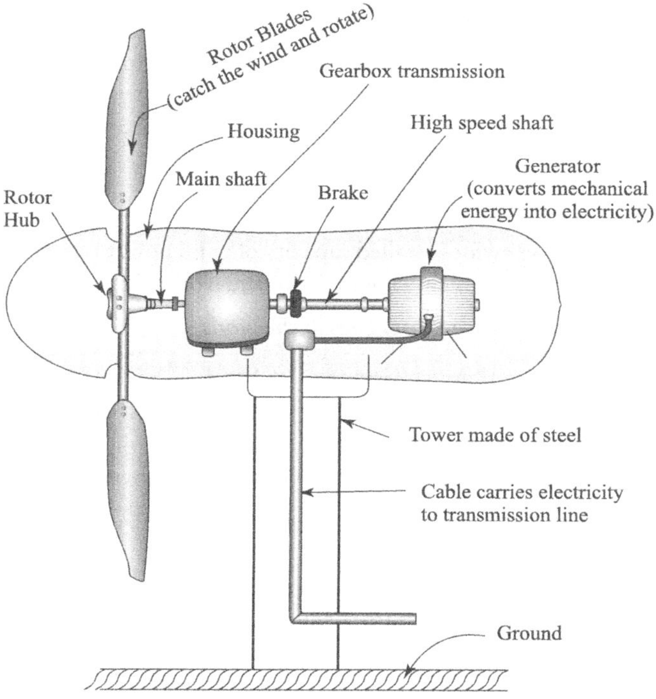


[A windmill essentially consists of a structure similar to large electric fan that is erected at some height on a rigid support. Inflow of wind activates rotor hub and blades. These hub and blades then spin the main shaft and gearbox. Generator in turn is spinned by gearbox, resulting in electrical output]

**Fig. 2.50** *How does a wind turbine work?*

(x) Wind power systems are specially useful for water pumping, battery charging, operating simple machinery, hybrid energy (wind/diesel, wind/photovoltaic) systems, etc., in remote areas.
(xi) The gestation period of installing wind turbines is short.

## (C) Limitations of Wind Energy

(i) Wind energy is variable, irregular, erratic, intermittent, unsteady and sometimes dangerous.
(ii) Location sites for wind farms are generally away from cities.
(iii) Design, manufacture and installation of wind turbines is very complex because of widely varying atmospheric conditions in which they have to operate.
(iv) Continuous whirling and whistling sound associated with rotation of blades of windmills can be very irritating.
(v) The output of a single windmill is quite small and cannot be used for commercial applications. Thus, a large number of windmills need to be erected to make a *wind energy farm*. This requires a very large open area and huge investment of money.
(vi) Cost of maintenance is very high because the towers and blades are exposed to rain, sun, storm and cyclones.
(vii) The location of some windmills are on the routes of migratory birds. The working of windmills in these locations cause bird deaths and decline in their populations. The wind energy industry is researching ways to reduce the impact of wind turbines on birds.
(viii) Initial cost for wind turbines is greater than that of conventional fossil-fuel generators per MW of installed capacity.

## (D) Wind Energy in India and the World

By the end of 2009, the worldwide installed capacity of wind power reached 157,889 MW. Top five wind-power producing countries in the world are listed below:

**Table 2.15** *Wind-power producing countries in the world*

| *Country* | USA | Germany | Spain | China | India |
|---|---|---|---|---|---|
| *Wind power [Installed capacity (MW)]* | 35,159 | 25,777 | 19,149 | 25,104 | 13065.37* |

The installed capacity of wind power as of 31st Dec. 2010

(i) In 2009–10, India's growth rate with respect to wind power was highest among other top four countries.
(ii) Wind power accounts for 6 % of India's total installed power capacity.
(iii) India generates 1.6 % of the country's power using wind.

The actual utilisation of wind power in India is low despite the high installed capacity. This is because policy incentives are for installation rather than operation of the plants. The Indian Government is still considering the addition of incentives for ongoing operation of installed power plants.

(iv) In India, additional 6000 MW of wind power capacity will be installed by 2012.

(v) Suzlon is an Indian-owned company and the leading manufacturer of wind turbines for the Indian market (with 52 % market share). It has made India the developing country leader in advanced wind turbine technology.
(vi) Tamil Nadu is the state with the most wind generating capacity (4906.74 MW) at the end of March 2010.
(vii) Maharashtra (2077.70 MW) is second only to Tamil Nadu in terms of generating capacity.

## 2.7.14 Tidal Energy

Periodic changes of water levels and the associated tidal currents, are due to the gravitational attraction of the sun and moon. At a particular location, the *magnitude of tide is the result of*

(i) the changing positions of the sun and the moon relative to the earth,
(ii) the effects of earth's rotation, and
(iii) the local geography of the sea floor and coastlines.

*Tidal power, or tidal energy,* is a form of hydropower that converts the energy of tides into electricity or other useful form of hydro power. Tidal power is a renewable energy resource because the earth's tides are practically inexhaustible.

### *(A) Power Generation Methods using Tidal Energy*

The following methods are useful:

**(i) Tidal Stream Generators (TSGs)** TSGs make use of the kinetic energy of moving water to power turbines. TSGs use moving water in the similar ways as wind turbines use moving air.

***Advantages*** Lower cost, lower ecological impact compared to tidal barrages.

**(ii) Tidal Barrage** They make use of the potential energy in the difference in height between low and high tides. They are essentially dams across the full width of a tidal estuary.

Motion of turbine blades help in the rotation of a coil of wire in a magnetic field of an electric generator. This induces a flow of electricity in the wire.

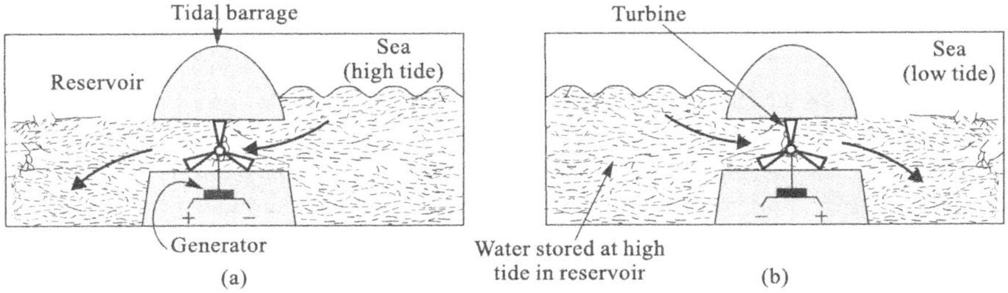


**Fig. 2.51** *Kinetic energy of moving water is used to power turbines for harnessing tidal energy using tidal barrage at (a) high tide, and (b) low tide*

***Drawbacks***

- Very high civil infrastructure costs
- Higher environmental impacts

- A worldwide shortage of viable sites

In the Gulf of Kutch on India's west coast, the Atlantis Resources Company will start construction in early 2012 to install a 50 MW tidal farm. This will be south Asia's first commercial-scale tidal power station.

***(B) Merits of Tidal and Wind Energy Resources*** A few important advantages of using tidal and wind energy are listed below:

- They are clean, renewable sources of energy.
- Consumption of tidal and wind energy does not create pollution.
- Use of tidal and wind energy reduces the dependence upon fossil fuels.
- They are highly efficient resources.
- Power generation using tides or waves does not create greenhouse gases.

***(C) Demerits of Tidal and Wind Energy Resources*** A few important disadvantages of using tidal and wind energy resources are given below:

- The initial construction cost for making the plant for harnessing tidal and wind energy is very high.
- The required technology is not fully developed for harnessing tidal energy.
- Ecological impacts relating to alteration of tides and waves is not fully understood.
- Waves are irregular in size, durability and direction; so they are a diffuse energy source.
- Appropriate waves and tides are highly dependent on location.
- Required technology for transporting electricity from the sea onto the land is yet to be discovered.

## 2.7.15 Advantages and Disadvantages of Various Sources of Energy

**Table 2.16** *Advantages and disadvantages of various sources of energy*

| | *Advantages* | *Disadvantages* |
|---|---|---|
| (1) Coal | • Widespread availability | • Mining of coal is dangerous to man and environment |
| | • Low cost of coal. | • Transportation of coal and its combustion in thermal power plants create pollution |
| | • Easier transportation to power stations | • Nonrenewable |
| | • Technology is known and used and supplying about 70% of energy requirement in India | • Domestic reserves of good-quality coal are very less in India |
| (2) Nuclear | • Low nuclear fuel requirements | • Generates very harmful and radioactive waste |
| | • Energy production does not generate greenhouse gases | • Nonrenewable because limited uranium reserves in the world; environmental concerns regarding digging of uranium mines in India |

(*Contd.*)

(*Contd.*)

| | *Advantages* | *Disadvantages* |
|---|---|---|
| | • Operating cost is about the same as coal-based thermal power plants | • Nuclear power stations are at risk from terrorist attack<br>• Time consuming and expensive development of nuclear power plants |
| (3) Wind | • Wind is free, renewable<br>• Wind power generation does not create greenhouse gases<br>• Wind turbines are relatively safe | • Number of wind turbines in the form of wind farms are needed to generate electricity; costly and requires large open area<br>• Availability of wind is variable<br>• Maintenance cost is also very high |
| (4) Hydro | • Hydro power does not create greenhouse gases<br>• Relatively safe<br>• Renewable | • Electricity production is variable<br>• Expensive |
| (5) Tidal | • Predictable<br>• Renewable<br>• No greenhouse gas emission | • Technological limitations for harnessing and distribution of energy<br>• Dam construction expensive and may harm plants and animals<br>• Tides happen only twice a day so electricity can be produced for that time only |
| (6) Geothermal | • Less costly<br>• Renewable, reliable<br>• No greenhouse gas emission<br>• Very useful for food processing sector and building sector | • Sites for building geothermal power stations are very limited<br>• While harnessing geothermal energy from underground, sometimes harmful gases also come up, which is difficult to control |
| (7) Solar | • Renewable<br>• Sun energy is free<br>• No greenhouse gas emission | • Construction of solar power station is expensive<br>• Electricity cannot be made during cloudy days or during night |

# Case Studies

## *Energy Resources*

### (i) Waste To Energy (WTE)

With the fast depletion of conventional resources and the growing awareness and concern regarding pollution created from their utilisation, there has been a major thrust in the recent past to identify and develop alternate energy resources.

Waste To Energy (WTE) is a proven environment friendly process that provides reliable electricity generation and is extensively used in Europe and other developed nations in Asia. Municipal Solid Waste (MSW) is a waste type that includes predominantly household waste (domestic waste) sometimes with the addition of commercial garbage or refuse that is generated by commercial establishments, industries, etc. Use of municipal solid waste as a fuel source is governed by moisture and energy content of the waste material. The thermal treatment of MSW results in the generation of 500–600 kWh of electricity per ton of MSW combusted.

In a case study of Eluru, AP, India (2010), it is estimated that by combusting the solid waste in Eluru Municipal Corporation, it is possible to generate nearly 3 MW of power.

**(ii) Electrifying Rural India (Husk Power System)**

In August 2007, HUSK Power Systems (HPS) lighted the first village from its power plant. As of August 2010, HPS has installed 60 mini power plants that provide power to 25,000 households in more than 250 villages.

The key advantages of HPS are the following:

(a) HPS electrifies rural India from its 100% biomass based power plant that uses discarded rice husks to generate electricity.

(b) One mini power plant of HPS serves about 400 households and replaces about 42,000 litres of kerosene and 18,000 litres of diesel per year.

(c) HPS power plants help in $CO_2$ sequestration.

(d) HPS has employed and trained more than 300 local people in rural India for running and managing its power plants.

(e) HPS plants are cheaper than the mega-size thermal power plants. The generation, distribution and installation cost of HPS is about $1/W.

[http://www/huskpowersystems.com]

## 2.8 LAND RESOURCES

### 2.8.1 Land as a Resource

*Land resources* are natural resources in the form of productive land. Land resources are essential for the survival and prosperity of humanity. These resources are also essential for the maintenance of all terrestrial ecosystemss.

The basic functions of land in supporting human and other terrestrial ecosystems are given below:

(i) Land is a store of wealth in its own.

(ii) Land is a storehouse of minerals and raw materials for human use.

(iii) Land helps in the production of food, fibre, fuel, etc.

(iv) Land is the biological habitat for many plants, animals and microorganisms.

(v) Land regulates flow of surface water and stores groundwater.

(vi) Land enables or hampers movement of people and animals between one place to another.

(vii) Land is a buffer, filter or modifier for chemical pollutants.
(viii) Land is co-determinant in the global energy balance and the global hydrological cycle, which provides both a source and sink for greenhouse gases.
(ix) Land is the physical space for settlements, industry and recreation.
(x) Land stores and protects evidence of past climates, archaeological remains from the historical or prehistorical record.

## 2.8.2 Texture, Structure and Composition of Soil

Soil is an uncemented aggregate of mineral grains and decayed organic matter with water and air occupying the void spaces between particles.

*Soil texture* refers to the size distribution of soil particles and the relative percentage of sand, silt and clay in a soil. Three sizes of particles are recognised in soil. These are tabulated below:

| *Soil Type* | Fine-Texture Clays | Medium-Textured Silt | Coarse-Textured Sand |
|---|---|---|---|
| *Particle Size* | < 2 μm | (2 – 50) μm | > 50 μm |

The relationship between particle size and class names is shown in Fig 2.52.

It is to be noted that a clay soil remains clay and a sandy soil remains sandy because the size of particles in the soil is not subject to ready change.

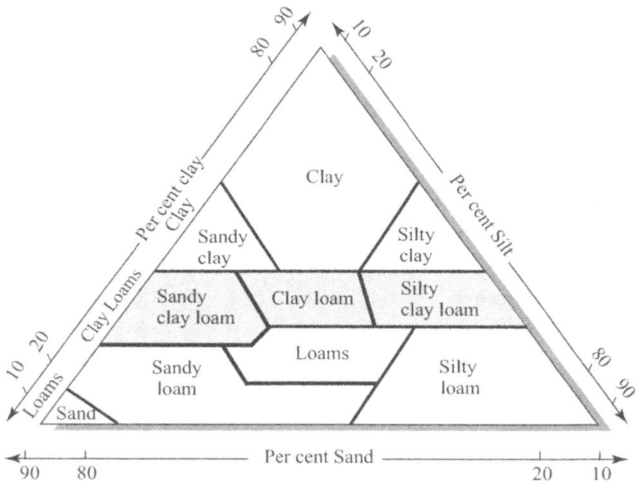


**Fig. 2.52** *Percentages of sand and clay in major soil textural classes: (a) Sandy clay contains more sand than clay. (b) Soils with the best structure for most crops have to retain water and dissolve mineral nutrients. These soils are various types of loams, viz. sandy loam, silty loam, etc.*

Soil texture controls the following properties of soil with respect to plant growth:
(i) aeration,
(ii) availability and movement of water,
(iii) content of plant nutrients, and
(iv) workability.

For instance, movement of water in various soil-textured types is illustrated in Fig. 2.53.

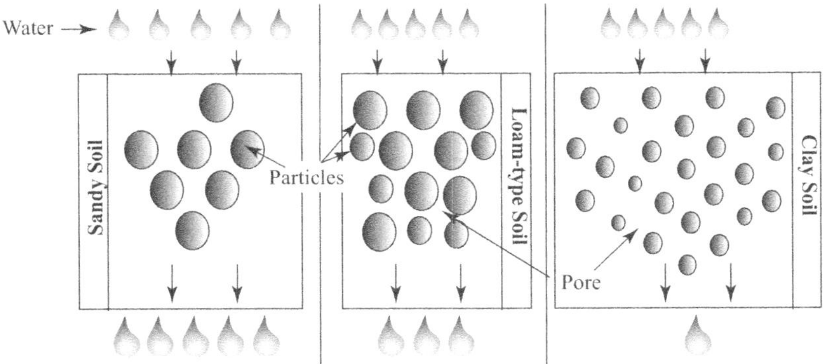


| | | | |
|---|---|---|---|
| **Particle size :** | Big | Big and small (mixture of Sand and Clay) | Small |
| **Number of pores :** | Few | More | Many |
| **Size of pore** | Large | Varying | Small |
| **Water drainage :** | Free and fast | Slow but still well drained | poor and slow, water tends to get trapped |

**Fig. 2.53** *Water drainage from various types of soils*

## (A) Soil Structure

*Soil structure* refers to the arrangement of soil particles into groups or aggregates. The *peds* are natural, fairly water stable aggregates; and *clods* are artificial, water unstable aggregates. Ploughing of wet clay soil makes the soil cloddy.

Soil structure is divided into the following groups:

(a) Types of soil structure.
 ***Examples:*** Platy, spheroidal, etc

(b) Class of soil structure.
 ***Examples:*** Fine, medium, coarse, etc.

(c) Grades of soil structure.
 ***Examples:*** Weak, strong, etc.

## (B) Soil Composition

Four major components of soils are mineral materials, organic matter, water and air. Approximate composition of soil is given below:

| ***Soil Constituent*** | Mineral Materials | Organic Matter | Water | Air |
|---|---|---|---|---|
| ***Volume Per Cent*** | 45 | 5 | 20–30 | 30–20 |

### (i) Mineral Materials

(a) Mineral materials are elements (Si, Fe, O, Mg, Al, Ca, Na and K), finely divided quartz ($SiO_2$), iron-silicates and aluminium silicates.

(b) They are derived from the underlying bedrock or from materials transported and deposited by surface run-off, wind flow, etc.
(c) Secondary minerals (viz., $Na^+$, $K^+$, $Ca^{2+}$, $Mg^{2+}$, $H^+$, $NH_4^+$) are held at the surface of all the silicate clays. These are not leached by water and are exchanged. Thus, they are available as plant nutrients.

**(ii) Organic Matter**

(a) Organic matter can be crop residues, weeds, grasses, bacteria, fungi, other microorganisms and animal manures.
(b) They come from the residues of plants and animals.
(c) *Functions*
- They provide food for microorganisms, other nutrients to plants. Thus, soil productivity is controlled by them.
- They provide a storehouse for nutrients.
- They increasing waterholding capacity of soil.

**(iii) Water** It is a good solvent for many nutrients which move into plant roots. Water is also important to maintain the proper form and position of leaves and new shoots for capturing sunlight and satisfactory growth.

**(iv) Air** Soil air encourages optimum rate of the essential metabolic processes of various organisms.

## 2.8.3 Land Degradation

In the face of the rapidly changing global situation, humanity is not paying attention to manage resources efficiently for sustainable future. Because of this land resources are under stress and the world is facing outcomes of declining crop production, increased land degradation, increased competition for land, etc.

Most persons are trapped in the downward spiral (Fig. 2.54) in which resource-limited farmers have shortage of land resources. They are even further degrading these limited resources by inappropriate farming practices in order to satisfy immediate subsistence needs. Decision-makers often face a situation where requirement is to increase production and reduce poverty, but at the same time resources must be conserved to prevent environmental degradation. Most of the time, short-term political

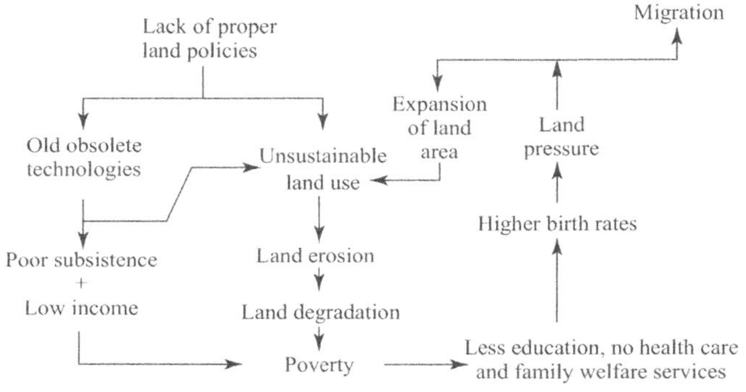


**Fig. 2.54** *Causes of land degradation*

gains have been made at the expense of long-term environmental damage. It is becoming more difficult for technological progress to keep up with the increasing demands generated by population explosion. This is partly a result of the extension of cropping to more marginal areas where risks of failures are higher and productivity is limited by physical factors. Externalities related to global change are also a major constraint to sustainable land management.

It is necessary to apply an integrated planning approach for the sustainable management of land resources (Fig. 2.55). This is because land in different regions varies greatly in productive potential. There is no universal technological solution for meeting human needs while protecting the terrestrial biosphere. This integrated planning approach requires establishment of improved land policies that will enable informed decisions to be made about land resources. Improved access to information and technology and the capacity to use them are essential for right decision making. This will help in sustainable land use and higher production per unit area.

**Fig. 2.55** *Sustainable land use*

***(A) Landslides*** A *landslide* means ground movement in offshore, coastal and onshore environments. This geological phenomenon includes deep failure of slopes, rock falls and shallow debris flows.

**Causes** When the stability of a slope changes from a stable to an unstable condition (due to either natural or human causes), landslides occur.

***(a) Natural Landslide*** Causes include groundwater pressure acting to destabilise the slope, earthquakes adding weight to barely stable slopes, earthquakes causing liquefaction, destabilising slopes, and volcanic eruptions. These causes are illustrated in Fig. 2.56.

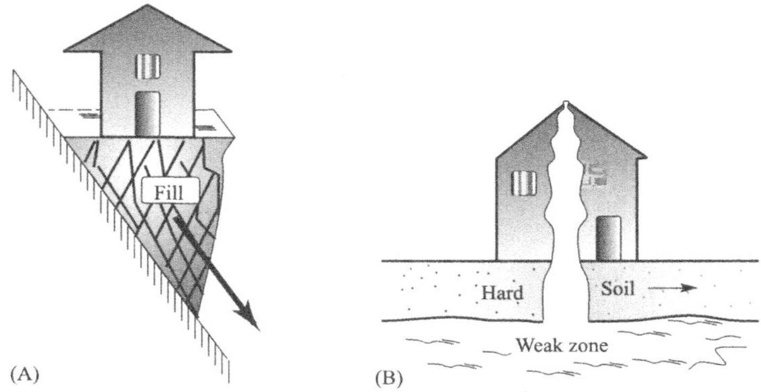


**Fig. 2.56** *Landslides as a result of translational ground movements. (A) Landslide caused by fill placement (B) Landslide caused by earthquake*

***(b) Human Induced Landslides*** Human activities aggravate landslides by destabilising already fragile slopes. Human causes include vibrations from machinery or traffic; blasting; the removal of deep-rooted vegetation that binds colluvium to bedrock in shallow soils; construction, and agricultural or forestry activities which change the amount of water which infiltrates the soil.

***(B) Soil Erosion*** *Soil erosion is the wearing away of soil or geological material from one point on the Earth's surface to be deposited elsewhere.*

Normally, soil erosion is a natural process which occurs over geological time scales. It occurs because of various physical forces such as rainfall, flowing water, wind, temperature change, gravity, etc.

By anthropogenic activity, rate of soil erosion significantly increases compared to the natural rate and it becomes a process of degradation. Tillage, mechanical influences, etc., are anthropogenic factors.

The natural factors of soil erosion are wind and water. Wind velocity and rain-splash dislodges soil particles that initially fly through the air (saltation). Slope wash causes overland flow. Wave motion in streams, rivers and lakes causes slumping of bank material. The various forces responsible for soil erosion are illustrated in Fig. 2.57.

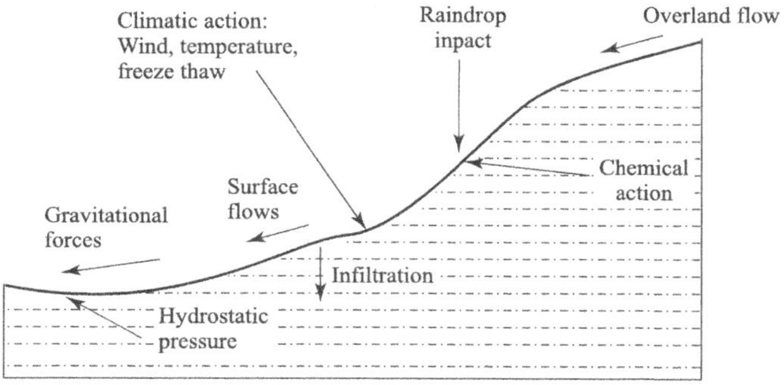


**Fig. 2.57** *Forces responsible for soil erosion*

**Table 2.17** *Consequence of soil erosion*

| ***On-site*** | ***Off-site*** |
|---|---|
| (i) Loss of productivity of soil | (i) Diffuse pollution of soil and water by contaminants and nutrients |
| (ii) Reduction of functional capacity of soil | (ii) Destruction of habitats, damage to property, infrastructure, etc |
| | (iii) Silting of lakes, reservoirs and river courses |

## *(C) Desertification*

*Desertification, or land degradation, means irreversible decline in the 'biological potential' of the land.* It is a process in which the soil loses its productivity and becomes infertile.

**(i) Causes** Desertification is caused by natural and/or man-made reasons. Some of these are climate change, deforestation, overcultivation, overgrazing, dam construction, mining, urbanisation, industrialisation, etc.

**(ii) International Efforts to Prevent Desertification** The following steps should be taken at international and national levels to control desertification:

- Sustainable usage of available groundwater resources
- Management of land use and farming techniques, etc; control of overgrazing
- Generation of public awareness, imposing ban on cutting of desert vegetation
- Plantation of ecologically suitable plants

## 2.9 CONSERVATION OF NATURAL RESOURCES

***(A) Need for Conservation of Natural Resources*** *Conservation* means wise use. We need to conserve natural resources so that present and future generations will be able to use natural resources to fulfill their needs of food, shelter and clothing.

There is an urgent need for the conservation of natural resources because overutilisation of natural resources has totally disturbed the ecological balance of the environment leading to pollution, extinction of species and disturbance in the food chain. This measurable degree of damage up to and including a state beyond use or repair is known as destruction.

***(B) Methods and Tools for Conservation of Natural Resources***
People belonging to different disciplines should take care of the following:

(i) Reduce usage of nonrenewable natural resources such as fossil fuels.
(ii) Preferably use renewable and environment-friendly energy resources like solar energy, wind energy, etc.
(iii) Avoid deforestation and conserve forest resources.
(iv) Help in afforestation by planting trees along the roadside, wastelands, etc., through social forestry programmes.
(v) Use biogas for cooking instead of fuel wood in rural areas.
(vi) Turn off unused appliances to reduce energy requirements.
(vii) Reduce usage of vehicles for energy saving avoiding pollution and conserving natural resources.
(viii) Preferably use ecofriendly products made through green technology.
(ix) Use less paper, reuse paper and recycle paper for saving forest and water resources.
(x) Donate money for conservation activities.
(xi) Reduce the use of chemical fertilisers. Use compost. It is a mix of dead plants and dirt which is buried underground for a few weeks. The compost can then be digged out for addition to soil and making it fertile and healthy.
(xii) Help governments to protect an entire wetland or a large section of tropical rainforest.

(xiii) Find more creative conservation solutions and then share ideas with many for timely action.

(xiv) Make a plan based on good natural resource information for the rational use of natural resources. For this, take help from satellite imagery and maps, GIS based analysis, real-world 3D modelling, statistical analysis of ecosystem services through software and many more.

## 2.10 SUSTAINABLE LIFESTYLES

*Sustainable lifestyle* or sustainable living is the application of sustainability to lifestyle choices and decisions.

Sustainable lifestyle attempts to reduce an individual's or society's use of the natural resources by shifting to a renewable energy-based, reuse/recycle economy with a diversified transport system.

***(A) Need for Sustainable Lifestyles*** The major cause of the continued deterioration of the global environment is the unsustainable patterns of consumption and production, particularly in the industrialised countries. Sustainable lifestyles and practices in agriculture, forestry, fishing, and manufacturing need to be maximised so that pollution and waste an be minimised.

### *(B) Relationship Between Population and Food Production*

For survival, people need to eat food. Thus, when the population increases, the food production will have to increase.

When there is a lack of land and water to grow food, there are large chances for a lack in food supply to meet the hunger need of the rising population. This leads to more starvation deaths which limits the population growth.

Impact on food production ($I$) is governed by population ($P$), Affluence ($A$) and Technology ($T$) as per the following relation:

$$I = f(P \times A \times T)$$

**(i) Population (*P*)** Keeping other factors constant, the more people there are, the greater is the impact on the environment, and the greater is the impact on food production capacity.

In many poor countries, rapid population growth directly results in land fragmentation. Often landholdings which are too small to provide a sufficient livelihood are sold to wealtheir landowners, making land distribution uneven and adding to the increase in landless labourers. As more land is in large holdings, the poor are forced to live on small holdings or in marginal areas.

Rapid population growth can lead to inappropriate farming practices, (like overuse and improper use of agrochemicals, water resources, etc.) resulting in severe land degradation and decline in food production.

When population growth is unchecked, it grows geometrically. However, food supply generally grows arithmetically because of a limited area of cropland. This leads to famine and rise in the death rate so as to balance population size and its food supply.

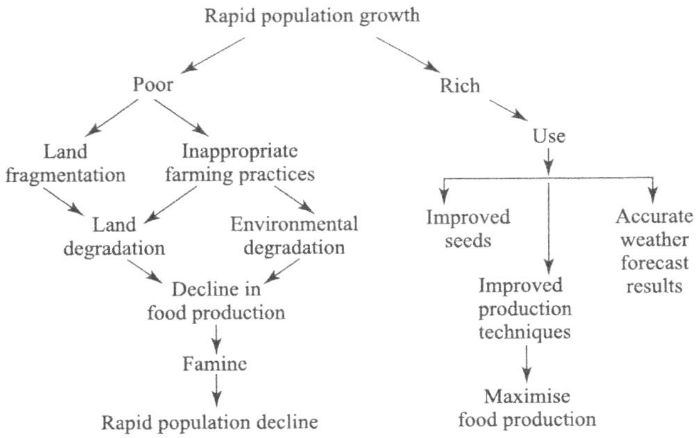


**Fig. 2.58** *Relationship between population and food production.*

**(ii) Affluence (*A*)** Lifestyle, income and social organisation determines the level of consumption of food. Poverty may prevent the adoption of a more appropriate technology that could halt or slow down environmental degradation. Affluent and awarded people can easily practice improved resource management technique for increasing crop yields.

**(iii) Technology (*T*)** The technologies in practice determine the extent to which human activities damage or sustain the environment. It also determines the amount of waste associated with a given level of consumption.

By using improved production techniques, improved seeds, accurate weather forecasts, etc., farmers can maximise agricultural food production.

## *(C) Equitable Use of Resources for Sustainable Lifestyles*

By equitable use of resources, it is possible to improve energy security while reducing the impact on climate. It is also possible to improve food and water security while reducing biodiversity degradation. This is because to generate economic value, it is a must to generate the resources on which it relies. Our prosperity is dependent on the prosperity of biodiversity (Fig. 2.59).

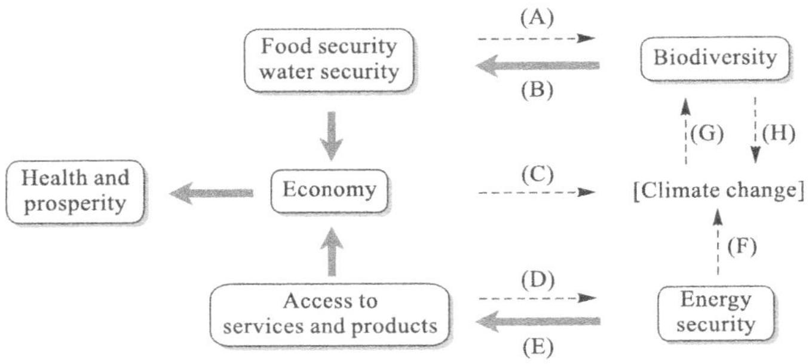


**Fig. 2.59** *How health and prosperity are linked with energy security and biodiversity. (Source: newanthropocene.wordpress.com)*

(A) Present approaches to secure food and water availability are leading to increasing natural environment fragmentation, pollution and unsustainable harvest. These approaches are harming biodiversity.

(B) Biodiversity provides the following valuable services:
   (a) Aerates and improves soil, water and air quality
   (b) Composes new, useful material
   (c) Decomposes spent material
   (d) Converts sunlight to chemical energy, etc.

   Thus, conservation of biodiversity is a must for food and water security.

(C) As economy improves, green cover on land is reduced, concrete cover increases, emission of greenhouse gases increases, protection from floods, cyclones, etc., decreases. All these leads to intensification of climate change effects.

(D) Energy availability governs the access of products and services. The demand for energy is rising day by day. As a result, fuel price is increasing, increasing price of consumables/personal transport/energy reliant services.

(E) As efficiency and security of energy improves, access to products and services likewise improves.

(F) When access to goods and services is dependent upon fossil fuels, it impacts climate change through increasing greenhouse gas emissions.

(G) Climate change results in heat and storm surge so it impacts all aspects of biodiversity, from distribution to damage.

(H) As the effects of food and water security increase, ecological services (such as green cover, carbon sequestration, storm surge protection) are reduced. This results in increasing the effects of climate change.

As is illustrated in Fig. 2.59, it is obvious that our wealth presently depends on undermining the wealth of natural resources and energy requirements. Thus, to have prosperous and sustainable societies, it is a must to improve (B) and (E) while reducing (A), (C), (D), (F), (G) and (H).

## 2.11 SUSTAINABLE WATER MANAGEMENT (SWM)

*The purpose* of *sustainable water management* is simply to manage our water resources while taking into account the needs of present and future users.

Sustainable water management is based on the following *principles* devised in 1992 during the International Conference on Water and the Environment (ICWE):

(i) Freshwater is a finite and valuable resource. It is essential for sustaining life, environment and development.

(ii) Users, planners and policymakers, all must participate in development and management of our water resources.

(iii) Women play a central role in the provision, management and safeguarding of water resources.

(iv) Economic and policy incentives must be given to manage water resources properly.

### (A) Problem or Factors Limiting the Adoption of SWM in India

**(i) Policy Failures and Institutional Weaknesses** Only government agencies are involved for water development projects (with full funding). There are policy failures on whether the demand for water was met by public, cost recovery and continued abuse of water resources.

**(ii) Competition for Water** Even presently, the urban, agricultural and industrial demands for water are greater than the available supplies.

**(iii)** Past projects did not fully address the health and environmental needs.

### (B) Action Plan for SWM

**(i) Right Planning** Environment, health, agriculture and other sectors must cooperate for the management of water resources. Each must understand the need of the other group and must be accountable for its actions. Policies should encourage demand management rather than continuing with the old policies of developing supplies of water resources.

**(ii) Optimisation of the Use and Distribution of the Current Water Supplies** This can be done through appropriate actions for water conservation. For this, all should join hands for

- (a) reducing wastage and leakage,
- (b) regulating demand,
- (c) using low-flow technologies,
- (d) reclaiming waste water, and
- (e) reusing water, etc.

**(iii) Right Budgeting** The needs of the ecosystem to maintain its functions must be included in water-balance budgets. The government should facilitate generation of sufficient data on the links between health and water quality.

**(iv) Right Decision-making Process** Government, industry, law and public, all should be involved in the decision-making process.

## 2.12 BIOGEOCHEMICAL CYCLE

A circulating pathway by which a chemical (an element or a molecule) moves through both biotic (*bio*) and abiotic (*geo*) compartments of an ecosystem is termed the *biogeochemical cycle*. The most important biogeochemical cycles are those of carbon, nitrogen, water, etc.

Biogeochemical cycles are essentially the continuous transport and transformation of materials in the environments and so these are also termed *material cycles*.

In natural systems, the flow of material and energy is circular. The wastes that living creatures excrete are degraded and become nutrients to living organisms. Likewise, when living creatures die, they too biodegrade to nutrients.

## 2.12.1 The Carbon Cycle

The *carbon cycle* is a complex series of processes through which all of the carbon atoms in existence rotate.

In the carbon cycle, plants absorb carbon dioxide from the atmosphere and through photosynthesis convert this $CO_2$ and water into oxygen and carbohydrates, which they need for growth. Animals breathe in this oxygen, eat the plants and use the carbon of carbohydrates to build their own tissues. These animals return carbon dioxide into the air, when they breathe and when they die, as the carbon is returned to the soil during decomposition. The carbon atoms in the soil may then be used in a new plant or small microorganisms. When we burn fossil fuels like oil, the carbon in the fuel combines with atmospheric oxygen to form carbon dioxide (Fig. 2.60). Every ton of $CO_2$ emitted from combustion of fossil fuels changes the carbon cycle for thousands of years as it upsets the carbon balance in the atmosphere. The current excessive levels of $CO_2$ lead to global warming, climate change, floods, droughts, etc.

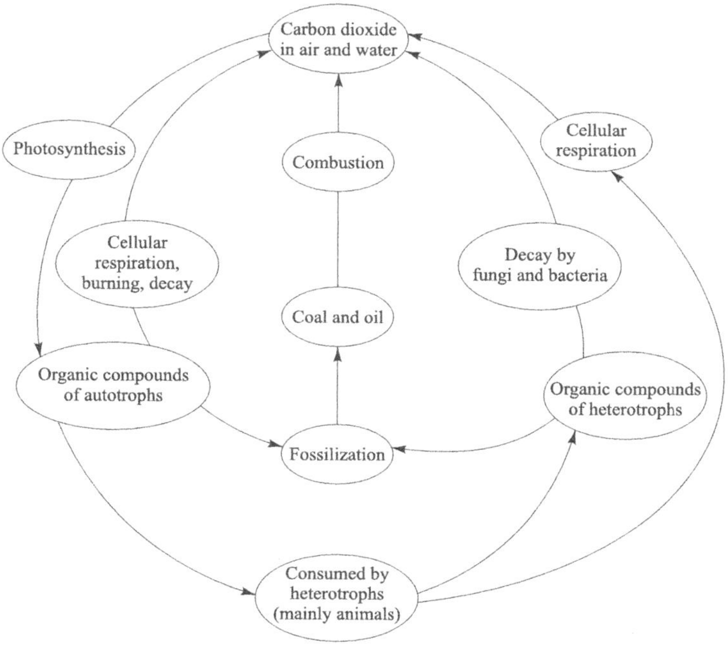


**Fig. 2.60** *The carbon cycle*

*Carbon exists in the nonliving environment as*

(i) Carbon dioxide in the atmosphere and by forming bicarbonates, it gets dissolved in water

(ii) Carbonate rocks (like limestone $CaCO_3$)

(iii) Deposits of coal, petroleum, and natural gas derived from once-living things

(iv) Dead organic matter, e.g. humus in the soil

*Carbon enters the biotic world through the action of producers (or autotrophs)*:

(i) Primarily *photoautotrophs* (like plants).
They use the energy of light to convert carbon dioxide to organic matter.

(ii) And to small extent, *chemoautotrophs* (like bacteria).
They do the same but use the energy derived from an oxidation of molecules in their substrate.

*Carbon returns to the atmosphere by*

(i) respiration (as $CO_2$),

(ii) burning or combustion of fossil fuels, and

(iii) decay (producing $CO_2$ if oxygen is present, methane if oxygen is not present).

However, the uptake and return of $CO_2$ are not in balance. The $CO_2$ content of the atmosphere is gradually and steadily increasing.

## 2.12.2 The Nitrogen Cycle

Nitrogen and its compounds are essential constituents of protein which is the building block of all living organisms. The ultimate source of nitrogen is atmospheric nitrogen but plants or animals are incapable of assimilating free nitrogen.

The process of converting atmospheric nitrogen to useful nitrogenous compounds by plants, passing it to animals and then the decomposition of these compounds to give back free nitrogen in the atmosphere is called the nitrogen cycle (Fig. 2.61). The percentage of nitrogen remains constant in the air because of the nitrogen cycle.

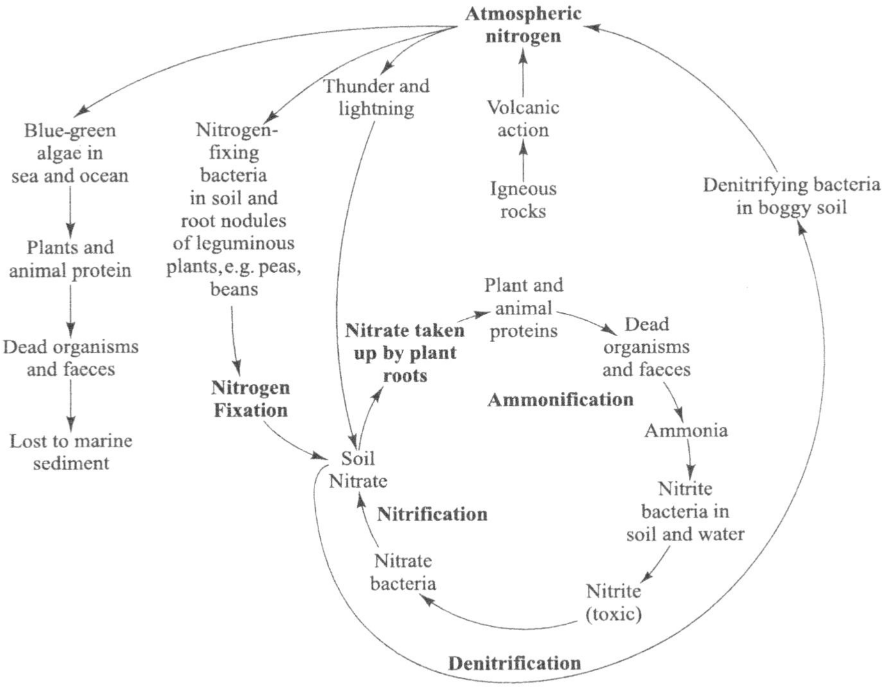


**Fig. 2.61** *The nitrogen cycle (detailed)*

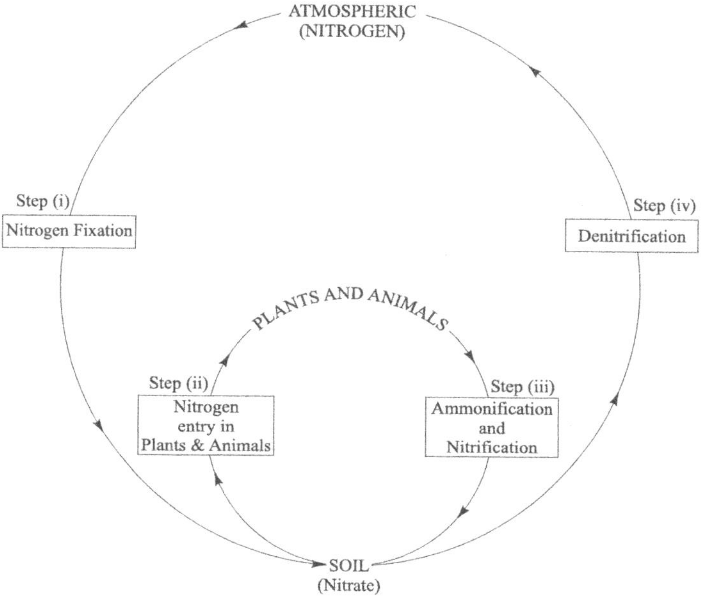


**Fig. 2.62** *The simplified nitrogen cycle*

The nitrogen cycle consists of following steps:

**Step (i) Fixation of Nitrogen** The process of conversion of free nitrogen of the air to useful nitrates is termed *nitrogen fixation*. This can be achieved by

(a) Microorganisms like *rhizobium* (a nitrogen-fixing bacteria) present in the roots of leguminous plants.
(b) Atmospheric thunder and lightning
(c) Addition of compost and fertilisers.

**Step (ii) Entry of Nitrogen in Plants and Animals** The nitrates in the soil are absorbed by plants as mineral salts. Plants convert the inorganic nitrates to organic proteins. The proteins from the plants enter the body of animals in the form of food.

**Step (iii) Ammonification and Nitrification Processes for Returning Nitrogen to the Soil** When plants and animals die, their bodies decompose. The organic proteins undergo a series of chemical changes, brought about by microorganisms, which convert the proteins back to inorganic nitrates.

In humans and animals, some proteins are broken down to ammonia and carbon dioxide, which forms urea and is excreated out in urine. Waste matter of animals and humans is also treated by microorganisms to convert it into inorganic nitrates.

**Step (iv) Denitrification** Denitrifying bacteria found in the soil convert nitrates of the soil to free nitrogen which escapes to the atmosphere and thus completes the cycle and maintains the atmospheric balance of nitrogen concentration.

***Impacts of Human Activities on the Nitrogen Cycle*** Human activities have altered and upset the nitrogen balance. The application of nitrogen fertilisers to crops has caused increased rates of denitrification and leaching of nitrates into groundwater streams, rivers, lakes, etc. The added nitrogen in these systems leads to eutrophication.

### 2.12.3 The Sulphur Cycle

Sulphur enters the atmosphere through both natural and human sources in the form of oxides of sulphur ($SO_x$). It reacts with rain and falls into earth as acidic sulphate ($SO_4^{2-}$) deposition. The sulphate is absorbed by plants as it is required for making amino acids, proteins, etc. Animals consume these plants, so that they take up enough sulphur to maintain their health. This is because sulphur is important for the functioning of enzymes and proteins.

A simplified version of the pathways, transformations, and chemical species in a sulphur cycle is illustrated in Fig. 2.63.

(i) Sulphate ($SO_4^{2-}$) is reduced to hydrogen sulphide ($H_2S$) by Sulphate-Reducing Bacteria (SRB).

(ii) Some sulphate is assimilated by organisms to form cell components such as amino acids and co-factors.

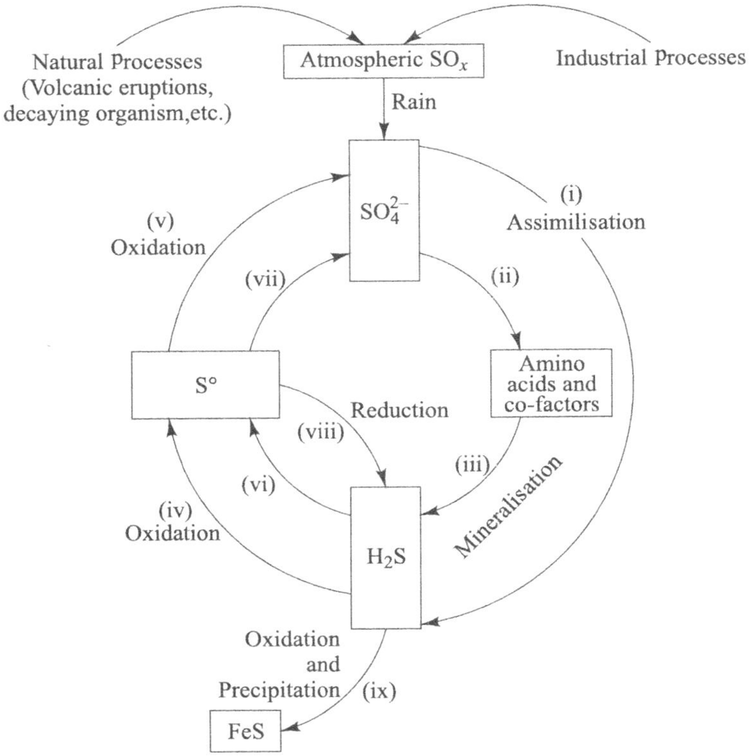


**Fig. 2.63** *The sulphur cycle*

(iii) Organic sulphur is converted to $H_2S$ upon mineralisation.
(iv) $H_2S$ is transformed to elemental sulphur ($S^o$).
(v) Sulfide oxidising bacteria convert $S^o$ into $SO_4^{2-}$.
(vi & vii) Anoxygenic phototrophic bacteria also convert $H_2S$ to $SO_4^{2-}$ via elemental sulphur.
(viii) Sulphur-reducing bacteria transform back the elemental sulphur to $H_2S$.
(ix) Some $H_2S$ complexes with iron to form black FeS precipitates, whose recycling is slow.

***Role of Human Activities in Sulphur Cycle*** Human activities such as coal mining, petroleum extraction and burning of coal and natural gas result in increase of sulphur content in the atmosphere.

$SO_2$ gets released as air pollutant. It reacts with moisture in the air and sulphuric acid is formed. It is responsible for acid rain which causes severe damage to natural and man-made ecosystems.

## Case Studies

### (i) Packaged and Processed Foods

(a) Canned foods with large amounts of sodium or fat.
(b) Breads and pastas made with refined white flour instead of whole grains. They have potassium bromate to increase their volume.
(c) Frozen fish sticks and frozen dinners contain excessive quantities of sodium.
(d) Packaged chips, candies, cakes and cookies have high calories.
(e) Processed meats such as ham, hot dogs, sausage, etc., contain sodium nitrate for maintaining colour and increasing shelf life.

**Commonly used preservatives used in packaged and processed foods and their harmful effects are listed below:**

- *Calcium, Potassium and Sodium Salts* These result in high blood pressure, kidney damage, worsening of heart-related diseases, calcification of tissues etc.
- *Sulphites* They worsen asthma.
- *Boric Acid* It causes cancer damage to kidneys and testis.
- *Sodium Benzoate* Responsible for allergies and liver problems.
- *Phosphoric Acid* Responsible for bone weakening, kidney stones and kidney damage.
- *Sodium Nitrate* It is known to worsen asthma and decrease lung function.
- *Potassium Bromate* When breads and other bakery products are not cooked long enough or not at a high temperature, a residual amount of this oxidising agent will remain in bread, which may be harmful if consumed.
- *Sodium Benzoate and Benzoic Acid* These are food preservatives found in sauces, fruit juices, jams and pickled products. If used over the permissible limit, regular intake of products containing these can cause allergic reactions and other side effects.

### Solutions

(a) Always read the label of pre-packaged food products for chemical content.
(b) High BP patients, kidney patients and heart patients should avoid too much of packaged food, particularly meat items that have high salt content.
(c) Choose fresh food over processed foods.

### (ii) Population in India

***(a) As per census 2011***

(i) Population of India is 1.2 billion.
(ii) It is almost equal to the combined population of the US, Indonesia, Brazil, Pakistan, Bangladesh and Japan (1214.3 million).
(iii) Absolute addition (181 million since 2001) is slightly lower than the population of Brazil, the fifth most populous country in the world.
(iv) Uttar Pradesh (with 200 million people) is the most populous state.
(v) Lakshadweep is the least populated at 64,429.
(vi) India now makes up 17.5 % of the world's population.

***(b) 586.5 million Females and 623.7 million males.***

**Table 2.18**

| *Decade* | *Population Growth Rate (%)* |
|---|---|
| 1951–1961 | 21.64 |
| 1961–1971 | 24.80 |
| 1971–1981 | 24.65 |
| 1981–1991 | 23.87 |
| 1991–2001 | 21.54 |
| 2001–2011 | 17.64 |

(vi) 60% of the population lives in the top seven states.
(vii) Uttar Pradesh and Maharashtra together now are home to more people than those in the United States of America.

## *Important Definitions*

- *Natural resources* are objects, materials, creatures, or any form of energy found in nature that can be used to perform any useful function.
- *Perpetual resources* are those natural resources that naturally perpetuate themselves and are not affected by human use.
- *Renewable resources* are those natural resources that have the inherent ability to renew or replenish themselves if given a reasonable amount of time.
- *Nonrenewable resources* are those natural resources that cannot be regenerated or renewed or replaced within a time framework.
- *Intangible resources* are those natural resources that are available in huge quantities, but at the same time can be destroyed easily.
- *Biotic resources* have originated from some living organism or have life.
- *Abiotic resources* are of nonliving origin.

- *Forests* help in production of timber, regulation of stream flow, control of erosion, recreation, provision of wildlife habitat, etc.
- *Deforestation* can be defined as the change of forest with depletion of tree crown cover of more than 90%. However, depletion of forest-tree-crown cover less than 90% is considered as *forest degradation*.
- *Timber extraction* is the removal of timber from forests.
- *Logging* is the work or business of felling and trimming trees and transporting the logs to a mill.
- *Mining* in the extraction (renewal) of metals and minerals from the earth.
- A *dam* is a huge and giant barrier constructed across a river to obstruct its natural flow.
- *Water resources* are sources of water that are useful or potentially useful to humans.
- Life is impossible without water. It is needed for health, ecosystem services, economic development, poverty reduction, protection of greenery, production of food and imparting of aesthetic beauty.
- *Water conservation* is the most cost-effective and eco-friendly way to reduce our demand for water.
- *Water-borne diseases* are illnesses caused by consuming water contaminated by pathogenic microorganisms.
- *Flood* is a temporary covering by water of land not normally covered by water.
- *Drought* may be defined as the deficiency of rainfall (relative to the statistical multi-year average for a region ) over an extended period of months or years.
- *Meteorological drought* is brought about when there is a prolonged period with less than average rainfall.
- *Agricultural drought* is a drought that affects crop production or the ecology of the range.
- *Hydrological drought* is brought about when the water reserves available in sources such as reservoirs, lakes or aquifers fall below the statistical average.
- Natural resources in the form of minerals are known as *mineral resources*.
- A mineral which can be extracted and processed at a profit is known as an *ore*.
- *Food security* is defined as the physical and economic access to sufficient, safe, nutritious food for maintaining a healthy and active life.
- *Undernourishment* means to receive less calories than needed.
- *Malnourishment* means lack of essential nutrients (like proteins, vitamins, lipids, minerals, etc.) in the diet.
- *Overnourishment* is due to overconsumption of saturated (animal) fats, sugar and salt in diet.
- *Overgrazing* can be defined as grazing plants (by livestock or wildlife) for extended periods of time or without sufficient recovery periods.
- *Agriculture* means the cultivation of soil and or production of crop plants or livestock products. It is synonymous with *farming*: the field-dependent production of food, fodder, and industrial organic material.
- The term *modern agriculture* depicts the push for innovation, stewardship and advancements continually made by farmers to sustainably produce higher quality products with a reduced environmental impact.
- *Organic farming* is based on development of biological diversity and the maintenance and replenishment of soil productivity through crop rotation, use of animal and green manures, and some forms of biological control of pests.
- *Sustainable agriculture* means meeting the food needs of the present generation without endangering the resource base for the future generation.
- *The Green Revolution* refers to the rapid increase in food production mainly through the use of High-Yielding Varieties (HYVs) of seeds, chemical fertilisers, pesticides, agriculture machinery and water.

- *Green Revolution* means better food-production methods or better agricultural practices to secure the world food supply.
- *Fertilisers* are substances (natural or synthetic) that are added to the soil to restore and enhance the soil fertility to improve the quality and quantity of plant growth.
- *Pest* is any organism that causes an economic loss or damage to the physical well being of human beings.
- *Risk (damage)* by pesticide is determined by both toxicity (hazard) and the likelihood of exposure.
- There is an increase in concentration of pesticides in successively higher trophic levels of a food chain or food web in a process known as bio-magnification.
- *Waterlogging* means too much water in the root zone of a plant so that roots cannot absorb enough oxygen to breathe.
- *Salinity* means accumulation of soluble salts in the soil due to over-irrigation.
- *Energy* is the ability to do work.
- *Conventional energy sources* are energy sources which are nonrenewable. However, *nonconventional energy sources* are energy sources which are renewable and ecologically safe.
- *Hydroelectricity or hydroelectric power* is the electricity obtained by harnessing the power of water flowing down from a high level.
- Petroleum and coal are formed from the fossilised remains of animals and plants, hence they are known as *fossil fuels*.
- *Coal* is defined as stratified rock, consisting of organic matter of fuel value derived from the partial decay and alteration of accumulated plant materials by the action of heat and pressure over millions of years.
- *Petroleum* is a complex mixture of paraffinic, olefinic and aromatic hydrocarbons with small quantities of organic compounds containing oxygen, nitrogen and sulphur.
- *Nuclear fission*: When an unstable nucleus of a heavy atom (like uranium–235) is bombarded with neutrons, the former splits up into two medium weight nuclei with the liberation of an enormous amount of energy.
- *Nuclear fusion*: When two lighter nuclei (like deuterium atoms) are heated to a very high temperature ($\sim 10^6$°C), they fuse together to form a heavy, more stable nucleus and an enormous amount of energy is liberated.
- *Solar energy* is the energy received by the earth from the sun that is converted into thermal or electrical energy. Solar energy influences the earth's climate and weather and sustains life.
- The term *biomass* is used for the dead plants and trees (e.g. wood, crop residue, etc.) and the waste material of living organisms (e.g. cattle dung, sewage, etc.).
- *Biogas* consists mainly of methane. It burns with a blue flame and its average calorific value is about 5300 kcal/$m^3$.
- *Green fuel* (or *biofuel*) is a type of fuel obtained from nonfood sources like green algae which is more environmental friendly than the widely used and quickly disappearing fossil fuels.
- *Hydrogen* is a very light gas, and its density is eight times less than that of natural gas. There are no significant problems with regard to transportation, dispensation as well as end use of hydrogen.
- *Oceans* cover more than 70% of the earth's surface making them the world's largest solar collectors. Just a small portion of the heat trapped in the ocean could power the whole population across the globe.
- *Geothermal energy* is a clean, renewable and environment-friendly energy source based on the heat inside the earth.
- Air in motion is know as *wind*. It is caused by the uneven heating of the surface of earth by the sun.

- *Wind* is a renewable energy source because the wind will blow as long as the sun shines.
- *Tidal power*, or *tidal energy*, is a form of hydropower that converts the energy of tides into electricity or other useful form of hydro power. Tidal power is a renewable energy resource because the earth's tides are practically inexhaustible.
- *Land resources* are natural resources in the form of productive land. Land resources are essential for the survival and prosperity of humanity. These resources are also essential for the maintenance of all terrestrial ecosystems.
- *Soil* is an uncemented aggregate of mineral grains and decayed organic matter with water and air occupying the void spaces between particles.
- *Soil texture* refers to the size distribution of soil particles and the relative percentage of sand, silt and clay in a soil.
- *Soil structure* refers to the arrangement of soil particles into groups or aggregates. The *peds* are natural, fairly water stable aggregates; and *clods* are artificial, water unstable aggregates. Ploughing of wet clay soil makes the soil cloddy.
- A *landslide* means ground movement in offshore, coastal and onshore environments. This geological phenomenon includes deep failure of slopes, rock falls and shallow debris flows.
- Soil erosion is the wearing away of soil or geological material from one point on the Earth's surface to be deposited elsewhere.
- *Desertification*, or *land degradation*, means irreversible decline in the 'biological potential' of the land. It is a process in which the soil loses its productivity and becomes infertile.
- *Conservation* means wise use. We need to conserve natural resources so that present and future generations will be able to use natural resources to fulfill their needs of food, shelter and clothing.
- The purpose of *sustainable water management* is simply to manage our water resources while taking into account the needs of present and future users.
- *Sustainable lifestyle* or sustainable living is the application of sustainability to lifestyle choices and decisions.
- By *equitable use of resources*, it is possible to improve energy security while reducing the impact on climate. It is also possible to improve food and water security while reducing biodiversity degradation.
- A circulating pathway by which a chemical (an element or a molecule) moves through both biotic (bio) and abiotic (geo) compartments of an ecosystem is termed the *biogeochemical cycle*.
- Biogeochemical cycles are essentially the continuous transport and transformation of materials in the environments and so these are also termed *material cycles*.
- The *carbon cycle* is a complex series of processes through which all of the carbon atoms in existence rotate.
- The process of converting atmospheric nitrogen to useful nitrogenous compounds by plants, passing it to animals and then the decomposition of these compounds to give back free nitrogen in the atmosphere is called the *nitrogen cycle*.

# EXERCISES

**Based on Natural Resources**

**1.** What are the different types of natural resources?

**2.** Distinguish between reserves and resources.

**3.** Give the classification of natural resources.

4. Discuss the associated problems of the unequal consumption of natural resources.

5. Why do we conserve natural resources?

6. Give in detail different types of renewable resources available and used in India and problems because of over-exploitation of them.

**Based on Forest Resources**

1. What is the importance of forest resources for economic and ecological wealth of a country?

2. Describe uses and over-uses of land and forest.

3. Describe the key benefits of forests.

4. How can we estimate and explain socio-economic and environmental impacts of deforestation?

5. What are the different benefits obtained from forest resources?

6. What are the causes of deforestation of India? Describe briefly its effects on the environment.

7. Write a short note on consequences of deforestation.

8. Describe in detail types of forest resources and the problems created by deforestations in India.

**Based on Water Resources**

1. Without water, life is impossible, Justify it.

2. Explain the water cycle and the problems that arise with disturbances to the water cycle.

3. Explain by drawing a sketch of the hydrological cycle.

4. Describe in detail the uses and over-uses of water resources in India.

5. Enlist different surface and groundwater sources and explain how over-exploitations of water resources exaggregate the problem of availability of safe drinking water

6. "Water shall be the source of the world's next big conflict." Explain.

7. What are the problems associated with the over-exploitation of water resources?

8. State and describe different sources of water.

**Based on Mineral Resources**

1. What are the environmental effects of using mineral resources?

2. What are the effects on environment due to extraction of mineral resources?

**Based on Food Resources**

1. What are the impacts caused by agricultural activities?

2. What is undernutrition and malnutrition?

3. State the biochemical effects of pesticides.

4. Write a short note on effects of modern agriculture.

5. What are the effects of modern agriculture on food resources?

6. Describe the relationship of population and food production.

**Based on Energy Resources**

1. State the merits and demerits of using geothermal energy.

2. Write a short note on wind energy. Also discuss its advantages and limitations.

3. What are green fuels?

4. State the advantages and limitations of using hydro energy.

5. What are the different types of energies which can be derived from the ocean? Explain briefly along with their advantages and limitations.

6. What are renewable and nonrenewable energy sources?

7. Explain the limitations and importance of solar energy.

8. What is the role of geothermal energy in India?

9. List the different nonconventional energy sources. What new thrusts are being given to renewable energy programme in India?

**10.** Explain in detail conventional (nonrenewable) and nonconventional (renewable) energy sources with environmental problems created by the use of each of them.
**11.** What are various sources of energy? Explain in detail any one nonconventional source of energy.
**12.** Describe the merits and demerits of nuclear power energy and discuss the major concerns regarding its use for electricity generation in India.

**Based on Land Resources**
**1.** Write a short note on desertification.
**2.** What are the causes of land degradation?
**3.** Discuss about the soil texture, structure and its composition.
**4.** What are the sources/factors responsible for soil degradation?
**5.** Discuss land use planning for minimising land-associated problems.

**Based on Conservation of Natural Resources**
**1.** What is the need of conservation of natural resources? Discuss the different tools used for conservation of natural resources.
**2.** Explain the different areas of environmental conservation to which people belong to different disciplines can contribute.
**3.** Write a short note on destruction versus conservation.

**Based on Sustainable Lifestyle and Sustainable Water Management**
**1.** Discuss the need for sustainable lifestyles.
**2.** Write about sustainable water management.

# OBJECTIVE TYPE QUESTIONS

**I. Fill in the Blanks**
**1.** Major purpose of most of the dams around the world is _________.
**2.** A fuel cell in order to produce electricity burns _________.
**3.** _________ declared that big dams are temples of modern India.
**4.** Energy and manure together are supplied by _________.
**5.** _________ became the first African woman to receive the Nobel Peace Prize for "her contribution to sustainable development, democracy and peace."
**6.** GH Brundtland is an international leader in _________ and public health.
**7.** _________ means the removal of minerals or metals from earth's soil.
**8.** Environmental impact assessment is a systematic process of predicting environmental, social and economic impacts of a proposed project prior to _________.
**9.** _________ means meeting the needs of the present without compromising the ability of future generations to meet their own needs.
**10.** When the sum total of nature's resources (natural capital) is used up faster than it can be replenished, _________ of environment occurs.
**11.** _________ is an example of on-site effect of agriculture on environment.
**12.** Climate change is an example of _________ effect of agriculture on environment.
**13.** _________ are major causes of childhood lead poisoning.
**14.** Intensive irrigation without drainage causes _________.
**15.** World Nature Conservation Day is recalled on _________.
**16.** Chipko movement in Tehri Garhwal region was initiated by _________ to conserve forests.

17. Most biotic resources are _________.
18. Overgrazing results in _________.
19. Green plants are also called the _________.
20. The disease caused by excess of fluoride in water is known as _________.
21. The largest portion of water is used for _________.
22. Deforestation takes place in mountain region due to _________.
23. World Water Day is recalled on _________.

## II. Match the following terms.

Match the terms of column I with appropriate terms of column II.

A.

| Column I | Column II |
|---|---|
| 1. Coal | (a) Earth |
| 2. Solar energy | (b) Lignite |
| 3. Nuclear energy | (c) Sun |
| 4. Geothermal energy | (d) Fission and fusion |

B.

| Column I | Column II |
|---|---|
| 1. Substandard housing | (a) Biodiversity loss |
| 2. Ecodesigning of houses | (b) Asthma |
| 3. Socially acceptable economic and environmentally compliant mining operation | (c) Reduced pollution |
| 4. Vehicular emission | (d) Sustainable mining |
| 5. Habitat fragmentation by roads or railway tracks | (e) $CO_2$, $NO_x$, particulate matter |

C.

| Column I | Column II |
|---|---|
| 1. Metal ores | (a) Aluminium |
| 2. Bauxite | (b) Mining |
| 3. Largest fresh water lake in the world | (c) Loss of biodiversity |
| 4. Deforestation | (d) Lake superior |

## III. Multiple Choice Questions

1. Biogas predominantly contains
   (a) $CH_4$ (b) $NH_3$
   (c) $SO_2$ (d) Ethane
2. Both power and manure are provided by a
   (a) thermal plant
   (b) biogas plant
   (c) nuclear plant
   (d) hydroelectric plant
3. Steam reforming is currently the least expensive method of producing
   (a) CNG (b) petrol
   (c) hydrogen (d) biogas
4. Peat, lignite, bituminous and anthracite are different types of
   (a) biogas (b) natural gas
   (c) nuclear fuel (d) coal
5. Identify the nonrenewable energy resource from the following:
   (a) Coal (b) Tidal power
   (c) Wind power (d) Solar energy
6. Power generation by which of the following is least polluting?
   (a) Coal (b) Natural gas
   (c) Gasoline (d) Crude oil
7. Extensive planting of trees to increase forest cover is called
   (a) afforestation
   (b) social forestry
   (c) deforestation
   (d) agroforestation
8. Deforestation generally decrease
   (a) drought (b) rainfall
   (c) global warming
   (d) soil erosion
9. Which of the following is a nonrenewable resource?
   (a) Forest (b) Wildlife
   (c) Coal (d) Water
10. Fossil fuel and mineral resources are
   (a) renewable (b) perpetual

(c) inexhaustive
(d) nonrenewable

**11.** Fluoride pollution mainly affects.
(a) teeth (b) heart
(c) kidney (d) brain

**12.** Material cycles go through.
(a) biosphere and lithosphere
(b) atmosphere and hydrosphere
(c) biosphere and hydrosphere
(d) all the four spheres

**13.** A forest is a complex community with
(a) birds and animals
(b) flora and fauna
(c) plants and trees
(d) all the above

**IV. Indicate True or False for the following statements.**

**1.** Hydro-power energy and biomass energy are renewable energy sources. True/False

**2.** The rural energy source in India is wood and animal dung. True/False

**3.** The largest production of hydro energy is carried out in Canada. True/False

**4.** Nuclear energy often causes thermal pollution. True/False

**5.** Power generation by hydro energy is least polluting. True/False

**6.** Deforestation is taking place only in developing countries. True/False

**7.** Population explosion is one of the reason for deforestation. True/False

**8.** Cleaning of forest for agriculture is deforestation. True/False

**9.** Cauvery water dispute is in between Karnataka and Tamil Nadu. True/False

**10.** Khetri (Rajasthan) is famous for copper mines. True/False

**11.** Mosquito is an example of Bacteria. True/False

## Answers to Objective Type Questions

**I. Fill in the Blanks**

1. electricity production
2. $H_2$
3. Pandit Jawaharlal Nehru
4. biogas plant
5. Dr. Wangari Muta Maathai
6. Sustainable development
7. Mining
8. Decision making
9. Sustainable development
10. Degradation
11. Soil erosion
12. Off-site
13. Lead based paints
14. Water logging
15. July $28^{th}$
16. Sunderlal Bahuguna
17. Renewable
18. Desertification
19. Producers
20. Fluorosis
21. Irrigation
22. Hydel projects and road construction
23. March, $22^{nd}$

**II. Matching the terms.**

**A.** 1. (b) 2. (c) 3. (d) 4. (a)
**B.** 1. (b) 2. (c) 3. (d) 4. (e) 5. (a)
**C.** 1. (b) 2. (a) 3. (d) 4. (c)

**III. Multiple Choice Questions**

| | | | |
|---|---|---|---|
| 1. (a) | 2. (b) | 3. (c) | 4. (d) |
| 5. (a) | 6. (b) | 7. (a) | 8. (b) |
| 9. (c) | 10. (d) | 11. (a) | 12. (d) |
| 13. (d) | | | |

**IV. True or False**

| | | | |
|---|---|---|---|
| 1. True | 2. True | 3. True | 4. True |
| 5. True | 6. False | 7. True | 8. True |
| 9. True | 10. True | 11. False | |

# 3

# ECOSYSTEM

## Learning Objectives

After studying this chapter, you should be able to

- define ecology and ecosystem
- describe the causes and basic types of ecological succession
- enumerate and explain different types of ecosystems
- explain the roles of producer, consumer and decomposer in an ecosystem
- enumerate and explain various types of food chains
- explain the flow of energy through the various components of the ecosystem
- describe grassland ecosystem, desert ecosystem, forest ecosystem and aquatic ecosystem
- define an ecological pyramid and its various types
- explain different models of energy flow in an ecosystem
- describe why a complex ecosystem is more stable than one with few species

## 3.1 INTRODUCTION TO ECOLOGY AND ECOLOGICAL SUCCESSION

The term ecology is derived from the Greek word *Oikologie*. Literally, *Oikos* means 'home or surroundings' and *logos* means 'study'. Thus, ecology is the study of nature.

Ecology can be defined as "*the study of interactions between an organism and its physical environment; the relationship between animals and plants and how one species affect another.*"

### 3.1.1 Classification of Ecology

Ecology can be classified

**(i) By Level of Complexity or Scope** For example, behavioral ecology, population ecology, etc.

**(ii) By Organisms under Study** For example, animal ecology, plant ecology, insect ecology, etc.

**(iii) By Biome under Study** For example, desert ecology, forest ecology, etc.

**(iv) By Geographic or Climatic Area under Study** For example, tropical ecology, polar ecology, etc.

**(v) By Spatial Scale under Study** For example, micro ecology, macro ecology, molecular ecology, global ecology, etc.

**(vi) By Phenomena under Investigation** For example, chemical ecology, evolutionary ecology, etc.

**(vii) By Technique used for Investigation** For example, theoretical ecology, quantitative ecology, etc.

**(viii) By Philosophical Approach** For example, conservation ecology, restoration ecology, applied ecology, etc.

**(ix) Ecology-involved Interdisciplinary Fields** For example, human ecology, agro ecology, etc.

## 3.1.2 Evolution of Ecosystems

An ecosystem is evolved through the following evolutionary processes:
Natural selection, life history, development, adaptation, populations and inheritance.

## 3.1.3 Ecological Succession

*Ecological succession is orderly changes in the composition or structure of an ecological community.* It is more or less predictable.

### *(A) Causes and Basic Types of Ecological Succession*

#### (i) Primary and Secondary Succession

(a) When the development begins on an area that has not been previously occupied by a community, the process is known as *primary succession.*
*Examples* A lava flow, a newly formed lake, or a newly exposed rock or sand surface.

(b) When the community development is proceeding in an area from which a community was removed, it is called *secondary succession.*
*Examples* It arises on cut-over forest, an abandoned crop, etc. These are the sites where the vegetation cover has been disturbed by nature or humans.

**(ii) Seasonal and Cyclic Succession** These are periodic changes arising from fluctuating species interactions or recurring events.

*Example* Vegetation changes which are not dependent on disturbance unlike secondary succession.

### *(B) Causes of Plant Succession*

(i) Changes in the soil caused by the presence of organisms there includes change in pH of soil by plants growing there, alterations of soil nutrients, etc.

(ii) Changes in the soil caused by external environmental influences includes soil changes due to erosion, deposition of silt and clays, changes caused by animals, etc.

(iii) Changes caused by climate factors are promoted by changes in temperature and rainfall patterns, specially global warming, floods etc.

## 3.2 ECOSYSTEM

*"An ecosystem is defined as a natural unit that consists of living and nonliving parts which interact to form a stable system."*

An ecosystem is generally an area within the natural environment in which physical (abiotic) factors of the environment, such as rocks and soil, function together along with interdependent (biotic) organisms, such as plants and animals, within the same habitat to create a stable system. It possesses all the characteristics required to sustain life. When we want to conserve species or to use natural resources in a sustainable manner, we need to focus on ecosystems. This is because an ecosystem is the minimal grouping of diverse organisms that interact and function together so as to sustain life.

The sizes of some ecosystems are illustrated below:

| *Ecosystem* | Bacteria | Pond | Desert | Ocean |
|---|---|---|---|---|
| *Size* | (1 to 100) $mm^2$ | (10 to 100) $m^2$ | > 100 $km^2$ | > 1000 $km^2$ |

**Example 1** *How is the sun the primary sustainer of life on the earth?*

*OR*

*Wshat sustains life in an ecosystem?*

**Solution** The energy from the sun

(i) enables plants to produce food through photosynthesis,

(ii) evaporates water and cycle it through the biosphere,

(iii) generates winds,

(iv) warms the atmosphere and the land,

(v) drives the climate and weather systems, and

(iv) powers the cycling of carbon, nitrogen and other matter.

### *Balanced Ecosystem*

The biotic (living) and abiotic (nonliving) parts of the ecosystem are in equilibrium in a balanced ecosystem. *Balanced ecosystem* means that the nutrients are able to cycle efficiently, and no community of organisms or natural phenomena is interrupting the flow of energy and nutrients to other parts of the ecosystem.

### 3.2.1 Structure of an Ecosystem

Structure of an ecosystem means:

**(i) The Composition of Biological Community** It includes species, their population, etc.

**(ii) The Quantity and Distribution of Abiotic Materials** It includes water, soil, nutrients, etc.

**(iii) The Conditions of Existence** It includes temperature, light, humidity, etc.

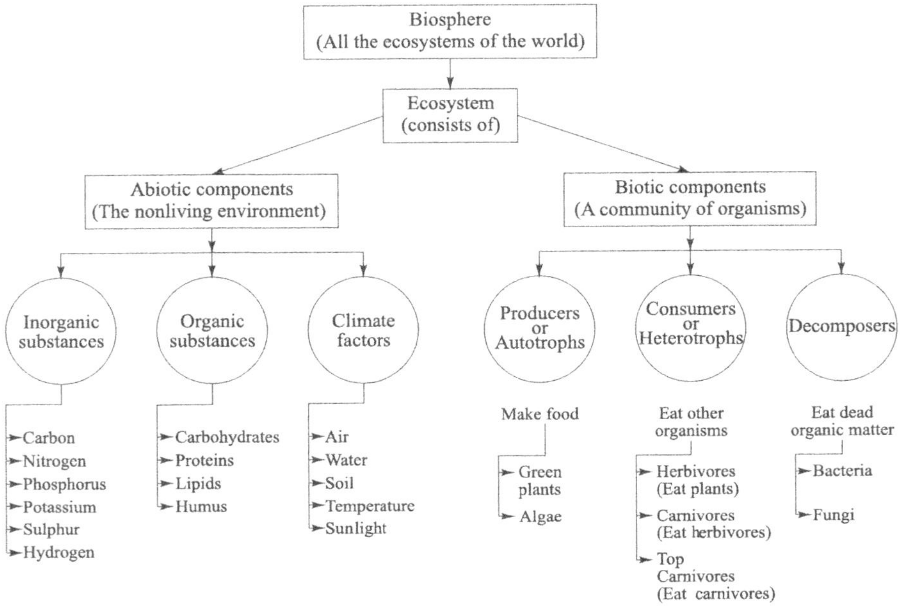


**Fig. 3.1** *Structure of an ecosystem*

### *(A) Abiotic Components*

All the nonliving components of the environment constitute the *abiotic components*. It includes:

- ***Inorganic substances*** which are involved in mineral (nutrient) cycles. *Examples*: C, N, P, K, S, H, etc.
- ***Organic substances*** present in the biomass or in the environment. They form the living body and influence the functioning of the ecosystem. *Examples*: Carbohydrates, proteins, lipids, humus, etc.
- ***Climate factors*** having a strong influence on the ecosystem.

**(i) Water** Plants and animals receive water from the soil and the earth's surface. Water is the medium by which mineral nutrients enter and are distributed in plants. For the survival of animals, water is necessary.

**(ii) Soil** Soil provides nutrients and water, a structural growing medium for organisms.

**(iii) Atmospheric Air** Within ecosystems, the atmosphere provides oxygen for respiration of organisms and carbon dioxide for photosynthesis in plants.

**(iv) Sunlight** Sunlight is necessary for photosynthesis. It is used to heat the atmosphere in ecosystems.

### *(B) Biotic Components*

All the living components of the environment constitute the biotic components.

Depending on their self-food producing capability, biotic components are of following types.

**(i) Producers or Autotrophic Components** Producers are self-nourishing organisms (so they are called *autotrophs*). They contain chlorophyll and are capable of converting carbon dioxide and water, in the presence of sunlight into carbohydrates through photosynthesis. In the process, they give out oxygen.

$$6CO_2 + 6H_2O + \text{Sunlight} \xrightarrow[\text{Producers}]{\text{Photosynthesis}} C_6H_{12}O_6 + 6O_2$$

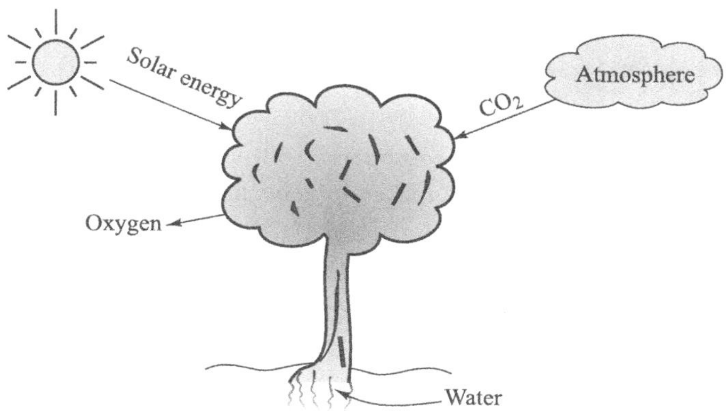


**Fig. 3.2** *Process of photosynthesis*

Autotrophs are of the following two types:

***(a) Photoautotrophs*** These are the producers who fix energy from the sun and store it in complex organic compounds.

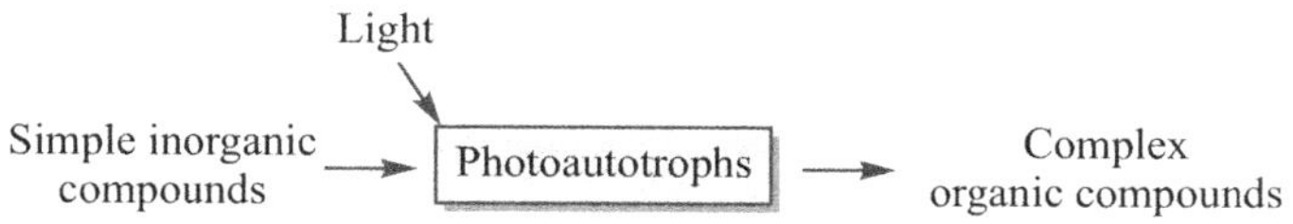


*Examples* Green plants, some bacteria, algae.

***(b) Chemoautotrophs (Chemosynthesizers)*** They are bacteria that oxidise reduced inorganic substances (typically ammonia and sulphur compounds) and produce complex organic compounds.

*Example* Nitrifying bacteria in the soil underground.

**(ii) Consumers (or Heterotrophic Components)** Consumers depend on producers to obtain their energy for survival. They utilise, rearrange and decompose the organic matter produced by autotrophs.

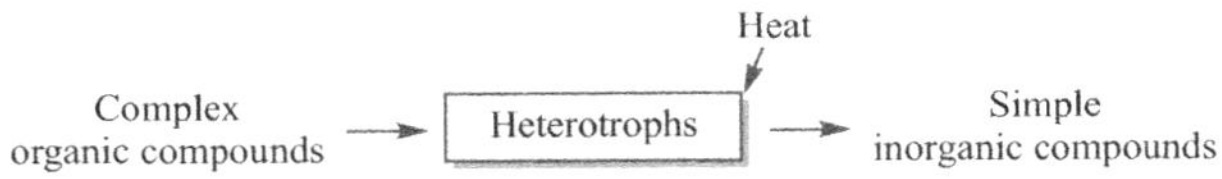


**Consumers** are classified as herbivores, carnivores and top carnivores depending on their food habits.

***(a) Herbivores (or Primary Consumers)*** They feed on green plants (autotrophs) to obtain energy for survival.

Seed-eaters are also known as *granivores*. Fruit-eaters are also known as *frugivores*.

*Examples* Grasshoppers, rabbits, goats, cows, horses, etc.

***(b) Carnivores (or Secondary Consumers)*** They feed on primary consumers.

*Examples* Lizard, fox, hawk, etc.

***(c) Top Carnivores (or Tertiary Consumers)*** They eat the flesh of both carnivores and herbivores and are not killed or eaten by other animals.

*Examples* Lions, tigers, vultures, etc.

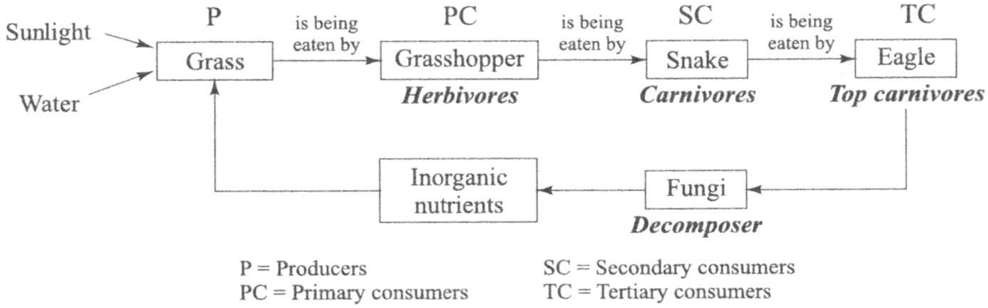


**Fig. 3.3** *Structure of an ecosystem*

**(iii) Decomposers** The decomposers are also known as *saprotrophs* (i.e. *sapros* = rotten; *trophs* = feeder). They feed on dead organic matter (from producers and consumers). They transform complex organic compounds back into simple inorganic substances like $CO_2$, $H_2O$, phosphates, sulphates.

*Examples* Bacteria, fungi, other microbes, etc.

Fallen leaves, parts of dead trees, and faecal wastes of animals are termed *detritus*. The consumers that feed on detritus are known as *detrivores*.

*Examples* Ants, termites, earthworms, crabs, etc.

Decomposers and detrivores are essential for the long-term survival of a community. Their vital role is to complete the matter cycle. Enormous wastes of plant litter, dead animal bodies, animal excreta, and garbage would collect on the earth without them. Furthermore, important nutrients would remain indefinitely in dead matter. The producers would not get their nutrients, and life would be impossible without detrivores and decomposers.

***Notes:***

- Humans act as primary consumers when they eat fruits and vegetables.
- Humans act as secondary consumers when they eat meat.
- Humans act as tertiary consumers when they eat the fish that eat smaller fish that eat the algae.
- Humans can also act as omnivores by eating both plants and animals.

## 3.2.2 Function of an Ecosystem

Major functions of an ecosystem are as follows:

(i) It regulates flow rates of biological energy. In other words, it controls the rate of production and respiration of the community.

(ii) It regulates flow rates of nutrients. In other words, it controls the production and consumption of minerals and materials.

(iii) It helps in biological regulation including:

*(a)* ***Photoperiodism*** (regulation of organisms by environment)

*(b)* ***Nitrogen-fixing organisms*** (regulation of environment by the organism)

## 3.2.3 Importance of an Ecosystem

An ecosystem provides number of services for the healthy survival of humans. For example:

(i) An ecosystem helps in water retention, thus facilitating a more evenly distributed release of water.

(ii) An ecosystem provides air and does its purification.

(iii) An ecosystem provides recreation for us via eco-tourisms facilitating the enjoyment of nature.

(iv) An ecosystem provides materials like minerals and food.

Importance of Ecosystem

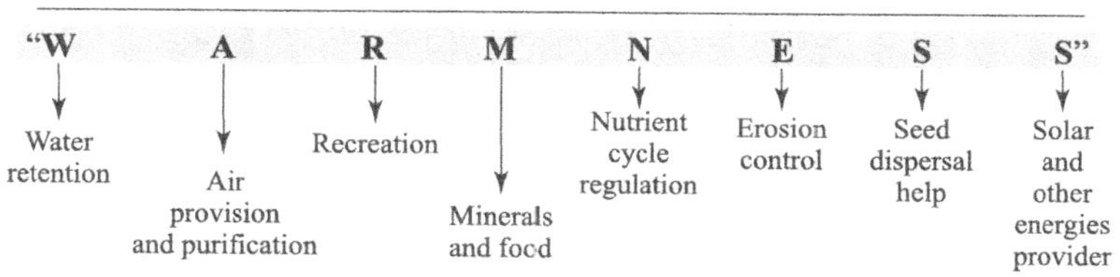


(v) An ecosystem regulates nutrient recycling and waste.

(vi) An ecosystem helps in erosion control, soil building and soil renewal.

(vii) An ecosystem helps in seed dispersal.

(viii) An ecosystem gives us solar energy (that accounts for 99% of the total energy used on earth). It also gives us

(a) Renewable energy like biofuels, and

(b) Non-renewable energy like fossil fuels.

(ix) Additional services provided by an ecosystem are the following:

(a) An ecosystem helps in the maintenance of the biogeochemical cycles like carbon cycle and water cycle. It also helps in the cycling of vital chemicals like sulphur, phosphorus, nitrogen and carbon.

(b) An ecosystem helps in natural pest and disease control.
(c) An ecosystem preserves genetic diversity.

## 3.3 FOOD CHAIN

*Food chain is a feeding hierarchy in which organisms in an ecosystem are grouped into nutritional (trophic) levels and are shown in a succession to represent the flow of food energy and the feeding relationship between them.* The directional flow of food energy from one organism to another is graphically represented by arrows. Food chain is just a sequence of organisms, in which each is food for the next.

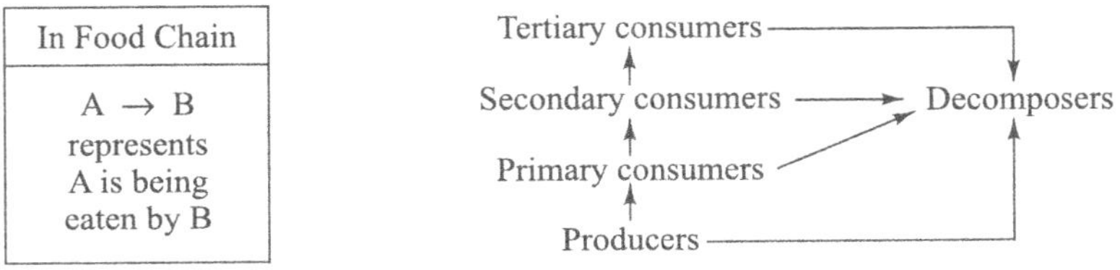


**Fig. 3.4** *A food chain*

Food chains overlap, because most consumers feed on multiple species and in turn, are fed upon by multiple other species. Thus, we have a complex network of interconnected food chains called a *food web*.

For example, a snake might feed on a mouse, a lizard, or a frog. In turn, the snake might be eaten by a bird or a badger.

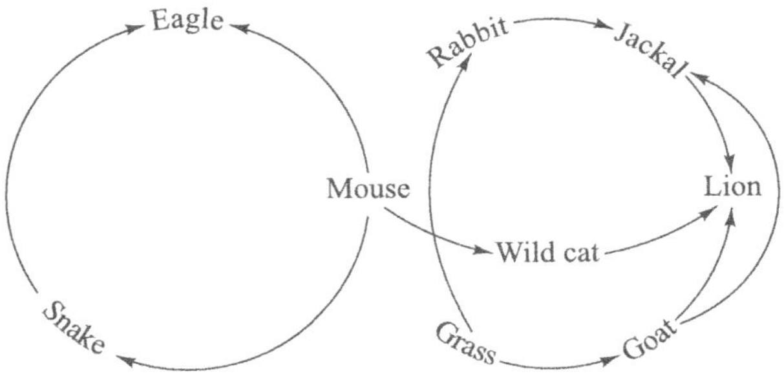


**Fig. 3.5** *Food web in a forest*

### *Types of Food Chains*

Food chains are broadly of the following two types:

**(i) Grazing Food Chain** The grassland and forest ecosystems follow this grazing food chain. Here, producers get energy from the sun and are grasses or green plants. They are subsequently grazed by animals.

***Examples*** (a) Grass → Grasshopper → Frog → Snake → Hawk
(b) Green plants → Goat → Wolf → Lion

**(ii) Detritus Food Chain** The estuarine and mangrove leaf ecosystems follow this detritus food chain. In this chain, the dead animals, dead plants and fallen leaves are consumed by detrivores and their predators.

***Examples*** (a) Dead plants → Soil mites → Insects → Lizards
(b) Dead organic matter → Bacteria → Protozoa → Rotifiers

**Example 2** *How is balance maintained in an ecosystem?*

**Solution** The food chains and other such interrelationships in ecosystems create a balance in the environment, called the ecological balance.

The components of the ecosystem are part of food chains and food webs. They do not try to modify the environment to suit their needs, rather they help in maintaining a balance in the ecosystem.

However, humans try to modify the environment to suit their needs. As they are also a part of these food chains and webs, modification of environment has upset the delicate balance which was maintained in the environment.

**Example 3** *What is the difference between a food chain and a food web?*

**Solution**

(i) Food chains follow just one path of energy as animals find food. Food webs show how plants and animals are connected in many ways to help them all survive.

(ii) A food chain is the hierarchy of consumption of food from sun to plant to herbivore to carnivore. It acknowledges only one single string of connected plants/animals. There are many food chains within a food web, and one creature is not necessarily at the top of the hierarchy. In a food web, one kind of prey may be eaten by several kinds of predators, and one predator may eat several different kinds of prey.

(iii) A food chain is very basic and doesn't show the full picture of an ecosystem. On the other hand, a food web refers to everything that goes on in the real world.

(iv) A food chain can be illustrated by a linear diagram. However, a food web can be illustrated by a complex diagram.

**Example 4** *Explain the significance of studying a food chain.*

**Solution**

(i) The knowledge of how species are inter-dependent in a food chain is necessary. It is also necessary to understand how natural and man-made environmental pressures affect ecosystems. These include mercury, DDT, etc., which are toxic chemicals and cause destructive pollution. They can alter or break the food chain. These also include nutrient pollution which shift whole ecosystems toward nutrient-hungry species. For example, crops, fertilizers, sewage and animal waste escape into lakes causing algae blooms at the expense of fishes.

(ii) The study of food chains and webs is critical for understanding the route by which pollutants gets bio-accumulated (i.e. concentrated) up the food chain.

(iii) By studying a food chain, we can understand how balance is maintained in an ecosystem.

## 3.4 ECOLOGICAL PYRAMIDS

### 3.4.1 Pyramid of Numbers

The *pyramid of numbers* represents the number of individuals at each trophic level. The shape of a pyramid of numbers can be upright, partly upright and inverted depending on the type of ecosystem.

#### *(A) Aquatic and Grassland Ecosystem*

In aquatic and grassland ecosystems, the number of producers are always more than that of primary consumers. Thus, the producer organisms remain in abundance near the base of the food chain and the consumers gradually decrease in number towards the apex. As a result, the shape of the pyramid is upright (Fig. 3.6 (a)).

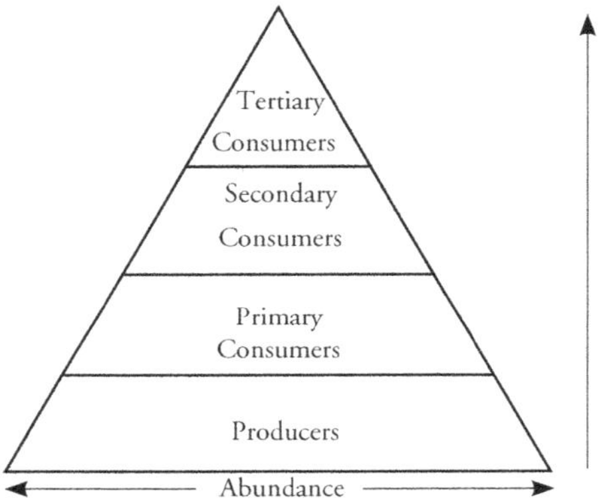


*(a) Upright pyramid*

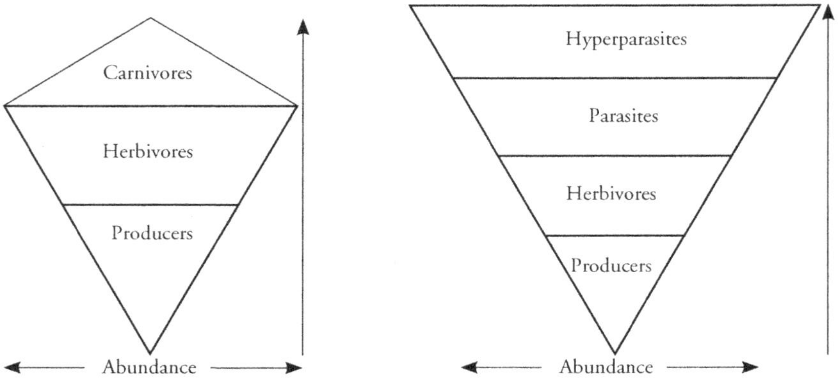


*(b) Spindle shaped*

*(c) Inverted pyramidal shaped*

**Fig. 3.6** *Pyramid of numbers*

#### *(B) Forest Ecosystem*

In a forest ecosystem, there are fewer number of producers that support a greater number of herbivores who in turn support a lesser number of carnivores. The shape of the pyramid of numbers is partly upright or spindle type (Fig. 3.6 (b)).

### (C) Parasitic Food Chain

In a parasitic food chain, one primary producer supports numbers of parasites which again support still more hyperparasites. The pyramid of numbers is inverted in shape because the producers are least in number and the predators are greater in number as we move up the food chain (Fig. 3.6 (c)).

## 3.4.2 Pyramid of Energy Flow (Flow of Energy in an Ecosystem)

Flow of energy in an ecosystem takes place through the food chain.

The main source of energy for most ecosystems is the sun. Solar energy is trapped by *producers.* They store it as carbohydrates, proteins and fats. When *primary consumers* eat the producers, the energy also moves up the trophic level. During this transfer, about 90% of the energy is lost as unusable heat to the environment.

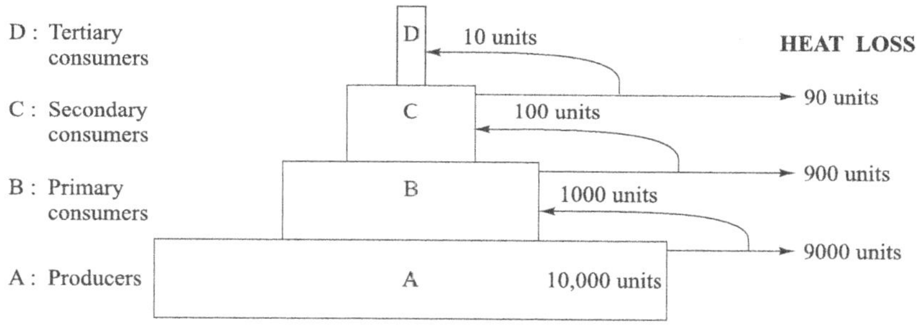


Fig. 3.7 *Pyramid of energy flow*

We have an upright pyramid of energy flow as we move up the trophic levels, and the amount of usuable energy available at each stage declines.

***Notes:***

1. The *ecological pyramid* is the graphical representation of the organism's position in the food chain. The base of the pyramid consists of the food-producer level and the successive levels make the tiers with the top carnivore or tertiary consumers forming the apex.
2. The size of each compartment in an ecological pyramid represents the amount of organisms (or item) in each trophic level of a food chain.
3. *Trophos* is a Greek word meaning nourishment.

Suppose the producer has 10,000 units of energy. When primary consumers eat the producer, they receive only 1000 units, and the rest 9000 units are lost as heat. Similarly, the secondary and tertiary consumers gets only 100 and 10 units respectively. The loss at each stage is simply released as heat into the environment.

*The flow of energy through the various components of the ecosystem is unidirectional and continuous.*

Unlike the nutrients which move in a cyclic manner and are reused by the producers after flowing through the food chain, energy is not reused in the food chain.

All organisms require energy for growth, maintenance, reproduction, locomotion, etc. The flow of energy in an ecosystem follows the laws of thermodynamics.

**(i) First Law of Thermodynamics** Energy can never be created or destroyed, but can be converted from one form to another.

**(ii) Second Law of Thermodynamics** Transformations of energy always result in some loss or dissipation of energy.

The trophic structure of an ecosystem is the pattern of energy flow among different organisms as illustrated in Fig. 3.8.

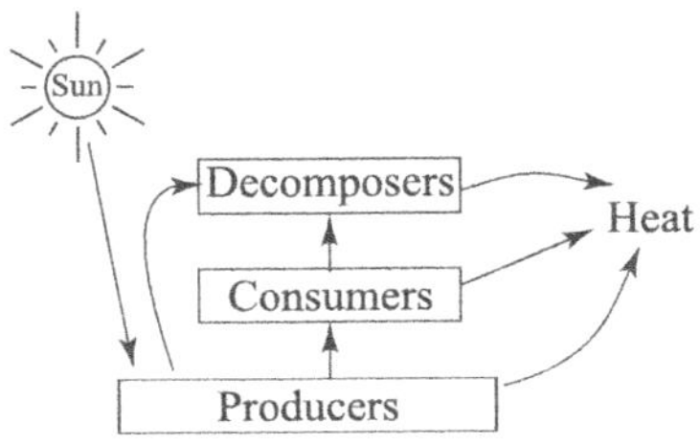


**Fig. 3.8** *The trophic structure of an ecosystem*

### 3.4.3 Models for Energy Flow in Ecosystem

The following models can be used for explaining the flow of energy through various trophic levels in an ecosystem.

***(A) Universal Energy-flow Model*** According to this model (as per E.P. Odum), as the flow of energy takes place, there is a gradual loss of energy at every level, thereby resulting in less energy available at the next trophic level as denoted by smaller boxes (for stored energy in biomass) and as indicated by narrower pipes (for energy flow). The energy not utilised (*NU*) is lost in excretion, locomotion, respiration (*R*). The rest of the energy is used for production (*P*).

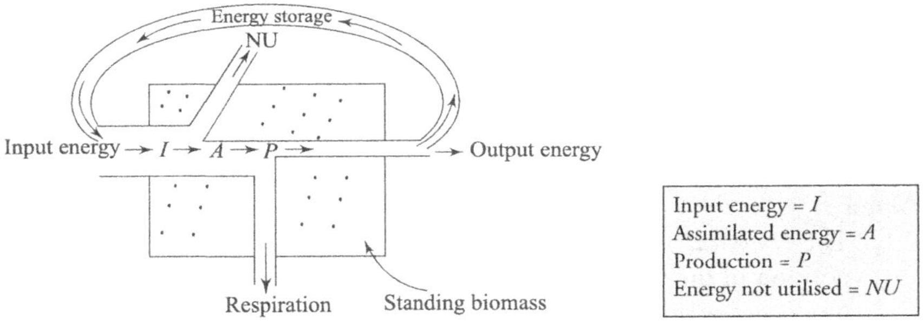


**Fig. 3.9** *Universal energy-flow model*

***(B) Single-channel Energy-flow Model*** According to this model, the flow of energy takes place in a unidirectional manner from producers to herbivores to

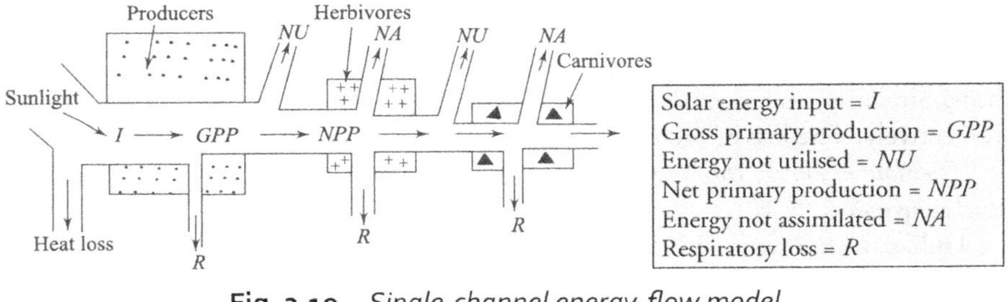


**Fig. 3.10** *Single-channel energy-flow model*

carnivores. Due to loss of energy at each successive trophic level in a grazing food chain, there is gradual decline in energy.

***(C) Double-channel or Y-shaped Energy-flow Model*** Both grazing and detritus food chain operate in the same ecosystem in nature. In a marine ecosystem, the grazing food chain predominates. In a forest ecosystem, the detritus food chain predominates. The grazing and detritus food chains are separated in space and time. A Y-shaped model of energy flow is used to show the passage of energy through these two chains.

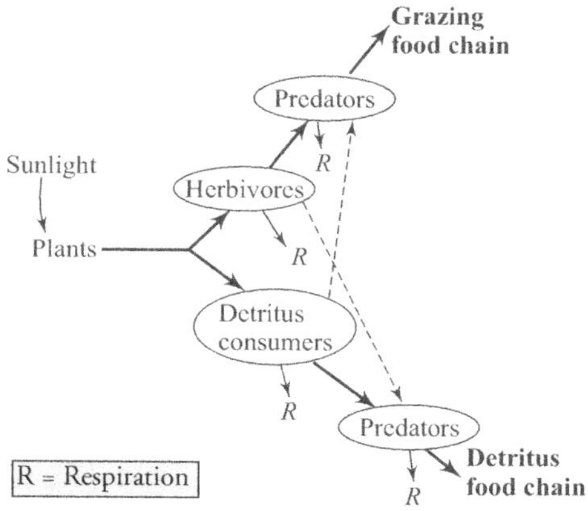


**Fig. 3.11** *Y-shaped energy-flow model showing linkage between the detritus and grazing food chains*

## 3.4.4 Pyramid of Biomass (Flow of Matter in an Ecosystem)

### *Estimation of Biomass*

**Step (1)** Collect (or trap) and weigh suitable samples at each trophic level.

**Step (2)** Total combined (net dry) weight (often, per unit area or volume) of all the organisms at each trophic level is *biomass*.

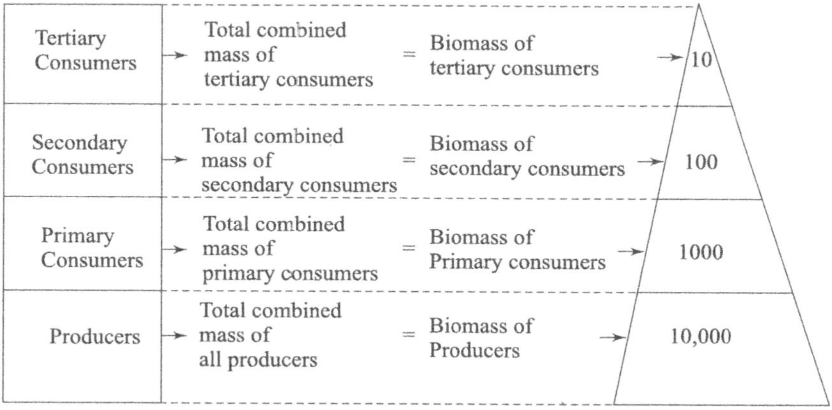


**Fig. 3.12** *Depicting biomass relationship graphically at successive trophic levels gives rise to a biomass pyramid*

The dry weight of all the matter contained in the organisms is known as *biomass*. Each tropic level contains a definite amount of biomass. As we move up trophic levels, biomass decreases drastically. There is 90 to 99 per cent loss of biomass at each level. This is known as the *pyramid of biomass*.

Reasons for the decrease of biomass as we move up trophic levels are the following:

(i) Only a fraction of the food taken in by a consumer is converted into body tissue. The remaining is stored as energy to be used by the consumer when needed.

(ii) Much of the biomass, especially at the producer level, is never eaten and goes directly to the decomposers.

**Example 5** *Why is the mass of water not usually included in biomass?*

**Solution** This is because the water content is variable and contains no usuable energy.

**Example 6** *What is net primary production?*

**Solution** Total amount of energy captured by producers is termed Gross Primary Production (GPP).

When energy lost due to respiration is subtracted from GPP, we get Net Primary Production (NPP).

$$\text{NPP} = \text{GPP} - \text{Respiration}$$

Net primary production is the amount of energy stored by the producers and potentially available to the consumers and decomposers.

**Example 7** *What is secondary productivity?*

**Solution** It is the rate at which consumers convert organic material into new biomass of consumers.

**Example 8** *"The pyramid of total biomass produced must resemble the pyramid of energy flow". Comment.*

**Solution** The above statement is TRUE.

This is because biomass can be equated to energy.

**Example 9** *Why can the pyramids of energy and yearly biomass production never be inverted?*

**Solution** Because this would violate the laws of thermodynamics.

## 3.5 TYPES OF ECOSYSTEMS

There are several ecosystems working at micro and macro levels in the world. The biosphere is the biggest ecosystem which combines all the ecosystems of the world. The world's smaller ecosystems are broadly divided into natural and artificial type ecosystems.

### 3.5.1 Natural Ecosystems

They operate by themselves under natural conditions without any interference by humans. Broadly they are subclassified into terrestrial and aquatic ecosystems.

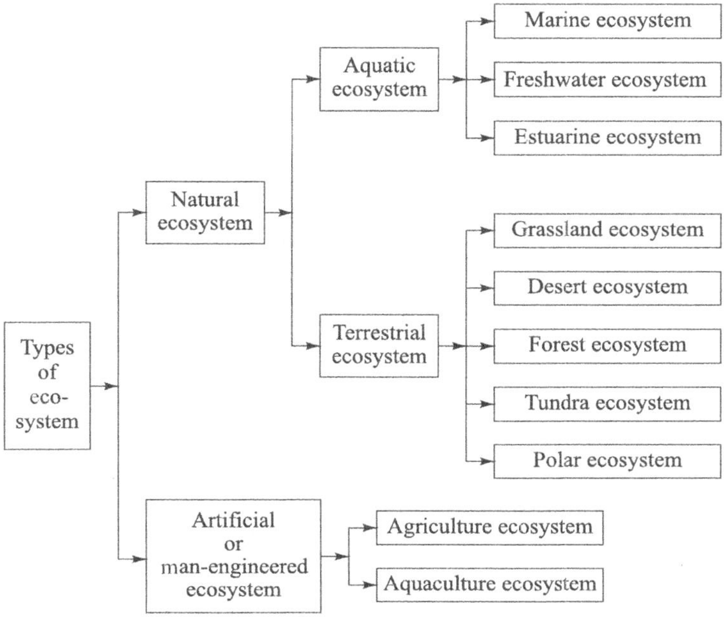


**Fig. 3.13** *Classification of ecosystems*

**(i) Terrestrial Ecosystems** They are known by the type of main vegetation in them. For example, grassland ecosystems have grass as the main vegetation.

**(ii) Aquatic Ecosystems** They are known by the type of habitat. They can be of estuarine, marine and freshwater types of ecosystems.

The freshwater ecosystems can be of standing freshwater ecosystems (or *lentic ecosystems*) or running freshwater ecosystems (or *lotic ecosystems*).

***Examples*** Ponds, lakes, etc., are examples of lentic ecosystems and rivers, springs, etc., are examples of lotic ecosystems.

## 3.5.2 Artificial Ecosystems

These ecosystems are controlled and manipulated by humans. These are created by humans in order to fulfill certain needs.

Broadly, they are subclassified into the following two types:

(i) Agriculture ecosystem
(ii) Aquaculture ecosystem

### *(A) Differences between Natural and Artificial Ecosystems*

| *Natural ecosystems* | *Artificial ecosystems* |
|---|---|
| (i) Polyculture systems | (i) Monoculture system |
| (ii) Stable ecosystems | (ii) Fragile ecosystems |
| (iii) Less productive in terms of yield of grains, milk, fish or meat | (iii) Highly productive as they are given increased supply of energy in the form of labour, extra nutrients, fossil fuels, fertilisers, pesticides, etc. |

(*Contd.*)

(*Contd.*)

| | |
|---|---|
| (iv) Pollution free | (iv) Generate lots of pollutants. |
| (v) *Examples:* Aquatic ecosystems and terrestrial ecosystems | (v) *Examples:* Agriculture ecosystems and aquaculture ecosystems |
| (vi) Functions:<br>• Air purification<br>• Water purification | (vi) Functions:<br>• To supply large quantities of grains, etc.<br>• To supply large quantities of fish, meat, milk, etc. |

### (B) Similarities between Natural and Arificial Ecosystems

(i) Both are open systems with no constraints of boundaries.
(ii) Both have all the essential components such as abiotic and biotic members.
(iii) Both permit constant interaction between biotic and abiotic components.

## 3.6 FOREST ECOSYSTEM

A *forest* is a community of trees, herbs, shrubs, and associated organisms that use oxygen, water and soil nutrients for their growth and reproduction.

A *forest ecosystem* is the organisms, soil, air and water associated with the forest.

A forest ecosystem is interdependent because every organism depends on every other living and nonliving elements of the system.

Fire, storms, drought, flood, death, disease, etc., are natural changes in a forest ecosystem. Harvesting, farming, trails, recreation and development, etc., are man-made changes in a forest ecosystem.

### Components of Forest Ecosystems

The different components of the forest ecosystems are the following:

**(i) Abiotic Components** The minerals present in the forest and all organic (litter, debris) and inorganic substances present in the soil and the atmosphere constitute the abiotic components.

**(ii) Biotic Components** All living components, viz. producers, consumers and decomposers, constitute the biotic components of the forest.

***(a) Producers*** Big trees, medium-sized bush, small herbaceous plants, or any vegetation of the forest is the producer, which performs photosynthesis.

***(b) Consumers***

*Primary Consumers* They graze over the primary producer, e.g. elephants, mongooses, squirrels, deer; birds and insects like flies, spiders, ants, etc.

*Secondary Consumers* They are the predators of primary consumers. They regulate the population size of primary consumers and thereby their grazing activity, e.g. jackal, fox, eagle, snake, etc.

*Tertiary Consumers* They feed on secondary consumers and are also known as top carnivores., e.g. lions, tigers, etc.

**(c) Decomposers** They have the ability to degrade all dead organisms to release nutrients into the soil which are again used by the producer. They remain confined to the soil of the forest floor.

***Examples*** Earthworms, bacteria, fungi, protozoa, nematodes, etc.

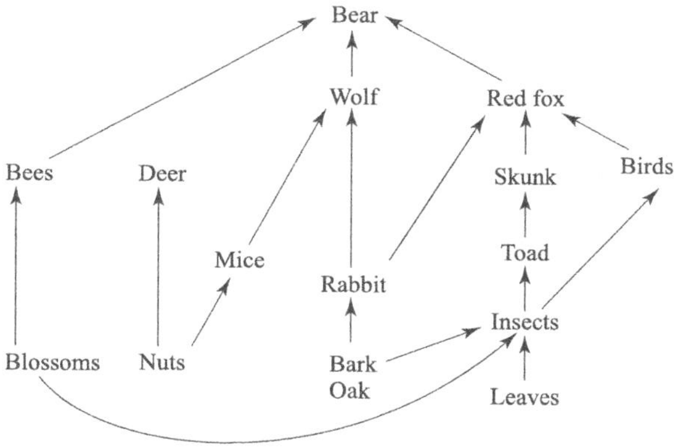


**Fig. 3.14** *Interconnection between food chain and food web in a forest ecosystem*

## 3.7 AQUATIC ECOSYSTEM

An *aquatic ecosystem* is an ecosystem located in a body of water.

Biotic and abiotic components (which are self-regulating and self-sufficient) constitute an aquatic ecosystem. About 70% of the earth's total surface is under the aquatic ecosystem. Broadly, an aquatic ecosystem is of the following three types: Freshwater ecosystem, marine ecosystem and estuarine ecosystem.

### 3.7.1 Pond Ecosystem (or Freshwater Ecosystem)

*The different components of a pond ecosystem* are as follows:

**(i) Abiotic Components** Oxygen, carbon dioxide, water, nitrogen, phosphorus, calcium, amino acids, etc., are abiotic components of a pond ecosystem.

**(ii) Biotic Components** They consist of the following:

***(a) Producers*** Some photosynthetic bacteria and the autotrophic green plants fix the solar energy with the help of nutrients obtained from the mud of the pond.

***(b) Consumers***

*Primary Consumers* They feed on the producers.
*Examples* Herbivores like zoo plankton and small invertebrates like copepod.

*Secondary Consumers* They feed on primary consumers.
*Examples* Small carnivores like small fishes.

*Tertiary Consumers* They feed on secondary consumers.
*Examples* Large fishes.

***(c) Decomposers*** They help in the release and recycling of nutrients. They decompose the organisms and are present at the base of the pond.
*Examples* Bacteria, fungi, etc.

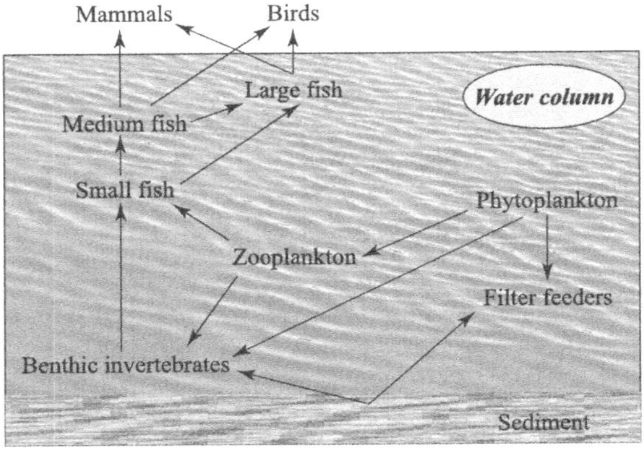


**Fig. 3.15** *How food chain and food web are interconnected in an aquatic ecosystem*

***Functions*** An aquatic ecosystem performs the following environmental functions:

- Recycle nutrients
- Purify water
- Recharge ground water
- Provide habitats for wildlife
- Attenuate floods
- Used for human recreation

### 3.7.2 Marine or Ocean Ecosystem

Oceans are gigantic reservoirs of water covering nearly 70% of the earth's surface. A marine ecosystem is different from a freshwater ecosystem mainly because of its salty water and also because the sea is deep, and the water is in continuous circulation.

A marine ecosystem can be divided into the following zones:

***(A) Littoral Zone*** It is the shoreline between the land and the open sea. Waves and tides have maximum effect in a littoral zone. Various regions of this zone are tabulated below along with the important organisms:

| *Region* | *Organisms* |
|---|---|
| (i) Rocky shore region | Starfish, barnacles, algae |
| (ii) Sandy shore region | Snails, clams |
| (iii) Bays | Algae |

Often photosynthetic bacteria are present below the algae. Moreover, by colonial coelenterates, coral reefs are also formed.

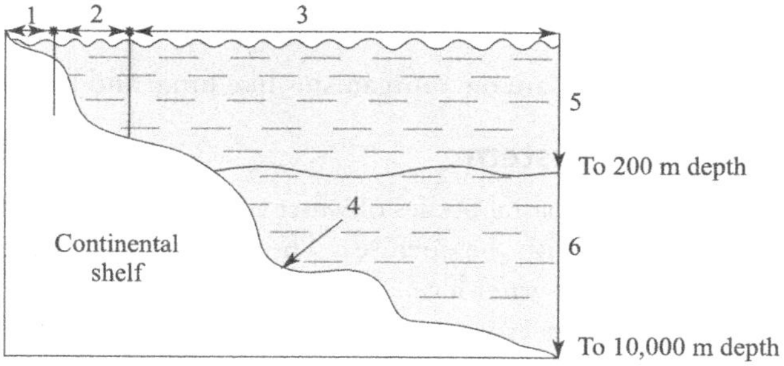


**Fig. 3.16** *Horizontal and vertical zonation in the ocean*

***(B) Neritic Zone*** This zone lies just above the continental shelf. The nutrients washed from land are found in this zone. Thus this zone is rich in species. The productivity of this zone is high because sunlight can penetrate through this zone. Zooplankton and phytoplankton are abundant here and support fishing grounds. Pollution also affects the neritic zone first.

***(C) Pelagic Zone*** The open sea constituting 90 per cent of the total ocean surface forms this zone. Phytoplanktons, zooplanktons, shrimps, jelly fish, fin, deep-water fishes and blue whale are found here.

Organisms of this zone are present below the light penetration zone and totally depend on the rain of detritus of upper regions for their nutrition.

### *(D) Benthic Zone*

The floor of the ocean constitutes this zone. It stretches from the edge of the continental shelf to the deepest ocean trenches.

Sponges, sea lilies, sea fans, snails, clams, starfish, sea cucumbers and sea urchins are found in this zone.

The components of marine ecosystem are the following:

**(i) Abiotic Components** High sodium, potassium, calcium and magnesium salt concentrations, variable dissolved oxygen content, light and temperature make a unique physico-chemical condition in marine water.

The size of marine populations is low because concentration of dissolved nutrients is less.

### (ii) Biotic Components

***(a) Producers*** Phytoplankton, seaweeds and mangrove vegetations are main producers in marine ecosystems.

***(b) Consumers*** Crustaceans, molluscs, fishes and other herbivorous which feed directly on producers are primary consumers.

Carnivorous fishes like herring, sahd and mackerel, etc., are secondary consumers.

Cod, haddock and other top carnivorous fishes are tertiary consumers.

**(iii) Decomposers** These are microorganisms like fungi and bacteria.

### 3.7.3 Estuarine Ecosystem

*Estuaries* are semi-enclosed coastal bodies of water connected on the one side with a river and on the other side with the open sea. Thus, estuarine is characterised as an ecosystem having fluctuating water level.

The organisms present in estuaries are known as *eurythermal* (which show a wide range of tolerance to temperature) and *eurysaline* (which show a wide range of tolerance to salinity).

Due to nutrient and energy inputs from both river water and sea water, estuaries are highly productive. They also offer high food potential for human beings. Deep-water fishes use estuaries as nurseries to bring up their younger ones.

The component of an estuarine ecosystem are the following:

**(i) Abiotic Components** A mixture of fresh and marine ecosystems.

**(ii) Biotic Components**

***(a) Producers*** Phytoplankton, benthic algae, sea grasses, seaweeds and marsh grasses.

***(b) Consumers*** Fishes, oysters, crabs, shrimp, etc.

***(c) Decomposers*** Bacteria and fungi.

### 3.7.4 Streams and River Ecosystems or Flowing-water Ecosystems

The water flows rapidly in mountain reaches. The lower reaches of rivers sustain phytoplankton, zooplankton, crustaceans, small fishes and big fishes. The river bottoms are covered by algae and bacteria.

The trophic levels in a river ecosystem are shown in Fig. 3.17.

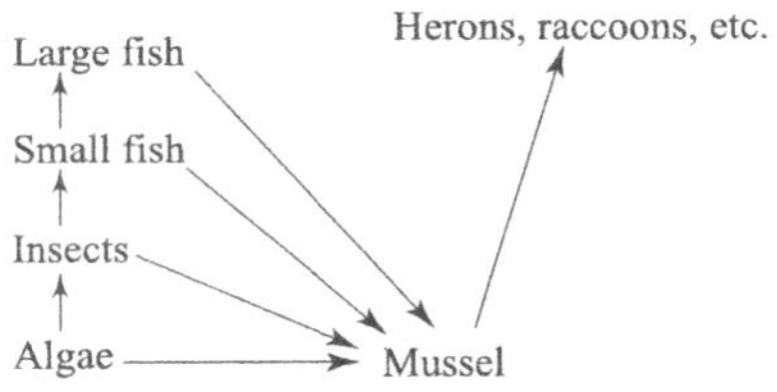


**Fig. 3.17** *River ecosystem*

## 3.8 GRASSLAND ECOSYSTEM

*Grasslands* are areas where the vegetation is dominated by grasses and other nonwoody plants.

*A grassland ecosystem* is a biological community that contains grasslands.

About 32% of the plant cover of the world is covered with grasslands. The most fertile and productive soils in the world have developed under grasslands. Generally, the natural species have been replaced by cereals (cultivated grasses).

Grasslands occur in regions too moist for deserts and too dry for forests. The annual rainfall in grasslands is usually seasonal. It ranges between 25 cm to 75 cm.

The principal grasslands include

(i) Steppes (Europe and Asia)

(ii) Prairies (Canada, USA)
(iii) Pampas (South America)
(iv) Veldts (Africa)

The dominant animal species in grassland ecosystems include large mammals in highest abundance and greatest diversity.

***Examples*** Horses, asses, antelope, herds of bison, etc.

## 3.8.1 Components of a Grassland Ecosystem

The components of a grassland ecosystem are briefly discussed below:

### *(A) Biotic Components*

**(i) Producer Organisms** Mainly grasses and a few herbs and shrubs contribute to primary production of biomass.

**(ii) Consumers** Three main types of consumers in a grassland are

***Primary Consumers*** They are herbivores feeding directly on grasses. These are grazing animals.

*Examples* Cows, buffaloes, goats, sheeps, deer, rabbits, etc.

***Secondary Consumers*** They are carnivores that feed on herbivores.

*Examples* Frogs, snakes, birds, foxes, lizards, etc.

***Tertiary Consumers*** They feed on secondary consumers.

*Examples* Hawks, tigers, lions, etc.

**(iii) Decomposers** They attack the dead or decayed bodies of organisms, and play an active role in their decomposition. In this decomposition process, nutrients are released for reuse by producers.

*Examples* Bacteria, fungi, Actinomycetes, etc.

***(B) Abiotic Components*** Abiotic components include inorganic and organic compounds present in the soil and aerial environment.

The essential elements like C, H, N, O, P, S, etc., are supplied by water; nitrates, sulphates, and phosphates are present in the soil and nitrogen is present in the atmosphere.

# 3.9 DESERT ECOSYSTEM

*Desert* refers to a region or landscape in which the rainfall is negligible; and annual rainfall is less than 250 millimetres. They occupy about 17% of the earth's surface.

Deserts are characterised by

(i) scanty flora and fauna,
(ii) hot days and cold nights, and
(iii) soils with abundant nutrients but little or no organic matter.

### *Structure and Functions of a Desert Ecosystem*

The structure and functions of biotic and abiotic components of a desert ecosystem are as follows.

**(i) Abiotic Components** Nutrients present in the soil and air are abiotic components. The organic substances are poorly present in the soil because of very low rainfall and high temperature.

**(ii) Biotic Components** Biotic components are producers, consumers and decomposers.

***(a) Producers*** In a desert, producers are mainly shrubs/bushes, some grasses and a few trees.
*Examples* Water-retaining plants adapted to arid climate or soil conditions (succulents), hard grasses.

***(b) Consumers*** They include animals which are capable of living in xeric conditions.
*Examples* Insects, reptiles, etc.

Some nocturnal rodents, birds and some mammals like camel, etc., are also found.

***(c) Decomposers*** In a desert ecosystem, decomposers are very few due to less vegetation and very low amount of dead organic matter.
*Examples* Bacteria and thermophillic bacteria.

## Important Definitions

- *Ecology* is the study of interactions between an organism and its physical environment; the relationship between animals and plants and how one species affect another.
- *Ecological succession* is orderly changes in the composition or structure of an ecological community.
- When the development begins on an area that has not been previously occupied by a community, the process is known as *primary succession*.
- When the community development is proceeding in an area from which a community was removed, it is called *secondary succession*.
- *Seasonal and Cyclic Succession* are periodic changes arising from fluctuating species interactions or recurring events.
- An *ecosystem* is defined as a natural unit that consists of living and nonliving parts which interact to form a stable system.
- *Balanced ecosystem* means that the nutrients are able to cycle efficiently, and no community of organisms or natural phenomena is interrupting the flow of energy and nutrients to other parts of the ecosystem.
- All the nonliving components of the environment constitute the *abiotic components*.
- All the living components of the environment constitute the *biotic components*.
- *Producers* are self-nourishing organisms (so they are called *autotrophs*). They contain chlorophyll and are capable of converting carbon dioxide and water, in the presence of sunlight into carbohydrates through photosynthesis. In the process, they give out oxygen.
- *Photoautotrophs* are the producers who fix energy from the sun and store it in complex organic compounds.
- *Chemoautotrophs (chemosynthesisers)* are bacteria that oxidise reduced inorganic substances (typically ammonia and sulphur compounds) and produce complex organic compounds.

- *Consumers* depend on producers to obtain their energy for survival. They utilize, rearrange and decompose the organic matter produced by autotrophs.
- *Herbivores (or primary consumers)* feed on green plants (autotrophs) to obtain energy for survival.
- *Top carnivores (or tertiary consumers)* eat the flesh of both carnivores and herbivores are not killed or eaten by other animals.
- The *decomposers* are also known as *saprotrophs* (i.e. *sapros* = rotten; *trophs* = feeder). They feed on dead organic matter (from producers and consumers).
- Fallen leaves, parts of dead trees, and faecal wastes of animals are termed *detritus*. The consumers that feed on detritus are known as *detrivores*.
- *Food chain* is a feeding hierarchy in which organisms in an ecosystem are grouped into nutritional (trophic) levels and are shown in a succession to represent the flow of food energy and the feeding relationship between them.
- Food chains overlap, because most consumers feed on multiple species and in turn, are fed upon by multiple other species. Thus, we have a complex network of interconnected food chains called a *food web*.
- The *pyramid of numbers* represents the number of individuals at each trophic level. The shape of a pyramid of numbers can be upright, partly upright and inverted depending on the type of ecosystem.
- Flow of energy in an ecosystem takes place through the food chain.
- The *ecological pyramid* is the graphical representation of the organism's position in the food chain. The base of the pyramid consists of the food-producer level and the successive levels make the tiers with the top carnivore or tertiary consumers forming the apex.
- The size of each compartment in an ecological pyramid represents the amount of organisms (or item) in each trophic level of a food chain. *Trophos* is a Greek word meaning nourishment.
- *First Law of Thermodynamics:* Energy can never be created or destroyed, but can be converted from one form to another.
- *Second law of Thermodynamics:* Transformations of energy always result in some loss or dissipation of energy.
- The dry weight of all the matter contained in the organisms is known as *biomass*. Each tropic level contains a definite amount of biomass. As we move up trophic levels, biomass decreases drastically. There is 90 to 99 per cent loss of biomass at each level. This is known as the *pyramid of biomass*.
- *Natural ecosystems* operate by themselves under natural conditions without any interference by humans.
- *Artificial ecosystems* are controlled and manipulated by humans. These are created by humans in order to fulfill certain needs.
- A *forest* is a community of trees, herbs, shrubs, and associated organisms that use oxygen, water and soil nutrients for their growth and reproduction. A *forest ecosystem* is the organisms, soil, air and water associated with the forest.
- An *aquatic ecosystem* is an ecosystem located in a body of water.
- A *marine ecosystem* is different from a freshwater ecosystem mainly because of its salty water and also because the sea is deep, and the water is in continuous circulation.
- *Estuaries* are semi-enclosed coastal bodies of water connected on the one side with a river and on the other side with the open sea. Thus, estuarine is characterised as an ecosystem having fluctuating water level.
- *Grasslands* are areas where the vegetation is dominated by grasses and other nonwoody plants. A *grassland ecosystem* is a biological community that contains grasslands.
- *Desert* refers to a region or landscape in which the rainfall is negligible; and annual rainfall is less than 250 millimetres. They occupy about 17% of the earth's surface.

## EXERCISES

1. **(a)** What are the different tropic levels of organisms in an ecosystem?
**(b)** Why is a complex ecosystem more stable than one with few species?
2. Explain in detail the different components of ecology.
3. Describe the concept of ecosystem and explain the relationship among its different parts using a schematic diagram.
4. How is an ecosystem evolved? Give component parts of an ecosystem.
5. Give a classification of ecology.
6. Give examples of aquatic and terrestrial ecosystems.
7. How is balance maintained in an ecosystem? Why is an ecosystem with a large population of one species considered to be not healthy?
8. "Decomposers are very important in an ecosystem". Give reasons for validity.
9. What do you understand by decomposers? Describe different types of decomposers explaining their functions.
10. Enlist types of ecosystems. Describe in detail the structure and functions of an ecosystem.
11. Narrate in detail energy flow in an ecosystem, Explain with one of the models of energy flow you have learnt in the class.
12. Explain the concept of food chain, food web and ecological pyramid.
13. What is an ecological pyramid? Describe the pyramid of mass and energy with a sketch.
14. Differentiate between food chain and food web.
15. Explain pyramids of number in parasitic food-chain energy flow with a neat sketch.
16. Explain the significance of studying food chains.
17. How are food chains and food webs interconnected? Explain this with an example of aquatic or terrestrial ecosystem.
18. Discuss the structure and function of a desert ecosystem.
19. Write a short note on 'marine ecosystem'.
20. Explain pond ecosystem.
21. Enumerate the aquatic ecosystems and describe the structure of a pond ecosystem stating its characteristic features.

## OBJECTIVE TYPE QUESTIONS

**I. Fill in the Blanks**

1. Herbivores are ______ consumers.
2. Autotrophic planktons are called ______ .
3. Dead plant parts and animal remains are called ______ .
4. Each stage in a food chain is called a ______ .
5. A food chain starts with ______ .
6. Producers produce ______ gas during photosynthesis.
7. Trophical rainforests occur in places where rainfall is more than ______ cm/ annum.
8. In terrestrial ecosystems, 1000 kg of vegetation can support (a) ______ kg of herbivores, which can support (b) ______ kg of carnivores.

9. Pyramid of ______ is always upright.

10. Tropical forests occurs in India in ______ .

11. The tundra biome occurs in ______ .

12. The concept of ______ was introduced by Charles Elton.

### II. Match the following terms.

Match the terms of column I with appropriate terms of column II.

A.

| Column I | Column II |
|---|---|
| 1. Wetlands | (a) African veldt |
| 2. Conifer forests | (b) Cactus |
| 3. Tropical rainforests | (c) bamboo |
| 4. Desert biome | (d) pine |
| 5. Grassland | (e) swamps |

B.

| Column I | Column II |
|---|---|
| 1. Wetland | (a) River |
| 2. Deltas | (b) Coral reefs |
| 3. Ponds | (c) seasonal |
| 4. Marine ecosystem | (d) Fan shaped |
| 5. Flowing water ecosystem | (e) Paddy field |

### III. Multiple Choice Questions

1. The most stable ecosystem is
   (a) ocean (b) forest
   (c) desert (d) mountain

2. Increase in fauna and decrease in flora would be harmful due to increase in
   (a) $O_2$ (b) $CO_2$
   (c) $N_2$ (d) S

3. The food chain in which micro-organisms breakdown dead producers is called
   (a) Predator food chain
   (b) Consumer food chain
   (c) Detritus food chain
   (d) Parasitic food chain

4. Which one of following is an abiotic component of the ecosystem?
   (a) Plants (b) Bacteria
   (c) Fungi (d) Humus

5. Which one is the correct food chain?
   (a) Phytoplankton → zooplankton → fish
   (b) Fish → zooplankton → phytoplankton
   (c) Zooplankton → phytoplankton → fish
   (d) Phytoplankton → fish → zooplankton

6. An ecosystem consists of
   (a) producers and consumers
   (b) producers, consumers, decomposers, and abiotic environment
   (c) producers and decomposers
   (d) consumers and decomposers

7. The main source of energy in an ecosystem is
   (a) mechanical energy
   (b) heat energy
   (c) solar energy
   (d) chemical energy

8. The graphical representation of the interrelation of producer and consumer in an ecosystem is termed the
   (a) food web
   (b) trophic levels
   (c) ecological niche
   (d) ecological pyramid

9. The shape of the pyramid of biomass for a pond or any aquatic ecosystem is
   (a) inverted (b) linear
   (c) upright (d) not certain

10. The shape of the pyramid of numbers for a parasitic food chain is
   (a) linear (b) inverted
   (c) upright (d) Not certain

11. The importance of an ecosystem lies in the
   (a) flow of energy
   (b) cycling of materials
   (c) both (a) and (b)
   (d) none of (a) and (b)

12. The food chain in an ecosystem helps to maintain
- (a) flow of energy in the ecosystem
- (b) passage of nutrients in the ecosystem
- (c) the feeding relationship in nature, thus biodiversity
- (d) all of the above

13. The interdependence of the living organisms among themselves and with the environment is called
- (a) ecosystem (b) ecology
- (c) chemistry (d) biology

14. Ecosystem consists of
- (a) abiotic components
- (b) biotic and abiotic components
- (c) biotic components
- (d) none of the above

**IV. Indicate True or False for the following statements:**

1. Green plants make high-potential-energy organic molecules from low-potential-energy raw material. True/False
2. Temperature affects the morphology, physiology and biochemistry of flora and fauna. True/False
3. The most important function of an ecosystem is gas regulation. True/False
4. Inorganic molecules do have C-C and C-H bonds. True/False
5. The most abundant element present both in humans and the ecosphere is oxygen. True/False
6. Grass → Mouse → Snake → Eagle is a typical food chain. True/False
7. Phytoplankton → zooplankton → small fish → large fish → Bacteria is a typical food chain in pond ecosystem. True/False
8. An ecosystem is defined as a natural unit that consists of living and non-living parts which interact to form a stable system. True/False
9. A food chain is just a sequence of organisms, in which each is food for the next. True/False
10. The dry weight of all the matter contained in organisms is known as biomass. True/False

## Answers to Objective Type Questions

**I. Fill in the Blanks**

1. primary
2. phytoplankton
3. detritus
4 trophic level
5. producer
6. oxygen
7. 200
8. (a) 100, (b) 10
9. energy
10 Kerala and Assam
11. Arctic zone
12. Ecological pyramids

**II. Matching the terms**

**A.** 1. (e) 2. (d) 3. (c) 4. (b) 5. (a)
**B.** 1. (e) 2. (d) 3. (c) 4. (b) 5. (a)

**III. Multiple Choice Questions**

| | | | |
|---|---|---|---|
| 1. (a) | 2. (b) | 3. (c) | 4. (d) |
| 5. (a) | 6. (b) | 7. (c) | 8. (d) |
| 9. (a) | 10. (b) | 11. (c) | 12. (d) |
| 13. (a) | 14. (b) | | |

**IV. True or False**

| | | | |
|---|---|---|---|
| 1. True | 2. True | 3. False | 4. False |
| 5. True | 6. True | 7. True | 8. True |
| 9. True | 10. True | | |

# 4

# BIODIVERSITY AND ITS CONSERVATION

## Learning Objectives

After studying this chapter, you should be able to
- define biodiversity and its conservation
- explain genetic diversity, species diversity and ecosystem diversity
- enumerate different biogeographic zones of India
- discuss the issues of food security and shelter security
- explain the major world food problems
- discuss the issues of social and economic security
- describe the rare species and threatened species
- explain poaching of wildlife

## 4.1 BIODIVERSITY

*Diversity* means the number and variety of species. *Biodiversity* is the diversity of plant and animal life in a particular region or in the world as a whole. It is often used as a measure of the health of biological systems. *The year 2010 was declared as the International Year of Biodiversity.*

Biodiversity represents the quality and characteristic features of life in an ecosystem. Being a combination of genes, species and the ecosystem itself, biodiversity can be considered at three levels: genetic diversity; species diversity and ecosystem diversity. These are briefly explained below.

### 4.1.1 Genetic Diversity

Living things contain in their cells, the basic instructions (which are called genes) for their own development. Many of these instructions result in physical characteristics that affect the way organisms interact with their environment. Variations in such characteristics within the same species give rise to *genetic diversity*. A significant level of variation must be present for a species to adapt to an ever-changing ecosystem. Domesticated species often have low levels of genetic diversity posing risks. A newly evolved virus or bacteria strand can invade a population of nearly identical organisms very rapidly. Thus, the protection that genetic diversity generally offers in wild populations is lost in such artificial selection or preferential breeding of crops and animals.

## 4.1.2 Species Diversity

It is a measure of the diversity within an ecological community that incorporates the number of species in a community and the evenness of species abundances.

Communities with more species are considered to be more diverse. Evenness measures the variation in the abundance of individuals per species within a community. Communities with greater evenness are considered to have greater species diversity.

(a) *Species Richness* Number of species per unit area
(b) *Species Evenness* Evenness of individuals in a species

**Table 4.1** *Species richness and evenness*

| | *Region I* | *Region II* | *Remarks* |
|---|---|---|---|
| Fishes of Type A | 5 | 20 | Region II has more species richness and species evenness, so it is more diverse than region I. |
| Fishes of Type B | 10 | 20 | |
| Fishes of Type C | 15 | 20 | |
| Fishes of Type D | 20 | 20 | |

***Evolution of Diverse Species in an Ecosystem*** Tropical ecosystems (like rainforests and coral reefs) are more biodiverse. Two or three times as many species live in the tropical ecosystems as live in temperate regions—and perhaps ten times as many species live in the tropics as live in the arctic.

Ecosystems near the equator have more species than do ecosystems nearer the poles. The tropics have been a diversity hot spot for more than 250 million years. This is because the tropics appear to be both cradles and museums of diversity.

As per biologists, diversity in a particular ecosystem is a result of local origination processes, local extinction processes and immigration processes,

[Diversity = Origination – Extinction + Immigration]

Thus, the tropics have high diversity because they have high origination, low extinction, high immigration or some combination of these processes.

To sum up, species diversity provides a quantitative idea of the number of species and the variety of species present in a particular region.

(i) *Cradle of diversity* means areas where new species are born and nurtured.
(ii) *Museum of diversity* means areas where old species are preserved despite having gone extinct in the rest of the world.
(iii) *Local origination* is a process which adds to diversity by generating new lineages. (*Lineages* means a series of organisms, populations, cells connected by ancestor/descendent relationships. It also means ancestry or family.)
(iv) *Local extinction* is a process which removes diversity by wiping lineages out in a particular area.
(v) *Immigration* is a process which adds to diversity through new migrant lineages from other areas.

### 4.1.3 Ecosystem Diversity

It indicates the variation in the structure and functions of ecosystems. It tells about trophic levels, energy flow, food and total stability of ecosystems. The ecosystems can be of various types as governed by the species composition and the physical structure. Following are a few examples:

(i) Terrestrial ecosystems
(ii) Aquatic ecosystems
(iii) Artificial or man-made ecosystems

## 4.2 VALUES OR BENEFITS OF BIODIVERSITY

Ecosystems and the species living in them are of enormous value to humankind. Benefits offered by biodiversity are briefly discussed below:

**(i)** ***Intrinsic Values*** Whenever something has a value for its own sake, we say, it has intrinsic value. It means, it does not have to be useful to humans to possess that value. For example, animals and plants have unique intrinsic value attached to them.

**(ii)** ***Agricultural Values*** Apart from wheat, maize and rice which fulfill about 60% of global food demands, thousands of plant species existing in nature could be utilised by humans as food.

**(iii)** ***Medicinal Values*** According to the World Health Organization, 80 percent of the world's people depend on natural-product-based medicine for healthy life.

**(iv)** ***Pest Control Values*** Selected species from nature have very important uses for controlling pests. For this, gene bank or natural biota must be preserved.

**(v)** ***Recreational and Aesthetic Values*** People enjoy fishing, camping, and other wildlife-related recreation activities. Contact with nature is ecologically and emotionally restorative.

**(vi)** ***Instrumental Values*** A species or individual organism has instrumental value if its existence or use benefits some other entity. Millions of people draw sustenance and income directly from forests, grasslands and fisheries.

**(vii)** ***Nonconsumptive Values*** Soil formation and protection of soil from erosion, carbon fixing through photosynthesis, etc., are a number of values that have a very important role in providing suitable conditions for living organisms.

**(viii)** ***Consumptive and Productive Values*** Fruits, vegetables, beverages and a wide array of food products are derived from biodiversity as they originate from living environment.

Medicines, fertilisers, fibres, rubbers and a number of other products obtained from or derived from the living environment have consumptive and productive values.

**(ix)** ***Ecological Values*** Solar energy absorption, nutrient cycling, air and water purification, soil formation, food production, waste disposal and a number of other ecological services depend on biodiversity. They are all very essential for humans, animals, plants, etc.

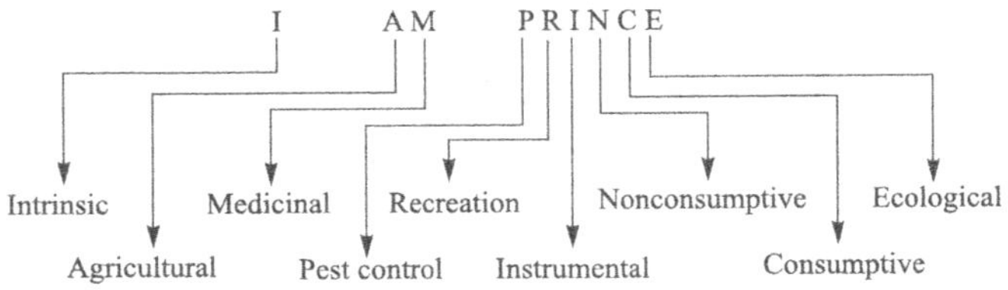


**Fig. 4.1** *Values of biodiversity*

## 4.3 BIOGEOGRAPHIC ZONES OF INDIA

For the planning of conservation of biodiversity at the state and national levels, classification of ecosystems is done based on biogeography.

The biogeographical classification uses the following four levels of planning:

(a) ***Biogeographic Zone*** It is a large distinctive unit of similar ecology, biome representation, community and species, e.g. the coasts, the islands, etc.

(b) ***Biotic Province*** It is the secondary unit within a biogeographic zone, giving weight to particular community.

(c) ***Land Region*** It is a tertiary set of units within a province. It indicates different landforms.

(d) ***Biome*** It is an ecological unit and is found in biogeographic zones or provinces.

### *Explanation of Characteristics of Biogeographic Zones of India*

A concise description of major biogeographic zones is outlined below:

(i) ***Trans-Himalayan Zone*** Climate is cold, vegetation is mountain type, animals found are sheep, goats and snow leopards.
Siachin, Leh and Srinagar are important places of this zone.

(ii) ***Himalayan Zones*** East, west, central and north-west Himalayas are the four biotic provinces of this zone. It has three climate zones and three vegetation zones. Elephants, apes, tigers, lions and bears, are main animals of this zone.
Arunachal Pradesh, and Jammu and Kashmir are important places of this zone.

(iii) ***North-East India*** Climate is highly moist because of heavy rainfall. Bamboo, citrus plants and banana, are important vegetation of this zone. The elephant is the main animal of this zone.
Cherrapunji, Imphal, and Shillong are important places of this zone.

(iv) ***North-West Desert Zone*** Summer is very hot and dry; winter is cold; rainfall is less. Ground vegetation and grasses are the main vegetation of this zone. The Indian Bustard (highly endangered species) is the surviving animal in this zone.
Jodhpur and Jaisalmer are important places of this zone.

(v) ***Gangetic Plain*** Moderate temperature and moderate rainfall is the characteristic climate. Many species of grasses, ground vegetables, fruits and flowering plants are found in this zone. Tigers, leopards, monkeys, apes and elephants are the main animals of this zone.
Lucknow, Allahabad, Patna and Kolkata are important places of this zone.

**(vi)** ***Semi-Arid Zone*** Moderate rainfall and temperature is the characteristic climate. Mixed deciduous, thorny and sal type of forests are found in this zone. Indore, Bilaspur, Jabalpur and Sambalpur are important places of this zone.

**(vii)** ***Deccan Penninsula Zone*** In this zone, the average annual rainfall is low. This zone extends from central India to south India.
Raipur, Hyderabad and Bangalore are important places of this zone.

**(viii)** ***Eastern and Western Sea Coasts*** In this zone, the climate is neither too hot nor too cold. Mangroves (estuarine ecosystem) are found in this zone. Thiruvanantpuram, Chennai and Mumbai are located in this zone.

**(ix)** ***Western Ghat Zone*** Rainfall is heavy, climate is neither too hot nor too cold. Different types of forests, viz. tropical moist evergreen, mangrove, mixed deciduous and temperate evergreen type, are found in this zone.
Pune and Surat are located in this zone.

**(x)** ***Islands Zone*** Climate is moist, neither too cold nor too hot.
Mangroves, beach forests and forests of tall trees are found in this zone.
Port Blair and Kavaratti are located in this zone.

## 4.4 HOT SPOTS OF BIODIVERSITY

*The areas on earth which exhibit high species richness as well as high species endemism are termed hot spots of biodiversity.*

To qualify as a hot spot, an area must satisfy the following criteria:

(i) It must support 0.5% of the global plant species.
(ii) It must have lost more than 70% of its original habitat.

There are 34 hot spots of biodiversity on a global level, out of which the following are present in India:

(a) The Western Ghats
(b) The Eastern Himalayas

About 1 billion people live in these hot spot areas. Many of these areas also suffer from overexploitation of land due to excessive agriculture, hunting, logging and climate changes. Thus, hot spots are in need of sincere conservation actions.

## 4.5 ENDANGERED AND ENDEMIC SPECIES

(i) *Endemic species* can be defined as those species that have very restricted distribution and are confined over relatively small ranges.
*Examples:* Lion-tailed Macaque, Nilgiri leaf monkey.

(ii) When there is no reasonable doubt that the last individual has died, the species is said to be *extinct.*

(iii) A species is *extinct in the wild* when exhaustic surveys in habitats have failed to record an individual.

(iv) A species is *critically endangered* when it is facing an extremely high risk of extinction in the wild in the immediate future.

(v) A species is *endangered* when it is not critically endangered but is facing a high risk of extinction in the wild in the near future.

(vi) A species is *vulnerable* when it is not critically endangered or endangered, but is facing a high risk of extinction in the wild in the near future.

Endangered species are provided with legal protection because their population decreases very rapidly. *Examples:* Tiger, Asian elephant, etc.

## 4.6 RARE AND THREATENED SPECIES

*Rare species*, although are not vulnerable or endangered, have a very small population in the world.

*Threatened species* are those species which may become extinct if not protected. They include the rare, vulnerable and endangered species.

***Examples*** Elephant, chinkara, Nilgiri tahr, Indian wild ass, lion-tailed macaque, tiger, cheetah, sloth bear, rhinoceros, etc.

## 4.7 THREATS TO BIODIVERSITY

In the last 150 years, the rate at which species are disappearing is about thousands per decade; while the natural extinction rate is only one or two species per decade.

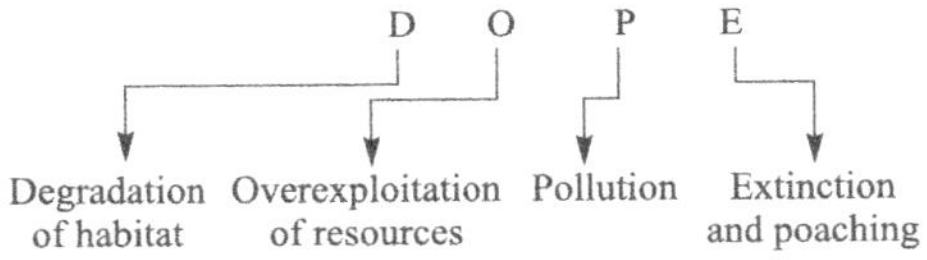


Fig. 4.2 *Threats to biodiversity*

Some of the main causes are as follows:

**(i)** ***Degradation of Habitat*** A habitat is place where living beings find food, cloth and shelter and a safe place to reproduce and bring up their offspring. Thus, loss of habitat is the greatest threat to the world.

**(ii)** ***Overexploitation of Resources*** A number of species like tigers, giant pandas, etc., are on the verge of extinction because of overexploitation of resources.

**(iii)** ***Pollution*** Pollution is responsible for global climatic changes and for the extinction of most species.

**(iv)** ***Extinction of Weaker Species due to Aggressive Non-native Species*** It is responsible for extinction of almost 50% of species on islands all over the world since 1800 AD.

**(v)** ***Poaching of Wildlife*** Poaching is the illegal killing of wildlife for sale in the international trade market. The animals are killed due to the following reasons.

- Some wildlife species are killed for consumption (eating).
- Elephants are killed to obtain their teeth for financial gains.
- Tigers/lions are killed to extract their skin to be sold for decoration of drawing rooms of some people.

We can stop poaching and conserve wildlife by

(a) Reporting poaching incidents to the concerned officers

(b) Encouraging effective wildlife legislation, and law enforcement

(c) Spreading awareness about the importance of wildlife

(d) Refusing to purchase products that have been illegally obtained from animals

## 4.8 HUMAN–WILDLIFE CONFLICTS

Human–wildlife conflict occurs when wildlife requirements overlap with those of humans, creating loss to both.

**(i) Consequence of the Human–Wildlife Conflicts** The conflicts has important consequences in terms of

(a) Wildlife conservation
(b) Micro and macro economy
(c) Safety and well-being for local population
(d) Food security

**(ii) Reasons of Human–Wildlife Conflicts** High human population growth rate which results in increasing demand for natural resources and the growing pressure for access to land.

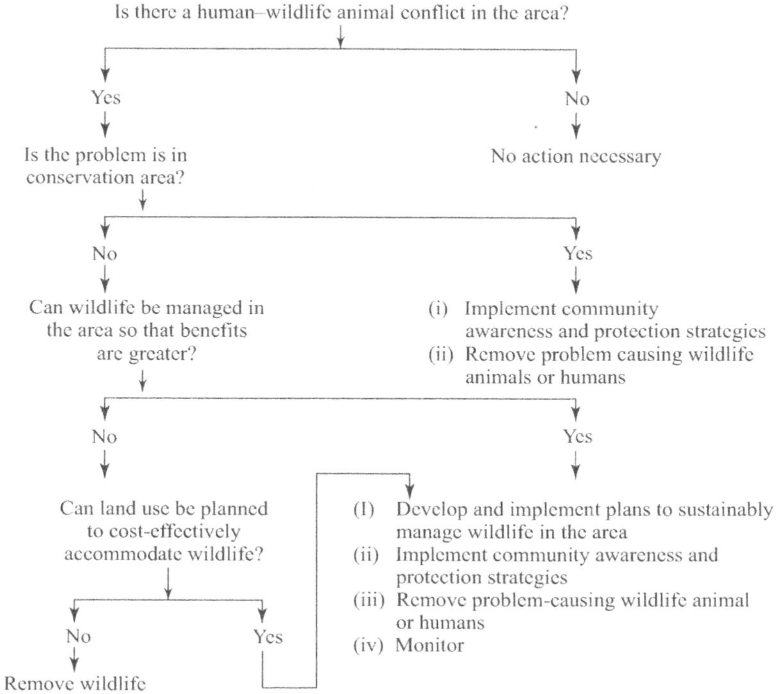


**Fig. 4.3** *A decision support process to determine appropriate management actions in areas with human conflict*

## 4.9 CONSERVATION OF BIODIVERSITY

***(A) Aims*** The protection, preservation, management or restoration of natural resources.

**(B) Objectives** As per the Ministry of Environment and Forests, Government of India, the objectives of conservation of biodiversity are

(i) To protect all critically endangered, endangered, and rare *species*
(ii) To protect natural *habitats* for preserving all varieties of old and new flora, fauna and microbes
(iii) To increase public awareness through media, government agencies, NGOs, etc, and implement strict restrictions on export of rare plants and animals
(iv) To reduce *pollution*
(v) To maintain ecological balance
(vi) To utilise the natural resources in a sustainable way

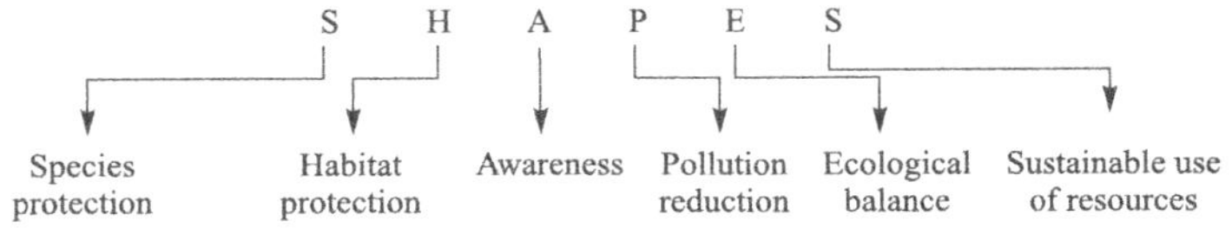


**Fig. 4.4** *Objectives of conservation of biodiversity*

## (C) Types

Conservation of biodiversity can be carried out in the following ways:

**(i) In-situ Conservation** It is the conservation of ecosystems and natural habitats, and the maintenance and recovery of viable populations of species in their natural surroundings, and in the case of domesticated or cultivated species, in the surroundings where they have developed their distinctive properties.

**(a)** ***Advantages*** Convenient and economical way as only supportive role is being played.
**(b)** ***Disadvantages*** It requires a large area for the complete protection of biodiversity. It implies a restriction of human activity and a greater interaction of wildlife with local residents near a reserve forest.

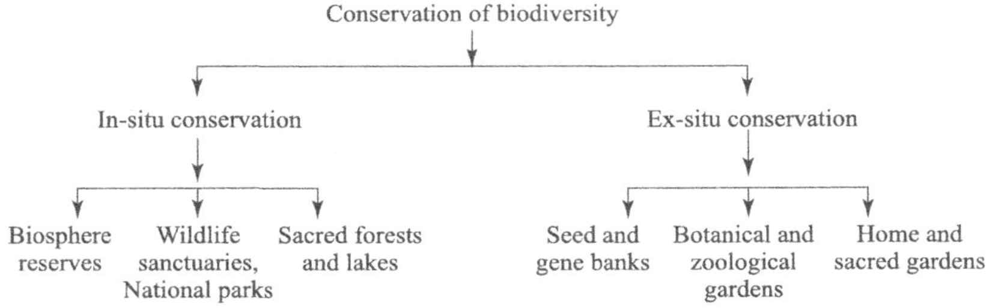


**Fig. 4.5** *Illustration of conservation methods of biodiversity*

**(ii) Ex-situ Conservation** It is the conservation of components of biological diversity outside their natural habitats. It is applicable for those threatened or endangered species whose population is so fragile and habitat fragmented that their survival in the wild is no longer possible.

Suitable locations for ex-situ conservation of animals and plants are zoological parks, and botanical gardens respectively.

## Important Definitions

- *Diversity* means the number and variety of species. *Biodiversity* is the diversity of plant and animal life in a particular region or in the world as a whole.
- Living things contain in their cells, the basic instructions (which are called genes) for their own development. Many of these instructions result in physical characteristics that affect the way organisms interact with their environment. Variations in such characteristics within the same species give rise to *genetic diversity*.
- *Species diversity* is a measure of the diversity within an ecological community that incorporates the number of species in a community and the evenness of species abundances.
  (a) *Species Richness:* Number of species per unit area
  (b) *Species Evenness:* Evenness of individuals in a species
- *Ecosystem diversity* indicates the variation in the structure and functions of ecosystems. It tells about trophic levels, energy flow, food and total stability of ecosystems.
- *Biogeographic zone* is a large distinctive unit of similar ecology, biome representation, community and species, e.g. the coasts, the islands, etc.
- *Biotic province* is the secondary unit within a biogeographic zone, giving weight to particular community.
- *Land region* is a tertiary set of units within a province. It indicates different landforms.
- *Biome* is an ecological unit and is found in biogeographic zones or provinces.
- *Endemic species* can be defined as those species that have very restricted distribution and are confined over relatively small ranges.
- When there is no reasonable doubt that the last individual has died, the species is said to be *extinct*.
- A species is *extinct in the wild* when exhaustic surveys in habitats, have failed to record an individual.
- A species is *critically endangered* when it is facing an extremely high risk of extinction in the wild in the immediate future.
- A species is *endangered* when it is not critically endangered but is facing a high risk of extinction in the wild in the near future.
- A species is *vulnerable* when it is not critically endangered or endangered, but is facing a high risk of extinction in the wild in the near future.
- *Rare species*, although are not vulnerable or endangered, have a very small population in the world.
- *Threatened species* are those species which may become extinct if not protected.
- Human–wildlife conflict occurs when wildlife requirements overlap with those of humans, creating loss to both.
- *In-situ conservation* is the conservation of ecosystems and natural habitats, and the maintenance and recovery of viable populations of species in their natural surroundings, and in the case of domesticated or cultivated species, in the surroundings where they have developed their distinctive properties.
- *Ex-situ conservation* is the conservation of components of biological diversity outside their natural habitats.

## EXERCISES

1. Write a brief note on biodiversity and ecosystem diversity.
2. Explain the evolution of diverse species in an ecosystem.
3. What do you understand by the term 'biodiversity'? Write briefly about the different kinds of diversity in organisms.
4. What is biodiversity? Write the major factors responsible for the loss of genetic biodiversity. How can it be conserved?
5. What are the values of biodiversity?
6. What are biodiversity hot spots?
7. What are endemic species? Name some endemic species of India.
8. What is poaching of wildlife? How can it be controlled?
9. How do biosphere reserves help in conservation of biodiversity?
10. Write a short note on conservation of biodiversity.
11. What are national parks?
12. What are wildlife sanctuaries?
13. What are the differences between endangered species and endemic species?
14. What are the differences between a national park, wildlife sanctuary and reserve forest?
15. What is the difference between in-situ and ex-situ conservation?

## OBJECTIVE TYPE QUESTIONS

I. **Fill in the blanks**

1. Biodiversity refers to the totality of genes, species and _______ of a region.
2. Biodiversity increase from the poles to the _______ .
3. Within a community, diversity is called _______ .
4. Western Ghats and _______ in India are among the 25 global biodiversity hot spots.
5. As per the latest (2011) quantitative evaluation done by the International union for conservation of Nature (IUCN), there are _______ critically endangered species of animals in India.
6. Sacred _______ and lakes are traditional protected areas.
7. Bharatput, Sultanpur Salim Ali, Kumarakom, Ranganthittu, Vedanthangal, kaundinya, Chilka lake, Mayani, Nal Sarovar are the top 10 famous _______ Sanctuaries of India.
8. _______ is the largest wildlife sanctuary of India.
9. _______ is the illegal killing of wildlife for sale in the international trade market.
10. The areas on earth which exhibit high species richness as well as high species endemism are termed _______ of biodiversity.
11. Biodiversity is the term popularised by the sociobiologist _______ .
12. Biodiversity is the foundation of healthy, functioning _______ upon which all life depends.

II. **Match the following terms.**

Match the terms of column I with appropriate terms of column II

A.

| Column I | Column II |
|---|---|
| 1. Pink head duck | (a) Endangered |
| 2. Himalayan brown bear | (b) Vulnerable |
| 3. Andaman wild pig | (c) Endemic |
| 4. Asiatic elephant | (d) Rare |
| 5. Black buck | (e) Extinct |

B.

| Column I | Column II |
|---|---|
| 1. Reserved forests | (a) Forests are regarded as most valuable for the conservation of forest and wild life resources |
| 2. Protected forests | (b) Forestlands are protected from any further depletion. |
| 3. Unclassed forests | (c) Other forests and wastelands belonging to both government and private individuals and communities |

C.

| Column I | Column II |
|---|---|
| 1. Black soil | (a) Tropical forests |
| 2. Arid zone | (b) Dwarfed plants |
| 3. Habitat loss | (c) Fragmentation |
| 4. Alpine zone | (d) Punjab |
| 5. Malabar region | (e) Deccan platean |

**III. Multiple choice Questions**

**1.** A biosphere reserve normally consists of
(a) one or several core zones
(b) a buffer zone
(c) a transition zone
(d) all of the above

**2.** Extinction of a weaker species by an aggressive alien species is the result of
(a) the domino effect
(b) habitat loss
(c) endemism of weaker species
(d) all of the above.

**3.** Variation of genes within the same species is
(a) species diversity
(b) genetic diversity
(c) biodiversity
(d) ecosystem diversity

**4.** The variety and the number of living organisms present in an ecosystem is called
(a) bioprospecting
(b) biopiracy
(c) biodiversity
(d) biogeography

**5.** Which of the following animals is endemic to India?
(a) Blue whale
(b) Snow leopard
(c) Asian elephant
(d) Red colobus monkey

**6.** Which of the following is a biodiversity hot spot in India?
(a) Sundland
(b) Eastern Himalayas
(c) Succulent karoo
(d) Mediterranean basin

**7.** Which of the following is an in-situ tiger reserve in India?
(a) Gulf of Myanmar
(b) Western Ghats
(c) Dudhwa
(d) Agasthyamalai

**8.** Which of the following is not a world heritage site of India?
(a) Manas Wildlife Sanctuary
(b) Kaziranga National Park
(c) Sundarbans National Park
(d) Simplipal

**9.** Which of the following is an endemic species found in Western Ghats, India?
(a) Brown Palm Civet
(b) Marsh Mongoose
(c) Indian Rhinoceros
(d) Flying squirrel

**10.** The concept of biodiversity hot spots was penned by the British ecologist

(a) F.P. Odum
(b) Norman Myers
(c) James Lovelock
(d) Rachel Carson

**11.** The major threat to biodiversity is due to
(a) habitat loss (b) pollution
(c) global climatic changes
(d) all of the above

**12.** Species with very restricted distribution over relatively small ranges is called
(a) extinct species
(b) endangered species
(c) endemic species
(d) none of the above

**13.** Biodiversity hot spots are also known as
(a) forest areas (b) desert areas
(c) land areas
(b) biologically rich areas with large percentage of endemic species.

**IV Indicate True or False for the following statements:**

**1.** Animals and plants are best protected in national parks. True/False

**2.** New approach to conservation is the establishment of biosphere reserves. True/False

**3.** The Tibetan antelopes *chiru* are killed for their wool, which is woven into the luxuary fabric shantoosh, threatening species survival. True/False

**4.** The dodo (bird) was first sighted around 1600 on Mauritius, an island in the Indian ocean. The dodo was extinct less than eighty years later. True/False

**5.** The Indian bustard or the great Indian bustard is a large bird with a horizontal body and long bare legs, being critically endangered by hunting and loss of its habitat. True/False

**6.** If all the plants of the earth die suddenly, all the animals will die due to deficiency of oxygen. True/False

**7.** The biogeographic zone is a large distinctive unit of similar ecology, biome representation, community and species. True/False

**8.** Biodiversity is the diversity of plant and animal life in a particular region or in the world as a whole. True/False

## Answers to Objective Type Questions

**I. Fill in the blanks**

1. ecosystem
2. equator
3. alpha diversity
4. Eastern Himalayas
5. 57
6. groves
7. Bird
8. Sunderbans, West Bengal
9. Poaching
10. hot spots
11. Edward Wilson
12 ecosystems

**II. Matching the following terms**

**A.** 1. (e) 2. (d) 3. (c) 4. (b) 5. (a)
**B.** 1. (a) 2. (b) 3. (c)
**C.** 1. (e) 2. (d) 3. (c) 4. (b) 5. (a)

**III. Multiple Choice Questions**

| | | | |
|---|---|---|---|
| 1. (d) | 2. (a) | 3. (b) | 4. (c) |
| 5. (a) | 6. (b) | 7. (c) | 8. (d) |
| 9. (a) | 10. (b) | 11. (d) | 12.(c) |
| 13. (d) | | | |

**IV. True or False**

| | | | |
|---|---|---|---|
| 1. True | 2. True | 3. True | 4.True |
| 5. True | 6. True | 7. True | 8. True |

# 5

# ENVIRONMENTAL POLLUTION AND ITS EFFECTS

## Learning Objectives

After studying this chapter, you should be able to

- explain the meaning of environmental pollution, water pollution, land pollution, noise pollution and air pollution
- describe the requirement of a nonpolluted environment
- identify the importance of efforts at individual levels to prevent environmental pollution
- describe the natural and man-made (synthetic) pollutant that cause air pollution
- suggest various remedial and control measures to minimise water pollution
- discuss the causes of land pollution and its control
- explain how industrial, agroproducts and pesticides deteriorate the soil
- describe the effects of air pollution on human health
- describe the measures used for controlling air pollution
- define what are solid wastes
- explain the various methods commonly employed for disposal of solid waste with their advantages and disadvantages
- describe the idea of solid waste management

## 5.1 INTRODUCTION

**(i) Environment** It is made up of air, water, land and biota. It is virtually everything that surrounds an organism. The surroundings may be living or nonliving. Each living organism constantly interacts with its environment and adapts to it.

**(ii) Pollutant** It is a material which is present in excess of the natural concentration and produces a bad effect upon the environment. For example, fertilisers having nitrates in them are added so as to increase plant growth. But an excessive nitrate concentration present in drinking water can be toxic, especially to children.

Thus, *anything or any substance, if present in undesirable concentrations and in the wrong place at the wrong time is a pollutant.*

**(iii) Source** It is the system (material or activity) which releases the pollutant.

**(iv) Receptor** It is something that is affected by the pollutant.

**(v) Sink** It is the store where the pollutant is received and stored for a long time.

**(vi) Pollution or Environmental Pollution** *It can be defined as an undesirable change in the physical, chemical or biological characteristics of our environment by the introduction of substances or energy by humans into the environment.*

Environmental pollution is liable to cause hazards to human health, harm to living resources, damage to structures, interfere with legitimate uses of environment and/or harm to ecological systems.

**(vii) Environmental Pollution Process** In it, a pollutant originates from a source. It gets transported by air or water or is dumped on land by humans. Some of the pollutant is assimilated (adsorbed) or chemically changed by the environment. The remaining pollutant builds up to concentrations which are harmful to the environment (Fig. 5.1).

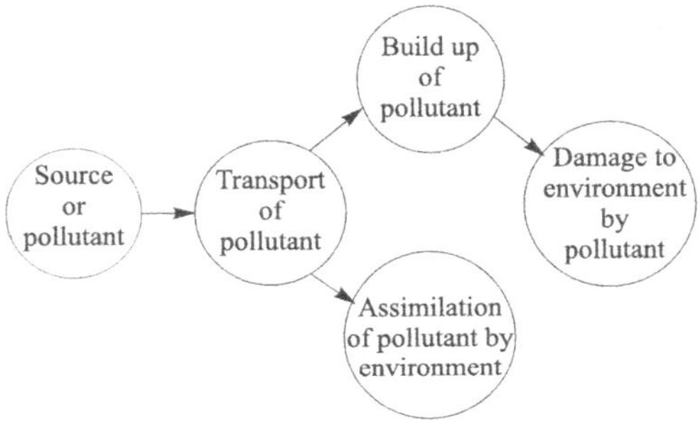


**Fig. 5.1** *Environmental pollution process*

## 5.2 REQUIREMENTS OF A NONPOLLUTED ENVIRONMENT

For the survival of humans, the requirements are air, water, food, shelter, forests and energy.

### 5.2.1 Harms of Polluted Environment

**(i) Air** Air quality is deteriorating day by day. Outdoor air pollution, whether in the form of visible haze or invisible ozone, and carbon monoxide, is a problem for nearly every country in the world.

Current emission of carbon dioxide is about 30 billion metric tons per year. It is expected to increase to around 60 billion metric tons per year by the middle of the century. In order to achieve climate stabilisation, an emission-reduction target of 50% to 100% by 2050 is needed from today's levels.

Indoor air pollution is the second leading environmental health threat in the world. It causes 1.6 million premature deaths every year and afflicts nearly half of the world's population, predominantly the rural poor. Indoor air pollution occurs in poor ventilation systems with accumulations of all sorts of contaminants and toxic substances in the air.

**(ii) Water** Without clean and usable water, nothing can survive. Global water consumption increased sixfold in the last century—more than twice the rate of population growth. Yet readily available fresh water is a finite resource. Moreover, water and populations are unevenly distributed across the globe. Arid and semi-arid regions receive only two per cent of all surface run off yet account for 40% of the global land area and about 50% of the world's poor live there. The existing freshwater resources are under heavy threat from overexploitation, pollution and climate change. There are 74 different kinds of pesticides which have been found in groundwater, used today as potential drinking water. Given these trends, equitably providing adequate water resources for agriculture, industry and human consumption is one of the biggest challenges of the $21^{st}$ century.

**(iii) Food** The United Nations estimates that agricultural output will have to rise 50% by 2030 to meet the increasing demand for food, because of

(a) Continued increase in the world's population
(b) Increased wealth generation from economic development
(c) Production of biofuels from food crops like corn and sugarcane

We have observed record-setting harvests over the last few years, and yet chronic hunger persists and has recently been increasing.

**(iv) Shelter** Cities are home to half of the world's people. They are the hot spots of consumption, production, and waste generation.

However, 50% of the world's people are poor, facing hunger and illness. Thus, sustainable development needs to be the focus area. It supports the concept of lifting populations out of poverty without endangering resources and the environment for future generations. At present, poor populations are suffering from the effects of unsustainable energy use such as adverse health impacts, deforestation, climate change and desertification.

**(v) Forests** Forests contain 70% of the world's biodiversity, provide vital ecosystem services such as soil protection and flood control, and support the subsistence livelihoods of up to 300 million poor people.

Between 2000 and 2005, roughly 13 million hectares of forest disappeared each year, mainly from the biologically rich tropical forests of the developing world. The intergovernmental panel on climate change estimated that deforestation contributes 15–20 per cent of global greenhouse gas emissions in 2007.

**(vi) Energy** On a yearly basis, just in the United States, power plants (that generate electricity), cause 3500 lung-cancer cases and over 35000 heart attacks. The situation is worse in underdeveloped or poor countries where the industrial and the private sectors do not follow consistent environment-protection guidelines.

### 5.2.2 Benefits of Nonpolluted Environment

A nonpolluted environment provides us clean air and water, generates less waste and helps in the conservation of habitat and biodiversity. The economic benefits include reduced expenses on health care, conveyance, and water and electricity bills.

## 5.3 PUBLIC HEALTH ASPECTS

Health is an outcome of the interactions between people and their environment. Disturbance of ecosystems and natural cycles, resource depletion, waste generation and pollution of natural resources affects human health.

Common cold, influenza, chicken pox, tuberculosis, silicosis, gout, black lung cancer, bronchitis and asthma are diseases caused by air pollution.

Cholera, typhoid, dysentery, minamata disease, hepatitis, intestinal disorders are diseases caused by water pollution.

Bolutism, viral food poisoning, fungal food poisoning, staphylococcal poisoning are food-borne diseases.

Teeth/gum diseases, rickets/osteomalacia, anaemia and avitaminosis are deficiency diseases.

Cancer is caused by harmful environmental exposure, tobacco smoking and alcohol consumption.

The key to attaining a good health is based on proper nutrition, safe drinking water availability, provision of maternal and child health care, immunisation against the major infectious diseases, prevention and control of locally endemic diseases, etc. (Fig. 5.2).

*Characteristics of a healthy person are*

(i) Absence of physical discomfort,
(ii) Cheerfulness,
(iii) Courage to face reality,
(iv) Enthusiastic and efficient ability to work,
(v) Self-control and self-confidence,
(vi) Stable mental attitude,
(vii) Efficiency, and
(viii) Freedom from disease.

**Fig. 5.2** *Important factors for attainment of good health*

| Memory aid |
|---|
| "A C C E S S E D" |

## 5.4 AIR POLLUTION

*Air pollution is the presence of substances in the air (which generally originate from human activities) in sufficient concentrations and sufficient time, to interfere with the comfort, health, safety or full use and enjoyment of property.*

### 5.4.1 Sources of Air Pollution

**(i) Point Sources** These are sources which cause direct release of air pollutants.

***Example*** The emission of gases from an industry through a chimney.

**(ii) Nonpoint Sources** These are sources which release substances which are capable of undergoing chemical reactions in the atmosphere to generate air pollutants.

***Example*** Photochemical smog

**(iii) Man-made or Anthropogenic Sources** These are sources which generate air pollutants by human activities.

***Examples*** Vehicular discharges, burning of fossil fuels, population explosion, etc.

- *Vehicular Discharges* A mixture of CO, $CO_2$, $NO_x$ and RH are emitted as exhaust gases from automobiles. They cause air pollution as such. Moreover, they also react with oxides of nitrogen in the presence of sunlight to produce highly toxic *photochemical smog.*
- *Burning of Conventional Fossil Fuels* Burning of coal, lignite, natural gas and combustion of petrol/diesel/CNG produces gaseous by-products like CO, $SO_2$ $NO_x$ which are toxic. They pollute the air and make it unfit for breathing.
- *Population Explosion* It creates the emission of greenhouse gases, global warming, destruction and loss of forest cover and wildlife, etc.

The classification of air-pollution sources is summarised in Fig. 5.3.

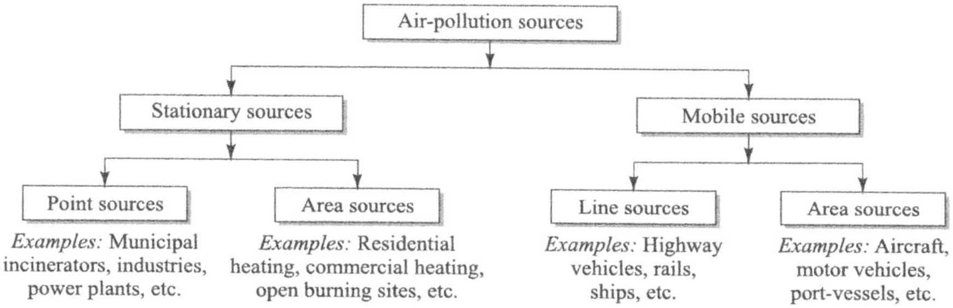


**Fig. 5.3** *Classification of air-pollution sources*

### 5.4.2 Air Pollutants

***Air pollutants** are gaseous, liquid or solid substances present in such concentration as may be or tend to be injurious to human beings or other living creatures or plants or the environment.*

***(A) Classification of Air Pollutants*** The air pollutants can be classified in the following three ways:

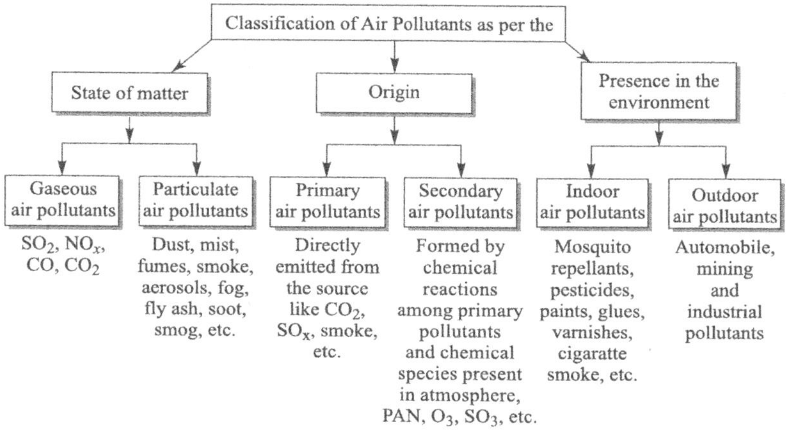


**Fig. 5.4** *Classification of air pollutants*

**(B) Air Pollutants for which National Ambient Air Quality Standards have been given in India** The Central Pollution Prevention and Control Board, New Delhi, has provided the standard for ambient air quality in India under the Act of 1981. Table 5.1 lists these standards.

The air-quality standards give the maximum limit of a particular pollutant permitted in the air around us (i.e. ambient air).

**Table 5.1** *Ambient air-quality standards in India [Concentration μg/m³]*

| | *Area Type* | $SO_2$ | *CO* | $NO_x$ | *SPM* |
|---|---|---|---|---|---|
| (i) | Sensitive areas like places with monuments, sanctuaries, tourist resorts | 30 | 1000 | 30 | 100 |
| (ii) | Residential and rural areas | 80 | 2000 | 50 | 200 |
| (iii) | Industrial and mixed use areas | 120 | 5000 | 120 | 500 |

## (C) Natural and Man-made (Synthetic) Air Pollutants and their Consequences

### (i) Carbon Monoxide (CO)

***Sources*** Forest fires, agricultural burning, incomplete combustion of fuels, tobacco smoking, automobile exhausts, etc.

***Effects*** Toxicity, blood poisoning leading to death, increased proneness to accidents.

### (ii) Sulphur Dioxide ($SO_2$)

***Sources*** Combustion of coal and petroleum products, sulphuric acid plants, powerhouses, metallurgical operations.

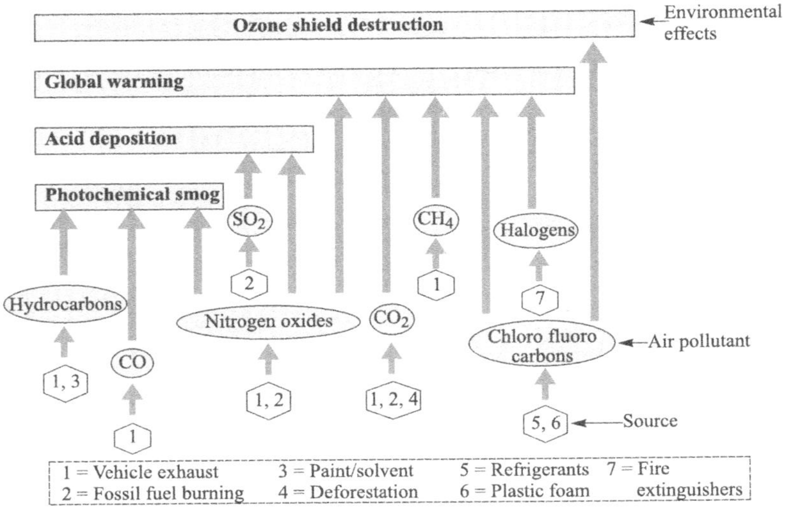


**Fig. 5.5** *Sources of air pollutants and their environment effects*

***Effects*** Irritation of throat and eyes, suffocation, aggravation of asthma and chronic bronchitis.

### (iii) Oxides of Nitrogen ($NO_x$)

***Sources*** Nitric acid plants, automobile exhausts, explosives and fertiliser industries, power stations.

***Effects*** Headache, respiratory irritation, impairment of lung defences, corrosion of teeth, loss of appetite.

### (iv) Mercury (Hg)

***Sources*** Mining and refining of Hg, industries linked with manufacture of medicinal products, pesticides which use organic mercurials.

***Effects*** Inhalation of Hg vapours cause toxic effects, highly toxic organo-mercurials may cause irreversible damage to brain and nervous system.

### (v) Lead (Pb)

***Sources*** Automobile emissions, electroplating waste, plumbing, lead paint industry, printing, etc.

***Effects*** Liver and kidney damages, mental retardation in children, abnormalities in fertility and pregnancy, gastrointestinal damage.

## 5.4.3 Measures Used for Controlling Air Pollution

The most effective method to control air pollution is to prevent the formation of pollutants or to reduce their emission at the source itself.

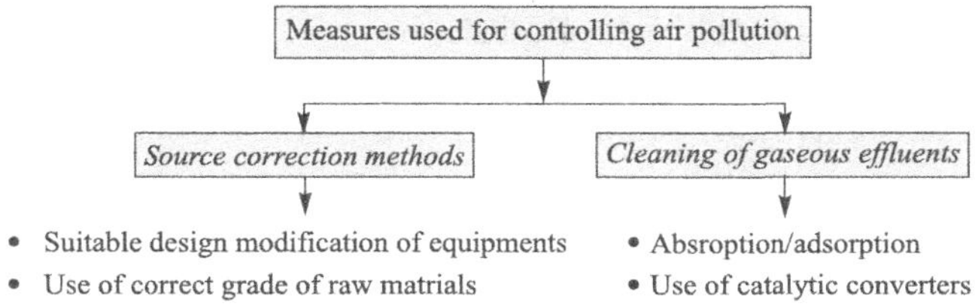


**Fig. 5.6** *Measures to check air pollutants*

**(i) Source Correction Methods** In case of industrial pollutants, the designing and development of plants may be so selected so as to have minimum emission of air pollutants.

***Examples***

(a) By suitable design modification of the tanks, evaporation from petroleum refineries can be minimised.

(b) Use of correct grade of raw material like low-sulphur oil and coal is recommended.

**(ii) Cleaning of Gaseous Effluents** These techniques control pollution by the removal of pollutants from the exhaust. These methods are used in combination with source correction methods.

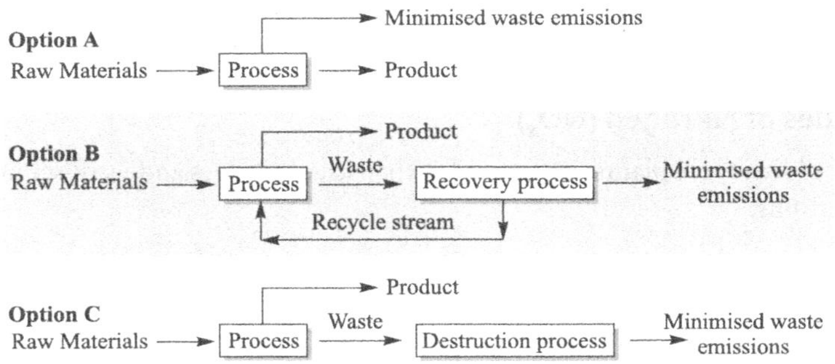


Fig. 5.7 *Options for controlling emissions at source*

***(a) For Gaseous Pollutants*** The gaseous pollutants are removed by absorption in a liquid, or adsorption on a solid. Catalytic converters are also used as they convert gaseous air pollutants into harmless gases.

***(b) For Particulates*** Following techniques are generally used for control of particulate emissions:

- Gravitational settling chambers,
- Cyclone separators,
- Fabric Filters,
- Electrostatic precipitators, and
- Wet scrubber, etc.

# Case Studies

**(i) Health Impact of Vehicular Pollution**

During 2007–2010, a cross-sectional study was carried out to investigate the health impact of vehicular pollution in Kolkata (India). This study was done among 932 male nonsmoking residents of Kolkata city and 812 age- and gender-matched rural subjects as control. The urban group included 56 motor mechanics, 78 bus drivers, 82 autorickshaw drivers, 188 street hawkers, 56 traffic policemen and 472 office employees compared with the rural control group.

The urban group had increased prevalence of hypertension, chronic obstructive pulmonary disease, reduced lung function, headache, asthma, and other respiratory diseases.

In essence, chronic exposure to vehicular pollution of Kolkata

(a) enchances cancer risk in the lungs,
(b) suppresses immunity,
(c) increases blood pressure, and
(d) reduces lung function.

[http: //www.theicct.org/workshops/India 2011/session/mkRey.pdf]

**(ii) Indoor Air Pollution** People in modern societies often spend many hours daily in the indoor environment. It is therefore very useful to find out the contribution of *Indoor Air Quality (IAQ)* to possible health outcomes at the household level.

The following important issues were found out based on empirical data collected from 5949 households from 35 wards of Delhi (NCT of Delhi):

(a) Many residents live in degraded indoor environmental conditions.

(b) The highest risks to health is from
- poor ventilation,
- exposure to Environmental Tobacco Smoke (ETS).
- lack of kitchen, and
- use of traditional fuels.

(c) Most prevalent health problems confronted by residents are
- cardiovascular diseases,
- lung cancer,
- asthma, and
- accute respiratory infections.

[http: //onlinelibrary.wiley.com/doi/10.1111/j.1600-0608.2011.0715.x/Full]

## 5.5 WATER POLLUTION

Pure water, like fresh air, is also becoming a scarcity. Whenever water has a bad taste or odour; oil or grease is seen floating on the surface of water; occurrence of outbreak of an epidemic happens; reduction in the number of fishes in water bodies is noticed; or offensive odours are emitted by water bodies, we say that the water is polluted.

*Water pollution is defined as presence of any foreign substance or energy in water in such concentration and for such duration that tends to degrade the quality of water so that humans, animals or any other organism cannot enjoy the beneficial qualities of water but the use constitutes a hazard.*

Water pollution is classified as freshwater pollution (it includes both surface and ground water pollution) and marine water pollution.

### 5.5.1 Sources of Water Pollution

**(i) Point Sources** When the cause and place of pollution is easily identifiable, it is known as a point source of water pollution.

***Examples*** Municipal and industrial discharge pipes.

**(ii) Nonpoint Sources** When the cause and place of pollution cannot be readily identified, it is known as a nonpoint source of water pollution.

***Examples*** Mining runoff and acid rain.

### 5.5.2 Water Pollutants

**(i) Organic Pollutants** They include oil, synthetic organic compounds, sewage and agricultural run-off, disease-causing wastes and oxygen-demanding wastes.

**(ii) Inorganic Pollutants** They include metals, metal compounds, organometallic compounds, mineral acids, inorganic salts, etc.

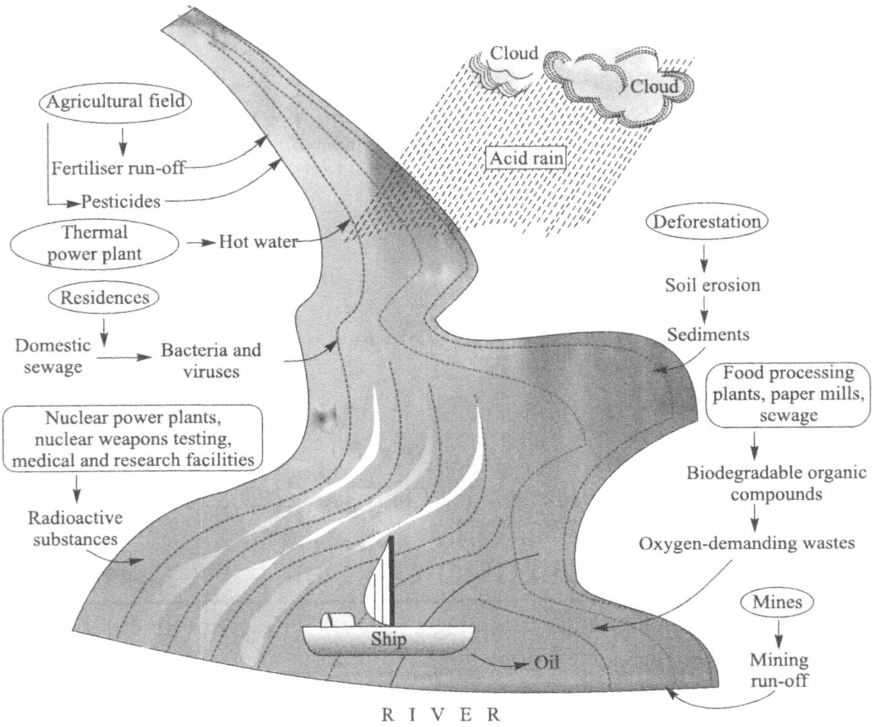


Fig. 5.8 *Sources of water pollution*

**(iii) Suspended Solids and Sediments** They comprise of sand, silt and minerals eroded from the land.

**(iv) Radioactive Materials** They include radioactive isotopes from nuclear reactors, nuclear power plants, research, industrial applications, agriculture and therapeutic as well as diagnostic medical applications.

**(v) Thermal Pollution** They include discharge of waste heat to water bodies by thermal and nuclear power plants.

## 5.5.3 Various Remedial and Control Measures to Minimise Water Pollution

(i) Enforce *zero-emissions laws* to protect water from pollutants and contaminants.
(ii) Create strict *rules and regulations* to prevent the continued pollution of water.
(iii) *Conserve* clean *water* supplies.
(iv) *Avoid littering* in any form.
(v) Support *clean agriculture* by preferably purchasing and consuming healthy organic foods. In the growing of organic food, no pesticide or other harmful contaminants are used.

(vi) Oppose coastal development and *preserve natural wetlands*, as they serve as nature's filter. The vast amount of various plant life, naturally occurring bacteria and algae and microorganisms help to filter destructive pollutants. Wetlands can be easily preserved through replanting efforts.

(vii) *Create awareness* in public through media, child education, etc., for preventing water pollution.

(viii) Treat waste water (from domestic and industrial houses) before disposal.

(ix) Prevent pollution caused by animals.

(x) Reduce urban/suburban run-off of lawn fertilisers and pesticides—stop use of chemical pesticides around your house and lawn.

(xi) Contribute some money to replace outdated municipal water-treatment plants.

(xii) *Stop deforestation*, save paper. A healthy forest acts like a sponge to soak up and clean rainwater and then supply it to nearby lakes and rivers.

(xiii) *Reduce pollution* from oil and petroleum liquids:
   (a) Maintain your vehicle more.
   (b) Drive your vehicle less.

(xiv) Reduce mercury emissions:
   (a) For new coal-fired power plants, use *coal-gasification* to filter out all mercury pollution.
   (b) For existing coal-fired power plants, use better *scrubber technology* to reduce mercury emissions by 90%.
   (c) Conserve electricity.

(xv) Insist that regulatory agencies force mining industries to
   (a) Use clean technologies for mining operations
   (b) Clean up long-abandoned but still-polluting mines
   (c) Stop mountaintop removal coal mining practice which often buries streams together

(xvi) As consumers and citizens, stop purchasing coal, metals, precious metals that are not produced by environment-friendly techniques.

(xvii) Use only green household cleaners and personal-care products.

(xviii) Fight global warming to avoid ocean acidification and rise in the ocean temperature.

(xix) *Increase tax* on chemicals, petroleum products, packaging so as to reduce their consumption. This helps in preventing water pollution.

(xx) Preferably use solar energy, wind energy, etc., because thermal power plants use fresh water for producing electricity. By this, fresh water will be available for other uses.

(xxi) Facilitate soaking of rain through permeable surfaces in parking lots, sports courts, driveways, sidewalks, etc. Impermeable surfaces increase rainwater run-off, resulting in more water pollution.

(xxii) Use less plastic bags, as they are easily blown around and end up polluting water bodies.

(xxiii) Promote *industrial symbiosis*. In it, the unusable waste from one company's industrial process become the input for another's. It helps in keeping effluents out of waterways, and keeping solid waste out of landfill.

(xxiv) Promote *green chemistry*. Eliminate the toxic components of an operation by using water-based solvents derived from corn, citrus fruits, soybeans and other biochemical sources.

## 5.5.4 Water Treatment

Using a network of sewers (i.e. underground pipes), sewage (i.e. waste water) from residential and commercial areas and industries is collected. This sewage is then transported through a sewerage system to treatment plants for purification.

*Waste-water treatment* is carried out in the following steps:

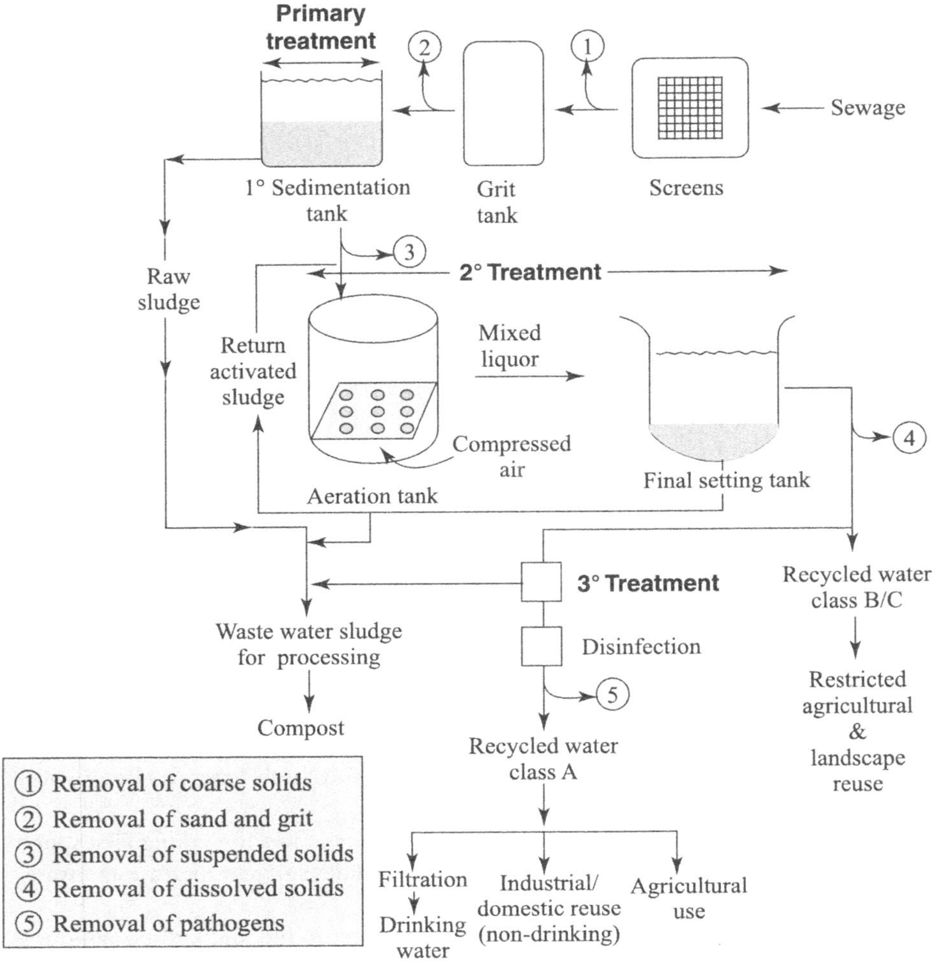


**Fig. 5.9** *Layout of a water-treatment plant*

**(i) Primary Treatment** It helps in the removal of suspended particles.

**(ii) Secondary Treatment** It helps in the aerobic decomposition of organic matter.

**(iii) Tertiary Treatment** It helps in the production of safe water, free from harmful chemicals and pathogenic bacteria. Figure 5.9 shows the layout of a water-treatment plant. The tertiary treatment consists of flocculation, activated carbon, reverse osmosis and/or chlorination methods to get pure water.

## Case Studies

### (i) River Water Pollution in India

According to a United Nations report released on March 22, 2010 on World Water Day:

(a) Contaminated and polluted water kills more people than all forms of violence including wars.
(b) In developing countries, 90% of waste water discharged daily is untreated.
(c) Sick water (thus obtained) is contributing to the deaths of some 2.2 million people a year from diarrhoeal diseases.
(d) At least 1.8 million children younger than five years of age die every year from water-related diseases.
(e) In India, about 80% of urban waste ends up in the country's rivers.
(f) The river water pollution is getting worse due to unchecked urban growth across the country combined with poor government oversight.
(g) In India, a growing number of water bodies are not fit for human use.
(h) The River Ganga (holy to India's 82% Hindu majority) is dying slowly due to unchecked pollution.
(i) Out of the 15 million Delhi residents, only 55% are connected to Delhi's sewage system. The remainder flush their waste, bath waters, etc., down pipes and into drains, and most of them empty into the Yamuna.
(j) Over 3 billion litres of waste (raw sewage industrial run-off and garbage) is thrown into the Yamuna river per day.
(k) On various clean-up efforts, nearly 20 billion rupees has been spent according to the Centre for Science and Environment.
(l) Recently samples taken from the Ganga River (near Varanasi) show that levels of fecal coliform (a dangerous bacterium that comes from untreated sewage) were some 3000% higher than what is considered safe for bathing.

[http: //www.gits4v.com/envo/envo4.htm]

### WATER-POLLUTION FACTS

- Asian rivers are the most polluted in the world. They have 20 times more lead than rivers in industrialised countries and three times as many bacteria from human waste as the global average.

- Pollution of freshwater is a problem for about 50% of the world's population. Each year there are about 250 million cases of water-related diseases, with roughly 5 to 10 million deaths.
- Each year, plastic waste in water and coastal areas kills up to 1 million sea birds, and 1 lakh marine mammals and countless fish.
- In one week, a typical cruise ship generates
  (a) > 8 tons of solid waste,
  (b) 10 lakh gallons of gray water from showers, sinks, dish washers and cloth washers,
  (c) 2.1 lakh gallons of sewage, and
  (d) 37 thousand gallons of oily bilge water.
- Around 80% of the pollution in seas and oceans comes from land-based activities.
- In India, almost 70% of surface-water resources have serious pollution problem.
- In India, domestic waste water is one of the major pollution sources. It causes heavy pollution in 14 major river systems as 50 million $m^3$ of untreated sewage is discharged into them each year. This also causes high incidence of water-related diseases.

### (ii) Water Pollution in Kolkata

In 2011, engineers of Kolkata Municipal Corporation suggested replacement of one-third of the total 5000 km pipeline of the water supply network. The main reason was corroded, age-old, worn-out water pipes. In some stretches, sewer lines run next to the water pipes. When the pressure in the pipes is low, often sewage seeps into the water and contaminates it. Because of unsafe drinking water, Kolkata is facing many health hazards, viz.

(a) Bacterial contamination is causing gastroenteritis, haemorrhage, septicaemia
(b) Many persons are suffering from irritable bowel syndrome
(c) Contaminated drinking water is also responsible for rapid spread of hepatitis A and E
(d) Malnutrition and poor growth is also linked to water-borne diseases

[http: //kolkatainfo.wordpress.com/2011/03/14/water pollution in kolkata/]

### (iii) Bottled Water

40% of the bottled water industry is branded. Key players are Parle Bisleri, Coca Cola India Inc., Kinley and Pepsico India Holdings Pvt. Limited. In 2010, the bottled water market generated $250 million in revenue. Bottled-water consumption is rising in India due to water shortage and health awareness. It is to be noted that both tap water and bottled water are unhealthy. The tap water is legislated to be 7 pH neutral. However, municipalities add huge quantities of chlorine in the water to kill all harmful bacteria, and this makes it highly acidic (unhealthy).

1. Water is taken from a source. (River, Stream, Aquifer, or a local water supply)

↓

2. The water is then processed and packaged in plastic bottles/containers

↓

3. The packaged product is then shipped to various locations

↓

4. Bottles are then received and stored at distribution facilities awaiting sale

↓

5. After consumption of water, most of the plastic bottles then either litter o ends up in land fills

**Fig. 5.10** *Flow chart depicting bottle water cycle*

According to a recent study on bottled water, carried out by the Ahmedabad-based Consumer Education and Research Society (CERS), an independent nonprofit institution with a sophisticated product-testing laboratory:

(a) In clear violation of norms, as many as 10 of the 13 brands had foreign floating objects.

(b) Two mineral water units in Bangalore on January 11, 2011 (during a surprise inspection) were bottling borewell water. They were also illegaly using several branded labels on the bottles to market the water.

(c) AM Enterprises were found selling water without an ISI mark from the Bureau of India Standards. They were found mixing mineral water with borewell water and selling it in cans to the public.

The public is indirectly raising the price of petrol and contributing to global warming and climate change by purchasing bottled water.

- According to a social activist R Srinivasan, more than 13,000 tankers carry water drawn from farmlands surrounding Chennai so as to generate profit from Chennai's accute water scarcity.
- According to the Pacific Institute, the manufacturers of plastic water bottles generated more than 2.5 million tons of carbon dioxide emissions and required 17 million barrels of oil (in 2007).
- About 47 million gallons of oil is required to make plastic bottles for storing water.
- Water when stored in earthen pots, becomes not only refreshingly cool and tasty, but is said to become bacteria free too.

Beverage majors are all measuring the water consumed for bottling every litre of their retailed drinks. Efforts to reduce water consumption is necessary for all companies in India because of limited availability of this natural resource. Hindustan Coca Cola Beverages uses 2.5 litres of water to make every 1-litre bottle of the carbonated drink. The company has 56 bottling plants in India.

Coca Cola's bottling plant in Kerala faced big problems after activists alleged the company was depleting local water supplies. The Palakkad bottling factory in Kerala was closed in 2005 after protests from residents and activists. A high-level state pancel concluded that the Coca Cola plant had caused water contamination, soil degradation and environmental degradation. They recommended a fine of 47 million dollars. A bill has been passed on February 25, 2011 allowing compensation claims in Kerala.

In drougt-prone Kala Dera near Jaipur (India), the focus of the protest is on the source of the packaged water and how bottling companies are grabbing underground water. The companies get their water free except for a small cess (for discharging the waste water). A Rs. 10 per litre bottled water has a raw material cost of just 0.02–0.03 paise. The companies siphon out, exhaust and export groundwater that once belonged to an entire village and buy it for themselves, forcing the villagers to pay for water that they used to be able to use as a community, free of charge. TERI advised Coca Cola in January to shut a bottling plant in Kala Dera.

The Tamil Nadu Water Supply and Drainage Board (TWAD) Workers Federation suggested that the government take over the sale of bottled drinking water. This is because the actual cost of purifying water is less than 5 paise per litre. So even if the government sells water at Rs. 2 a litre, they can generate a revenue of Rs 5,000 crore.

[www.gits4u.com/water/water16.htm]

***Hazards of Bottled Water***

(i) The global annual amount of plastic used to produce water and carbonated water bottles is about 3 million tons. Most of this plastic ends up in landfills.

(ii) Each bottle requires nearly 5 times its volume in water to manufacture, creating wastage of valuable resource.

(iii) The global annual amount of carbonated bottled water consumed outside its country of origin is more than 20 million tons. This results in huge transportation and environmental damage.

(iv) More than 70 per cent of bottles are not recycled. These are responsible for increased soil pollution as they end up in landfill sites.

(v) The bottled water industry has less stringent testing policies than governmental agencies which require rigorous testing of tap water. Thus, bottled water is not safer than tap water.

(Source: www.infographicslibrary.com)

(vi) More than 20% of tested bottled water brands contain chemical contaminants at levels above strict state health limits

(vii) About 40% of all bottled water is tap water but it is very expensive.

**Table 5.2** *Comparison of bottled water and tap water*

| *S.No.* | *Attribute* | *Bottled water* | *Tap water* |
|---|---|---|---|
| (i) | Required to provide source | No | Yes |
| (ii) | Tested for *E.coli* bacteria | No | Yes |
| (iii) | Required to produce quantity reports | No | Yes |
| (iv) | Can still be distributed when tap water quality standards are not met | Yes | No |

(Source: www.lifegivingwater.com.au/health.php)

## 5.6 LAND POLLUTION OR SOIL POLLUTION

*Soil pollution can be defined as the introduction of substances, biological organisms, or energy into the soil that lead to a change in the quality of soil so that plant growth and animal health is adversely affected.*

### 5.6.1 Causes of Soil/Land Pollution

Soil pollution is caused due to direct and indirect sources. The direct sources harm the soil much more than the indirect sources. Examples of direct causes are poor management of solid and liquid domestic/industrial/agricultural waste, waterlogging, soil erosion, salination, disposal of medical wastes, etc. Examples of indirect causes are acid rain and disposal of radioactive substances.

The main reasons of soil pollution are briefly described below:

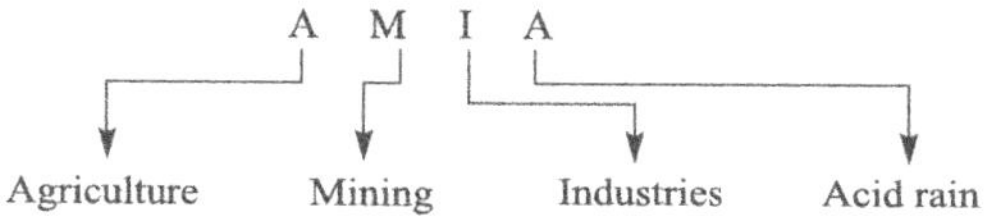


**Fig. 5.11** *Soil pollution causes*

#### (i) Faulty Agricultural Practices

***(a) Unskilled Irrigation*** *Waterlogging* may occur when the drainage system of the agricultural field is not maintained scientifically. Waterlogging closes the passage of air to the soil, stops the growth of soil organisms and makes the soil barren.

***(b) Shifting Cultivation*** In it, the forest is burnt to use the land for cultivation. However, this practice exposes the soil for soil erosion.

***(c) Injudicious Use of Chemical Fertilisers*** Use of inorganic fertilisers increases the nutrient contamination. The microbes of the soil reduce the nitrogen to nitrite ions which enter the animal body through food or water. It is directly absorbed in the bloodstream and oxidises the oxyhaemoglobin (the $O_2$ carrier) to methemoglobin. The latter cannot carry oxygen any more; so ultimately the animal dies.

***(d) Pesticides*** Pesticides are chemicals which are used by farmers to protect their crops. Large amounts of pesticide in the soil interfere with the soil's metabolic process. Pesticides kill many nontargeted beneficial soil organisms such as earthworms. Thus, the soil becomes infertile.

Organo-chlorides (like Dichloro Diphenyl Trichloroethane, DDT) are second generation pesticides. They are nonbiodegradable substances. They accumulate and magnify in the food chain and interfere with the calcium metabolism of birds. As a consequence, birds lay fragile, thin-shelled eggs.

High concentration of pesticides get accumulated in fatty tissues of prey organisms. When predators eat these prey organisms, they also get killed.

Thus, pesticides lead to poisoning of the ecosystem.

**(ii) Mining** For mining, trees are cut down. Loss of greenary results in land degradation, drought and desertification. This is illustrated in Fig. 5.12.

**(iii) Solid Wastes from Homes and Industries** Chemical, petroleum, and metal-related industries, dry cleaners and gas stations produce hazardous waste such as oils, battery metals and organic solvents. These hazardous wastes contaminate soil and water resources.

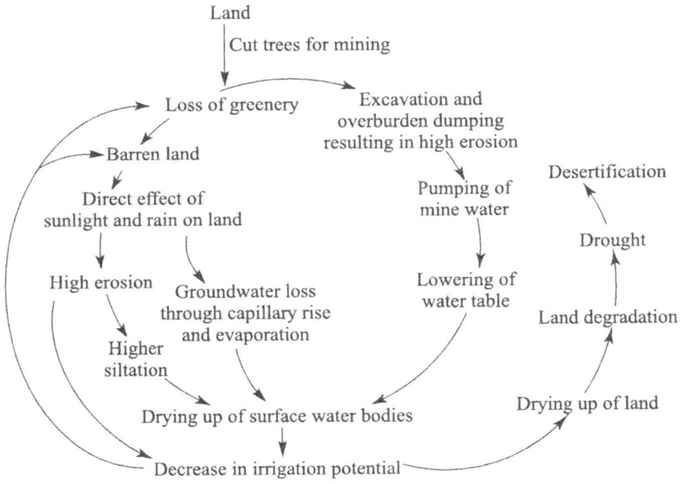


**Fig. 5.12** *Land-degradation cycle due to mining and related activities*

**(iv) Acid Rain** It converts neutral soil to an acidic one.

## 5.6.2 Effects of Soil Pollution

The harmful effects of soil pollution are briefly described below:

(i) Fluorosis occurs as a result of consumption of fluoride containing maize and jawar *crops*. The fluoride is absorbed by the crops from the fluoride-contaminated soil.

(ii) Emission of toxic gases (from dumped solid wastes on land) are detrimental to health. The unpleasant smell and spread of insects cause inconvenience to people.

(iii) Poisoning of the ecosystem takes place by soil pollution.

(iv) *Contamination* of underground and surface drinking *water* takes place by soil pollution.

(v) Reduction in the fertility of *soil* takes place by soil pollution.

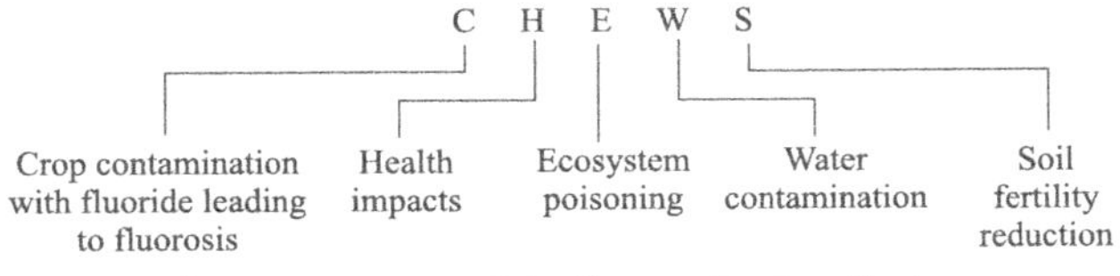


**Fig. 5.13** *Harmful effects of soil pollution*

## 5.6.3 Control of Soil Pollution

The land pollution can be controlled by the following methods:

(i) Polluted soil can be *treated* by *bioremediation*. It uses microorganisms (yeast, fungi or bacteria) to break down, or degrade, hazardous substances into less toxic or nontoxic substances (such as $CO_2$ and $H_2O$). Proper treatment of liquid wastes from industries and mines must be done.

(ii) The principles of three Rs, namely, *Recycle*, *Reuse* and *Reduce*, help in minimising the generation of solid waste. For example, use of biofertilisers and natural pesticides help in minimising usage of chemical fertilisers and pesticides.

(iii) Proper disposal methods must be employed. For example, composting of biodegradable solids and *incineration* of nonbiodegradable solids should be done.

(iv) Planned *afforestation* helps in preventing soil erosion.

(v) Formulation and effective implementation of stringent pollution-control *legislation* also helps in controlling soil pollution.

(vi) Faulty *sanitation practices* must be improved.

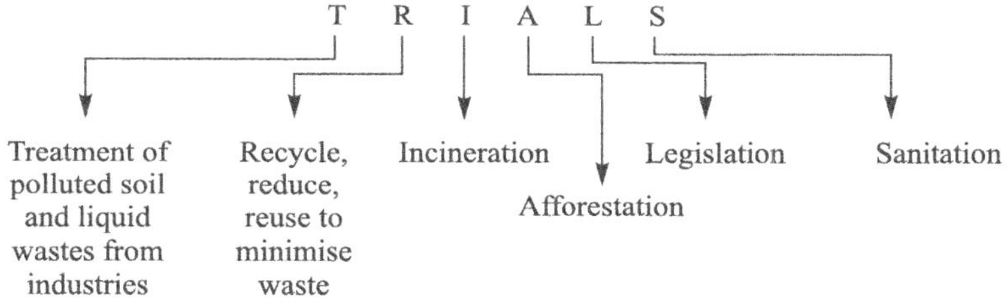


**Fig. 5.14** *Control of land pollution*

# Case Studies

## (i) Soil Pollution in Patancheru Industrial Area, Hyderabad, Andhra Pradesh, India

The Patancheru industrial development area is in Hyderabad, which is the fifth largest city in India. In this area, soil environmental hazards are increasing at an alarming rate because of rapid industrialisaiton and indiscriminate disposal of waste materials without adequate knowledge of toxic pollutants and their control. Studies have shown that the soil from resudential areas is moderately contaminated with Cr, Ni and Pb. The area needs bioaccumulators to remove toxic metals.

[Res. J. Environ. Earth Sci:, 3(3): 214-220, 2011; http: //mawellsci.com/print/rjees/v3-214-220.pdf]

## (ii) Heavy Metal Contamination in Different Vegetables

[http: //scialert.net/fulltext/? doi=rjet.2011.162.179 and org = 10]

Heavy metals such as cadmium, copper, lead, chromium, and mercury are very harmful because of their long biological half-lives, nonbiodegradable nature and their potential to accumulate in different body parts. Plants take up heavy metals by absorbing them as deposits on the parts of the plants exposed to the air, from polluted environment as well as from contaminated soil.

Recent findings of heavy metal contamination in vegetables:

(a) Waste water contains a substantial amount of toxic heavy metals.

(b) Waste-water irrigation results in the accumulation of heavy metals in soil and vegetables.

(c) Excessive accumulation of heavy metals adversely affects the food quality and safety.

(d) Unwashed vegetable contain high concentrations of toxic elements resulting from the accumulation of air-borne heavy metals.

(e) Heavy metals get concentrated/accumulated in edible portions of leafy or root crops than storage organs or fruits.

(f) Accumulation of heavy metals in fruits is low (unlike vegetables). This is because a large proportion of heavy metals absorbed by trees are stored in other organs like leaves.

(g) Consumers of heavy-metal-contaminated vegetables are at health risks.

(h) Consumption of vegetables grown on urban waste dumpsites is dangerous to human health.

(i) The transportation and marketing of vegetables in contaminated environments elevates the levels of heavy metals in vegetables through surface deposition.

(j) To reduce health risks, vegetables should be washed properly before consumption. This is because on washing, a significant amount of aerial contamination from the vegetable surfaces is removed.

## 5.7 MARINE POLLUTION

When the salt content of a water body is equal to or more than 35 parts per thousand (ppt), then it is known as a *marine water body*.

***Examples of Marine Water Bodies:*** Seas, oceans, brackish water, salt marshes, etc.

***Marine pollution*** *is defined as the direct or indirect discharge of matter or energy by humans into marine water bodies that is harmful to living organisms, hazardous to human health, hinders marine activity, adversely affects sea-water quality and reduces its amenities.*

Thus, marine pollution is harmful and is caused by human activities. Damages or disturbances caused by earthquakes, volcanic eruptions, tsunamis, etc., are not considered marine pollution.

*Oceans are ultimate sink for most of the waste we produce.* This is because any waste material which is released into the river system (and many sewage systems around the world) will exit into the ocean. Marine pollution is also known as *ocean pollution*.

**(i) Land Sources of Marine Pollution** Radioactive substances, toxic chemicals, solid waste, thermal pollution, sewage, nutrients, pathogens, etc., are some of the land-based pollutants which cause maximum loss to coastal and marine ecosystems.

**(ii) Offshore Source of Marine Pollution** They include release of 'produced water' from the oil-bearing strata with the gas and the oil at the time of production. 'The produced water' together with waste-drilling chemicals and mud is discharged into the ocean.

From offshore operations, '*oil spills*' are caused by equipment malfunctioning, pipeline breakage, etc.

From submarine oil deposits, '*natural seepage of petroleum hydrocarbons*' also contributes to marine pollution.

## 5.7.1 Sources and Effects of Marine Pollution

The major sources of marine pollution and their effects are briefly described below:

### *(A) Sewage*

All over the world, untreated (or undertreated) sewage is discharged into the oceans. It contributes pathogens and nutrients to oceans.

**Effects** Nutrients cause algae blooms (eutrophication) in coastal waters. Decomposing of algae depletes the water of oxygen, killing other marine animals. Algal bloom releases toxins that can kill fish and poison people.

Eutrophication has created enormous dead zones in several parts of the world, including the Baltic sea and the Gulf of Mexico.

Pathogens contaminate coastal swimming areas and seafood, spreading typhoid, cholera, etc.

### *(B) Toxic Chemicals*

Chemicals like DDT, heavy metals, etc., which have toxic effects are known as toxic chemicals. They come from industrial discharge, waste-water discharge from cities; seepage from landfills and pesticides from home use, farms, forests, etc.

**Effects** Toxic chemicals cause disease in coastal marine life especially near industries or major cities. They contaminate seafood. Disease and reproductive failure is caused by fat-soluble toxins that biomagnify in predators. By eating contaminated seafood and animal fats, cancer, damage to the immune system, behavioural problems, and reduced fertility etc., are caused in humans.

### *(C) Thermal Pollution*

Discharge of cooling water is at about 10°C above the temperature of the receiving water from power stations.

**Effects** Heat can kill corals and other temperature-sensitive sedentary species. It can displace other marine life. Increase in water temperature causes a lowering of dissolved oxygen levels.

### *(D) Oil*

Oil pollution in marine water is caused by maritime accidents, wars, offshore oil drilling, oil-tanker accidents, conventional shipping, tank cleaning at sea, oil exploration, etc. It has been estimated that about 80 to 200 litres of oil is spilled in the ocean per second.

**Effects** A drop of petroleum oil spread over marine water surface becomes more dark and viscous due to slow evaporation of lighter volatile fractions. The oil is polymerised to highly viscous tar balls by sunlight and oxygen, causing litter. Besides this, an oil film floating on water inhibits sunlight from entering the sea, thereby prohibiting photosynthesis of aquatic plants and simultaneous oxygen production.

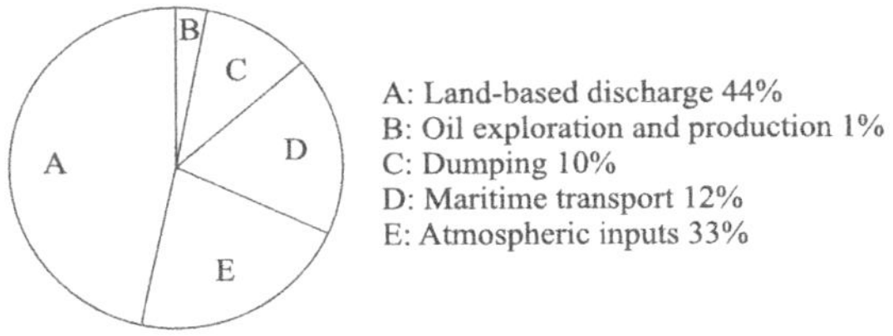


**Fig. 5.15** *Sources of marine pollution (Safe-environment.wordpress.com//)*

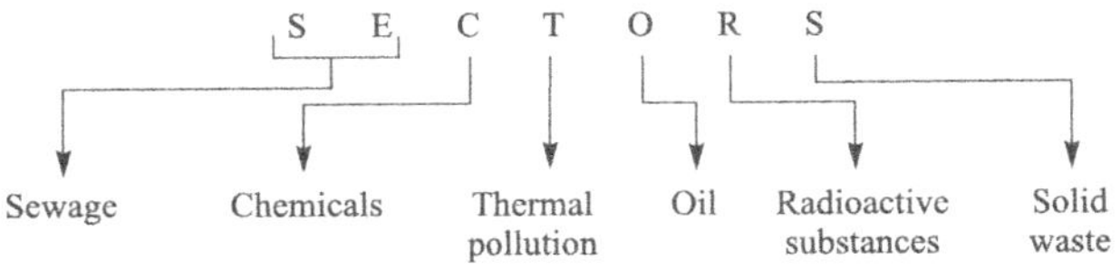


**Fig. 5.16** *Marine pollutants*

The fur of marine mammals and feathers of diving birds get soaked with oil, displacing the air present there, interfering with their buoyancy and natural insulation required for maintenance of body temperature. Once these mammals or birds are soaked, they either get drowned or die due to loss of body heat. From oils, hydrocarbons also enter the human body through the food chain and cause biomagnification.

### *(E) Radioactive Substances*

The oceans are a sink for radioactive waste from the production of nuclear weapons, electricity, testing of nuclear weapons and nuclear accidents.

**Effects** Radioactive substances get concentrated in predators and shellfish, which are eaten by humans.

Radioactive substances cause disease in marine life, and are a great threat to civilisation. It has been predicted that the survival ability of the human species is reduced because several generations are exposed to radiation.

### *(F) Solid Wastes*

Much of the solid wastes are composed of plastics like polythene containers, plastic sheets, nylon ropes and nylon nets with variable sizes and shapes.

**Effects** Plastics are nonbiodegradable. They remain in the oceans for 200 to 400 years and come back to the shore and continue littering beaches and coasts.

Plastics eaten by seabirds causes reduction in appetite, sensation and development of stomach and intestinal ulcers. Fish, turtles, seals and seabirds can become entangled in discarded plastic fishing nets which trap and slowly strangle them. Plastic waste is often mistaken for food by marine animals. Plastic bags have been found blocking the breathing passages and stomachs of many marine species including turtles, dolphins, whales, etc.

### 5.7.2 Control of Marine Pollution

Marine pollution can be controlled through the following measures:

(i) ***Countries*** should improve existing *sewage-disposal facilities.*

(ii) ***Industries*** should adopt zero waste *technologies.* They should discharge waste after treatment with best available technology. It will help in the minimisation of release of toxic or damaging substances.

(iii) ***Agricultural and aqua-cultural*** *practices* should integrate recycling, waste treatment and proper handling for minimum discharge of pollutants.

(iv) ***Bioremedial controls*** should be used in which bacteria and other microorganisms are employed to alter and break the complex toxic chemical into harmless products and $CO_2$.

(v) ***Governments*** should develop and enforce *legislation* relevant to the management of residential and tourism development in the coastal zone.

(vi) ***Legal*** permit should be issued or denied based on detailed *Environmental Impact Assessment (EIA)* or *Environmental Impact Statement (EIS).*

(vii) ***Individual humans*** should ensure that their houses have sewage disposal systems, such as *septic tanks.*

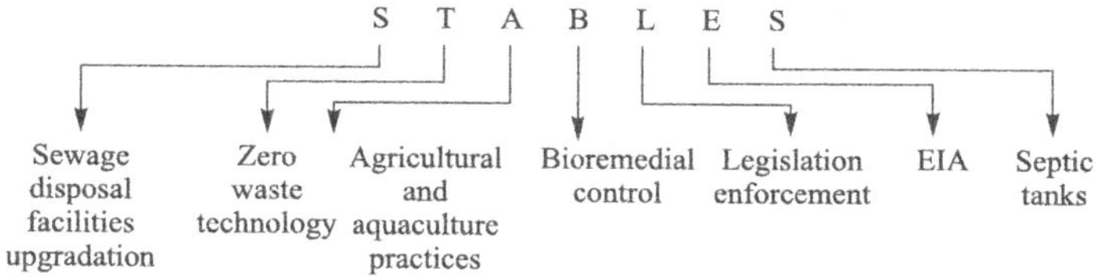


**Fig. 5.17** *Control of marine pollution*

# Case Studies

**(i) Effects of Thermal Power Plant on Marine Ecological Habitat**

Marine backwaters including estuaries and tidal creeks face tremendous stress from disposal of industrial, agricultural and domestic wastes. These water bodies possess self-purification capacity through diurnal tidal cycle and constant ecological integrations.

Major ecological concerns due to thermal discharges from power plants are sublethal damage to biological communities, adverse impact on regeneration process in native biota, alternation in primary and secondary production, etc.

Periodic observations are necessary to estimate the exact response of biotic communities to changing temperature, pH, salinity, enabling retaining a better quantity of creek and associated mangrove habitats.

[Jagtap et al; Journal of Environmental Biology, March 2011, www.job.co.in]

### (ii) Ocean Pollution

There are eighty-seven cities and towns located within 50 km of India's 8,000 km coastline. In these coastal areas, nearly 250 million people live and dump 5.5 billion litres of waste water into the sea every day. Less than 0.5 billion litres of waste water is treated. Municipal waste water constitutes the single largest source of marine pollution. The three states of Maharashtra (45%), West Bengal (26%) and Tamil Nadu (9%) account for the bulk of waste water flushed into our coastal seas.

Fifteen locations that degrade our coastal ecosystems with untreated effuents, contaminated with metals, chemicals and raw sewage are Hooghly, Paradip, Visakhapatnam, Kakinada, Bhimavaram, Ennore, Pondicherry, Cuddalore, Tuticorin, Kochi, Mangalore, Goa, Thane creek, Mahim, Versova and Bassein (Mumbai), Hazira.

Common effluent-treatment plants for small-scale industries, monitoring, local-scale solutions, costs, benefit-sharing and empowerement are possible solutions for ocean pollution.

[RahulGoswami;http://infochangeindia.org/agenda/coastal-communities/ocean-pollution.html]

## 5.8 NOISE POLLUTION

Vibrations transmitted through an elastic medium (air, water or solids), with frequencies in the approximate range of 20 to 20,000 hertz, capable of being detected by ears is known as *sound*.

*Noise* is a sound that is undesired, unexpected, unpleasant or loud. It is undesired in that it interferes with sleep, rest, recreation, work or communication. The word *noise* comes from the Latin word *nausea* meaning sea sickness.

***Noise pollution** is defined as environmental noise or an unwanted sound that is annoying, distracting, or physically harmful. Harms include hearing loss, stress, sleeplessness, etc. Noise pollution is also known as **sound pollution**.*

There are about 25,000 hair cells in our ear which create waves in our ear, corresponding to sound in the environment, as a response to different levels of frequencies. With increasing intensity/pitch/loudness of sound, the cells get destroyed decreasing our ability to hear high-frequency sounds.

*Decibel (dB)* is used as a measure of sound intensity level or sound pressure level. It is named after Alexander Graham Bell, the inventor of the telephone.

$$\textit{Intensity level (dB)} = 10\log_{10}\left[\frac{\text{Intensity measured}}{\text{Reference intensity}}\right]$$

$$\Rightarrow \quad dB = 10\log_{10}\left(\frac{I_m}{I_o}\right) \tag{5.1}$$

As intensity varies with the square of pressure, intensity level can also be replaced as *Sound Pressure Level (SPL).*

Thus,

$$\text{SPL (dB)} = 10\log_{10}\left(\frac{\text{Pressure measured}}{\text{Reference pressure}}\right)^2$$

$$\Rightarrow \quad \text{SPL (dB)} = 20\log_{10}\left(\frac{\text{Pressure measured}}{\text{Reference pressure}}\right)$$

$$\Rightarrow \quad \text{SPL (dB)} = 20\log_{10}\left(\frac{P_m}{P_o}\right) \tag{5.2}$$

Note that decibels is a logarithmic scale. A change from source (A) of sound (intensity measured = 10 watts/m$^2$) with another source (B) of sound (with intensity measured = 1000 watts/m$^2$) causes a change of 20 dB of noise:

$$dB_B - dB_A = 10\left[(\log I_m^B - \log I_o) - (\log I_m^A - \log I_o)\right]$$

$$= 10[(\log(1000) - \log I_o - \log(10) + \log I_o]$$

$$= 10\left[\log\left(\frac{1000}{10}\right)\right] = 10 \times 2 = 20$$

Thus, *a change of 20 dB represents a 100-fold increase in loudness.*

Apart from loudness, the frequency or pitch of the sound also determines whether it is harmful or not. A modified scale called decibel–A [dB(A)] takes pitch into account. Average noise levels of some sources are summarised in Fig. 5.18.

For humans, normal level of tolerance is 80 dBA. Sound level above or below this is considered as noise pollution.

Most of the electronic vehicles and motors emit sound above 80 dBA level.

Amplified rock music is 120 dBA.

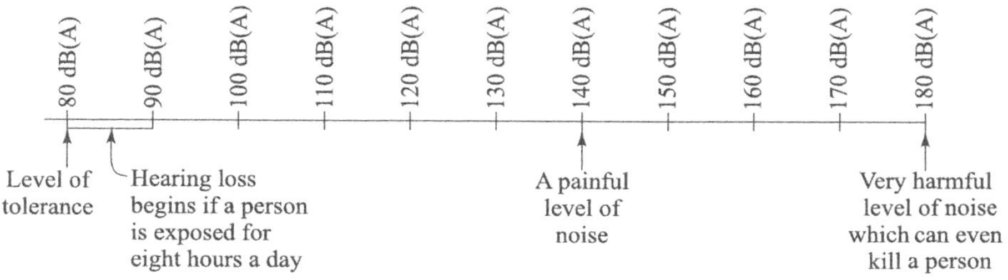


**Fig. 5.18** *Harmful levels of noise*

***Legislation*** In 2000, the Government of India notified Noise Regulation Rules under the Environment (Protection) Act of 1986. Sectorwise limits to noise levels are summarised on next page:

| Sector | Industrial | Commercial | Residential |
|---|---|---|---|
| Limit | 75 dB | 65 dB | 55 dB |

Further, use of public address systems after 10 p.m. and before 6 a.m. is not permitted.

Rules also establish zones of silence within a radius of 100 m of schools, courts, hospitals, etc.

## 5.8.1 Sources of Noise

Source is the equipment or process directly responsible for sound generation.

The major sources of noise are summarised below:

**(i) Transportation Sources** Railways, road traffic and air traffic.

**(ii) Industrial Sources** Noise is generated in mostly all industrial activities such as power generation, processing, product fabrication and product assembly.

**(iii) Public Address System Sources** Use of loudspeaker at any occasion like marriages, functions, festivals, etc.

**Table 5.3** *Average noise levels of some noise sources*

| *Source* | *Noise level in dB(A)* |
|---|---|
| Threshold of audibility/ hearing | 0 |
| Conversation–quiet | 20–30 |
| Conversation–Face to face | 60 |
| Classroom teaching | 55–60 |
| Home appliances | 65–75 |
| Road Traffic–Medium | 70–80 |
| Road Traffic–Heavy | 80–90 |
| Inside cinema hall | 85–95 |
| Horns of vehicles | 90–105 |
| Rail engine at 15 m | 97–105 |
| Loudspeakers | 100–120 |
| Threshold of pain | 130 |
| Jet engine at 25 m | 140 |
| Diwali crackers | 125–160 |
| Bomb explosion | 190 |

**(iv) Agricultural Machine Sources** Use of tractors, tubewells, farm machines for agriculture.

**(v) Defence Equipment Sources** Shooting practices, wars, bomb explosion, etc.

**(vi) Household Sources** Mixer-grinder, lawn mowers, food blenders, vacuum cleaners, etc.

**(vii) Other Sources** Rock concerts, barking dogs, construction equipments, etc.

## 5.8.2 Effects of Noise Pollution

Noise affects human health in the following ways:

**(i) Physical Effects** Damage to ear drum, temporary impairment of hearing, permanent deafness.

**(ii) Physiological Effects** Muscular strain, headache, eye strain, decreased colour perception, nervous breakdown, pain in heart, etc.

**(iii) Psychological Effects** Emotional disturbance, depression, fatigue, frustation, irritation, reduced efficiency, etc.

### 5.8.3 Control of Noise Pollution

Noise pollution can be controlled by reducing noise at the source, interrupting the path of noise and/or protecting the receiver.

**(i) Noise Control at the Source** It is most effective to eliminate noise at the source.

***Examples***

(a) Reduction of noise generated by mechanical vibration of a machine by damping or isolation of the vibration by applying a damping material (rubber) to the vibrating components.
(b) Reduction of impact force by optimising the impact distance and covering either or both impact surfaces by rubber.
(c) Modification of manufacturing design like enclosing the engine parts within proper noise-insulating material.

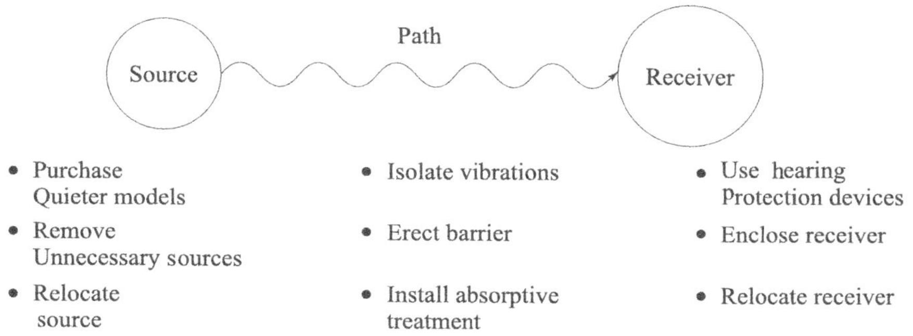


**Fig. 5.19** *Noise control*

**(ii) Noise Control at Path** When the source cannot be made quiet, noise can be controlled by modifying the path.

***Examples***

(a) Attenuation of noise by moving noise source away from sensitive area
(b) Suppression of noise from automobiles using silencers
(c) Reduction of noise around residential areas by planting trees in the form of green belt
(d) Reduction of transmission of noise using acoustic screens and barriers
(e) Enclosing noisy machines in isolated buildings

**(iii) Noise Control at Receiver** If source and/or path control do not work, control at the receiver should be explored.

(a) Use of Hearing Protection Devices (HPD) like ear plugs, ear muffs, etc. They reduce the level of noise (by 10 dB to 55 dB) entering the outer and middle ears before it reaches the inner ear.
(b) Enclose receiver.
(c) Relocate receiver.

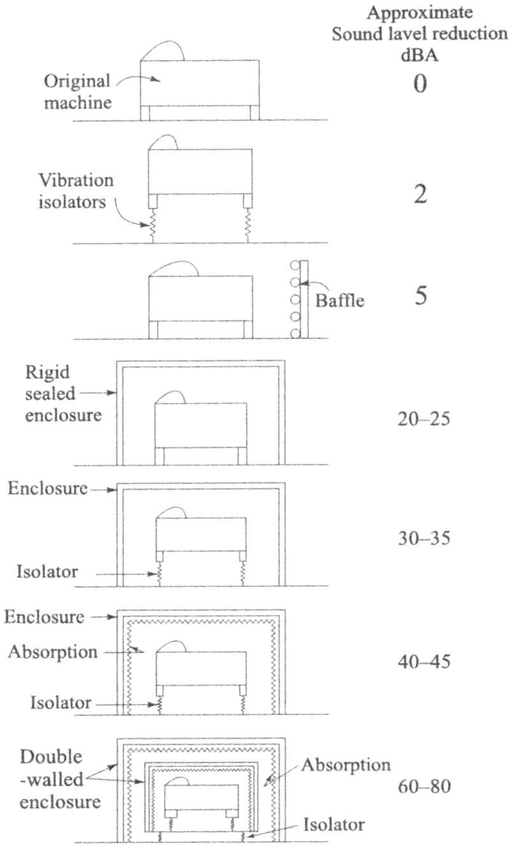


**Fig. 5.20** *Comparison of noise-reduction methods applied to a machine*

# Case Studies

### (i) Solution of Noise Pollution: Sustainable Urban Transport

The transportation sector is growing fast in India due to rapid urbanisation. It has resulted in overcrowded roads and traffic noise pollution. The major pollutants are gasoline-driven vehicles. In most congested parts of Delhi, the city experiences noise pollution beyond the accepted prescribed limit. The city also faces problems due to excessive air pollution of more than 2000 tonnes per year. The traffic problems are heterogeneous movement of traffic, inadequate road capacity, poor mass-transportation facilities and high intensity of private vehicles on the road.

The transport sector is also responsible for the greenhouse effect. It is one of the biggest threats to our society and sustainability. Negative impacts of greenhouse effects include the melting of most of the polar ice and rise in sea level resulting in population displacement, destruction of low-lying urban infrastructure, inundation of arable land, contamination of freshwater and spread of diseases such as malaria, dengue, etc.

Carbon dioxide, methane and nitrous oxide gases are generated by vehicular traffic. These gases and chlorofluorocarbons (CFCs) are the primary greenhouse gases contributing to global warming.

Owing to environment-friendly concepts, energy requirements, possible consequences of energy scarcity of oil-based transport and damages caused by it on the environment, there is an urgent need of sustainable transport.

**Sustainable transport** refers to any means of transport with less impact on the environment and which promotes healthy lifestyles. It includes walking, cycling, green vehicles and transport systems that are fuel efficient.

For metros like Kolkata we need sustainable transport. This is because, around 13 million inhabitants of Kolkata suffer from respiratory problems such as lung cancer and asthma. These problems are caused by pollution from the chaotic transport system of the city.

[Sarkar et al., current science, Vol.100, No. 9, 10 May 2011
http: //www.ias.ac.in/currsci/10 may 2011/1349-pdf]

### (ii) Noise and Quality of Life

The noise levels in the work zone of a thermal power plant are greater than the prescribed standards. The thermal power plant cannot achieve its targeted demand because the quality of life of all the workers is not maintained.

The noise levels of a steel plant even affect far-field locations and thus disturb the quality of life of the rural settings.

An increase in the ambient noise results in deteriorated indoor noise levels and speech intelligibility in classrooms.

In different mining complexes, the mechanised mining operations create excessive noise both in the work areas as well as in the surrounding areas. The noise generation greatly disturbs the quality of life of the inhabitants residing near the mining areas.

Workshops of many areas have maximum noise pollution. To avoid occuptional hearing loss, older workers of workshops should be regularly changed to work in less-influenced areas. Ear protective less-influenced devices should be worn by all workshop workers.

[Kerketta et al., Internationsl Journal of Environmental sciences, Vol.1/ No. 7, 2011 (http: //:publishing.co.in/jes vol 1 no 1 2010/E1JES 2124.pdf)]

## 5.9 THERMAL POLLUTION

*Thermal pollution may be defined as the degradation of water quality by any process that changes ambient water temperature.*

### 5.9.1 Causes of Thermal Pollution

Causes (or sources) of thermal pollution are briefly described below:

**(i) Coal-fired Power Plants** River water is used for cooling the condenser rods of coal-fired thermal power plants. When water used as a coolant is returned to the river, its temperature is high, which lowers down the dissolved oxygen of water and affects ecosystem composition.

**(ii) Nuclear Power Plants** Large amount of heat along with toxic radionuclides are discharged into nearby water streams by nuclear power plants. Radiation leakages are also responsible for increasing the temperature of water bodies. Nuclear experiments and nuclear explosions are also responsible for thermal pollution.

**(iii) Domestic Sewage** Normally, the municipal water sewage has a higher temperature than normal water. When domestic sewage is discharged into lakes, rivers, etc., it causes thermal pollution.

**(iv) Industrial Effluents** Textile, sugar, paper, pulp and various other industrial effluents when discharged into lakes, rivers, etc., cause thermal pollution.

**(v) Deforestation** When shade-providing trees are cut down, water temperature rises.

## 5.9.2 Effects of Thermal Pollution

The harmful effects of thermal pollution are described below:

**(i) Reduction in DO** Elevated temperature typically decreases the level of dissolved oxygen in water. This can harm aquatic animals.

**(ii) Change in Quality** With rise in temperature, the density, viscosity and solubility of gases in water decreases.

**(iii) Damage to Biological Activity** Above 37°C, biological activity of enzymes of aquatic flora and fauna gets severely damaged.

**(iv) Interference with Reproduction Capability** Temperatures higher than 9 to 10°C interfere with reproduction capabilities of certain fishes.

**(v) Increase in Metabolic Activity** At increased temperatures, metabolic activities such as oxygen uptake, food intake and mobility of fishes are increased.

**(vi) Shortening of Longevity** The lifespan shortens due to increased metabolic activity. For example, a crustacean lives for 108 days at 7.8°C. But if the temperature of water is increased to 21°C, it can live for only 29 days.

**(vii) Increased Vulnerability to Diseases** With temperature rise, the vulnerability or susceptibility to diseases also increases.

**(viii) Increased Mortality Rate** At higher temperatures, the mortality rate of fish and all other aquatic organisms increases.

**(ix) Decreased Biodiversity** Adjacent to thermal discharge, some fish species will avoid stream segments or coastal areas because they may not be used to the warmer temperatures. However, the more adapted organisms might move in. An increased metabolic rate may result in fewer resources which lead to compromise in the new and old food chains. Biodiversity can decrease as a result.

**(x) Malnutrition** High temperatures can lead to the denaturing of life-supporting enzymes. It means, within the quaternary structure of the enzymes, hydrogen bonds and disulphide bonds break down. In aquatic organisms, these reduced activities of enzymes can cause problems such as the inability to break down lipids, leading to malnutrition.

**(xi) Ecological Effects of Cold Water** Elimination of native fish species, and drastic alteration of macroinvertebrate fauna has been observed by releases of unnaturally cold water from reservoirs like dams.

### 5.9.3 Control Measures of Thermal Pollution

Heated water from power plants, petroleum refineries, pulp and paper mills, steel mills and chemical plants can be cooled down for controlling thermal pollution by using cooling ponds, cooling towers, etc.

(i) ***Cooling ponds*** are man-made bodies of water which help in reducing the temperature of water by evaporation, convection and radiation (Fig. 5.21).

(ii) ***Cooling towers*** transfer waste heat to the atmosphere through evaporation and/or heat transfer (Fig. 5.22).

(iii) ***Cogeneration*** is a process for recycling waste heat for domestic and/or industrial heating purposes.

(iv) ***Stormwater management facilities*** absorb urban run-off or direct it into groundwater, such as bioretention systems. Otherwise, urban run-off can have significant thermal impact during summers on small streams, as stormwater passes over hot parking lots, roads and sidewalks.

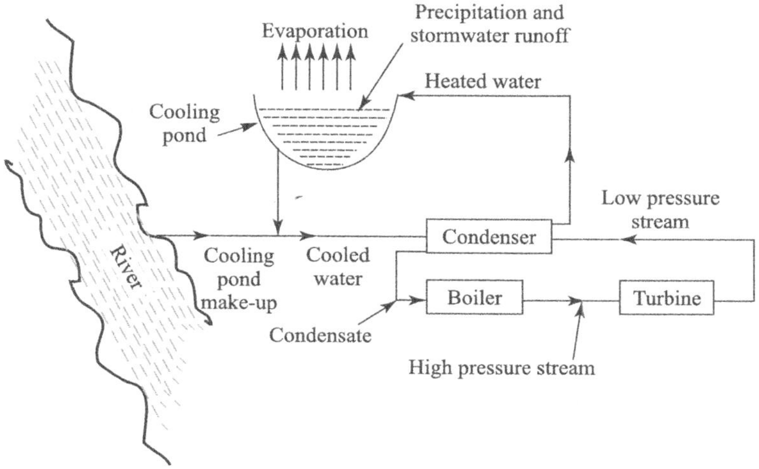


**Fig. 5.21** *Power plant with recirculating cooling pond*

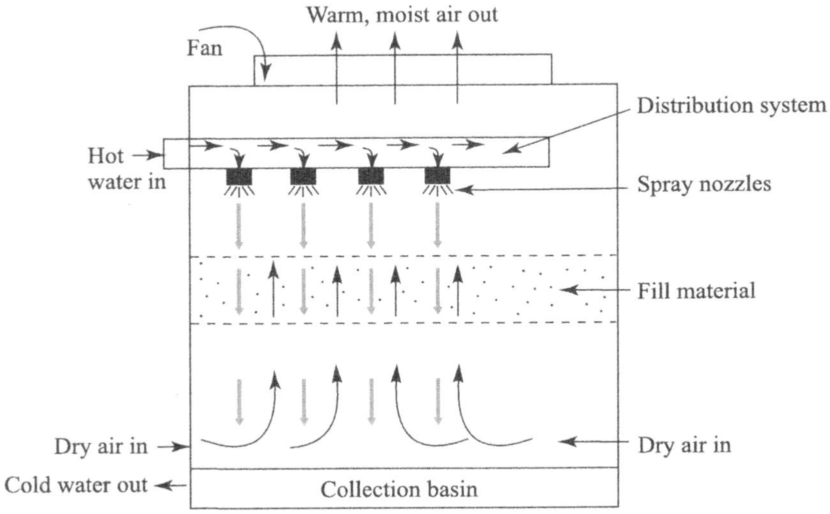


**Fig. 5.22** *Counterflow type cooling tower*

(v) ***Afforestation*** By planting trees along streams and shorelines, thermal pollution can be controlled. If these trees and tall plants are not there for providing shade, the water warms by as much as 10°C. Even removal of vegetation far away from a lake can speed up the erosion of soil into water, making it muddy. Muddy water absorbs more energy from the sun than clear water does, resulting in further heating, Afforestation controls erosion, keeps water clearer and thus, cooler.

# Case Study

## Thermal Pollution

**Thermal pollution** is the act of altering the temperature of a river, lake, ocean or any other natural water body. Environmental consequences of thermal pollution include decrease in biodiversity, creation of an environment hospitable to alien aquatic species, etc. Waste heat discharged to natural waters typically reduces the dissolved oxygen content. The resulting higher water temperatures specifically raises the metabolic rate of aquatic organisms.

Higher enzyme activity occurs causing plants and animals to take in greater amounts of nutrients and $CO_2/O_2$. These metabolic changes can alter the balance of species composition, species migration, etc.

Use of cooling ponds, cooling towers, productive use of the heated water for a secondary industrial process or space heating are some of the methods of reducing impacts of warm-water thermal discharges.

The Turkey Point nuclear power station (South Florida) removes water from Biscayne Bay and then passes it through condensers. The output water with an increased temperature of up to 15°C above ambient is then pumped back into the receiving water. The cooling water is discharged into a shallow soft-bottomed area. The water temperature around the outfall pipe at Turkey Point is 30–35°C. It is estimated that about 9.3 hectares of turtle grass has been destroyed around this outfall because of a temperature increase of 5°C.

Some of the impacts of thermal pollution on the local marine community surrounding the outfall pipes of coastal power stations are

(i) Death of marine animals like sponges, molluscus, crustaceans (e.g. crabs)
(ii) Replacement of turtle grass by blue-green algae
(iii) Change in community structure around the outfall pipe

[http://www.chengapedia.de/vsegine/vlu/vsc/en/ch/16/uc/vlus/thermal pollution/case studies/thermal/thermal turkey.vscml.html]

## 5.10 HAZARDOUS WASTES

*Hazardous wastes* can be defined as useless, unwanted and discarded materials that may pose a threat to human, plant or animal life.

OR

"A waste or combination of wastes which because of the concentration, quantity or physical, chemical or infectious characteristics may cause, or significantly contribute to increase in mortality or an increase in serious irreversible, or incapacitating reversible illness, or pose a substantial present or potential hazard to human health or the environment when stored, transported, disposed off, improperly treated or improperly managed."

OR

"Waste that possessess reactivity or ignitability or corrosivity or toxicity."

(i) *Reactive waste* is a waste which is not stable at normal conditions. It can cause explosive reactions or liberate toxic fumes, gases or vapours. It can react violently with water.

 ***Examples*** Dry picric acid, sodium metal, etc.

(ii) *Ignitable waste* is a waste which is easily ignited and burns vigorously at or below 60°C.

 ***Examples*** Paint thinner, solvents (ethers, acetone, toluene)

(iii) *Corrosive waste* is generally a waste in liquid physical form that can corrode metal and which has pH less then 2 or greater than 12.5.

 ***Examples*** Battery acid, drain openers, $H_2\ SO_4$, HCl, NaOH, lime, etc.

(iv) *Toxic waste* is a waste which releases toxic materials on leaching in excess of the permissible concentration. These toxic substances are harmful or fatal when ingested or absorbed.

 ***Examples*** Nickel–cadmium batteries, lead batteries, mercury batteries, rat poison, antifreeze, etc.

A waste is considered ***hazardous*** (as regards human toxicity) if

(i) It has an oral LD 50 toxicity $\geq$ 50 mg/kg
(ii) It has an inhalation LC 50 toxicity = 2 mg/kg

***LD 50*** is the lethal dose of the toxic waste at which 50% of experimental animals die as a result of dermal penetration or oral ingestion.

***LC 50*** is the lethal ambient concentration of the toxic material in mg/L of air causing 50% mortality to test rats during 4 hours of inhalation.

Toxicity can be acute or chronic. The adverse effects of a substance which result either from a single exposure or from multiple exposures within a day are described as ***acute*** toxicity.

The adverse health effects from repeated exposures, often at lower level, to a substance over a long time period (months or years) are described as ***chronic toxicity***.

## 5.10.1 Environmental Problems and Health Risks Associated with Hazardous Wastes

For a hazardous waste to pose a risk, a material/plant/animal/person must be exposed to it in a way that can cause harm.

Exposure may occur through air, drinking water, soil or food. When the waste spilled onto the soil includes volatile organic chemicals, these can evaporate. Direct inhalation by on-site workers or humans who live extremely near the site can cause harm. Evaporated volatile organic compounds could settle onto corn plants and trees and other vegetation. Animals may eat the contaminated vegetation or grain. Later, humans may eat the contaminated vegetation, grain, or meat. Soil is a direct exposure source for plants and animals on site. Spilled waste could percolate into groundwater or run off into surface water. The most likely exposure route for those living near such sites is the groundwater that is used for drinking.

(i) Arsenic, cyanide, pesticides, etc., are *toxic substances* which can adversely affect the health of living organisms exposed to them.

(ii) Strong acids and strong alkalis are *corrosive substances* which can cause severe injury at the point of contact, i.e. the skin, eyes, lungs, or mouth.

(iii) Petroleum distillates, many organic solvents, ethyl ether are ignitable or *flammable substances* which can easily catch on fire. They are thus fire hazards.

(iv) If ammonia and chlorine bleach are mixed together, they will react to form toxic fumes, which can adversely affect humans. If ethers are left sitting in their containers for years, they can form explosive peroxides.

(v) Battery acids can corrode metal.

(vi) Strontium-90 and Iodine-137 are slow decaying and produce hazardous effect on human life.

(vii) *Radioactive wastes* generate radiations which can have somatic or genetic effects on living cells.

The *somatic effects* are caused on exposed individuals. The cell damage caused may result in cancer or leukemia. The *genetic effects* are transmitted from the exposed individuals to their descendants.

The radiation-induced changes in the genes may result in gene mutations, chromosome aberrations and changes in the number of chromosomes. Such changes can result in mild or lethal abnormalities in the offsprings.

(viii) As, Pb, Cd and Hg are *inorganic toxic chemicals*, they act as biological poisons even at parts per billion (ppb) levels.

When coal, leaded gasoline and other fossil fuels are burnt, these toxic metals enter the atmosphere. Low pH caused by acid rain or the generation of $CO_2$ increases the transportability and hence availability of these metals by rendering them more soluble. These chemicals also leach out from mines and landfills. Then they contaminate land. In soil and sediments, these toxic elements accumulate in organic matter and are taken up by growing plants, thus entering the food chain. They accumulate in organs and tissues to toxic levels in the body since they are poorly excreted by humans.

(ix) Pesticides (DDT), Polychlorinated Biphenyls (PCBs) are *hazardous organic chemicals*. They slowly degrade in the environmental. They are fat soluble; so they accumulate in the food chain. They cause immediate toxicity as well as long-term effects such as carcinogenicity and mutagenicity.

(x) In the course of curing health problems, the health-care sector produces huge amounts of *biomedical waste* which is hazardous to all those who come in contact with it,

Bacteria (tuberculosis) and viruses (HIV, Hepatitis B and Hepatitis C) can be transmitted by contact with an infected patient or contaminated body secretion/fluid.

Progressive increase in hospital infection rate, and increasing resistance to wide variety of antibiotics are the pointers to the way in which poor hospital waste management can contribute to ill health. There is also risk of pollution of water, air and soil.

## 5.10.2 Different Methods for Treatment of Hazardous Waste

There are two reasons for treating hazardous waste: one is to reduce the volume of the waste; the other is to reduce its toxicity.

Hazardous waste can be treated by various methods (Fig. 5.23). These and many other are briefly described below:

### (A) Physical Treatment Methods

(i) A liquid is separated from a solid using a membrane in *filtration*. The liquid goes through membrane pores. The solid particles are retained on the surface of the membrane, and its volume is much reduced. If enough hazardous material is filtered out, the liquid coming through the membrane (the filtrate) may no longer be hazardous.

(ii) To separate a mixture of liquids, *distillation* is sometimes done. The mixture of liquids is heated. Lower temperatures drive off the more volatile substance, leaving behind those with higher boiling points. Depending on the chemicals, the waste remaining may or may not be hazardous. Volumes of some components can be reduced too.

(iii) To remove hazardous material from a solution, *aeration* can be done. Contaminated waste solution is sprayed downward through a packing material in a tower, through which air is blown upward carrying away the

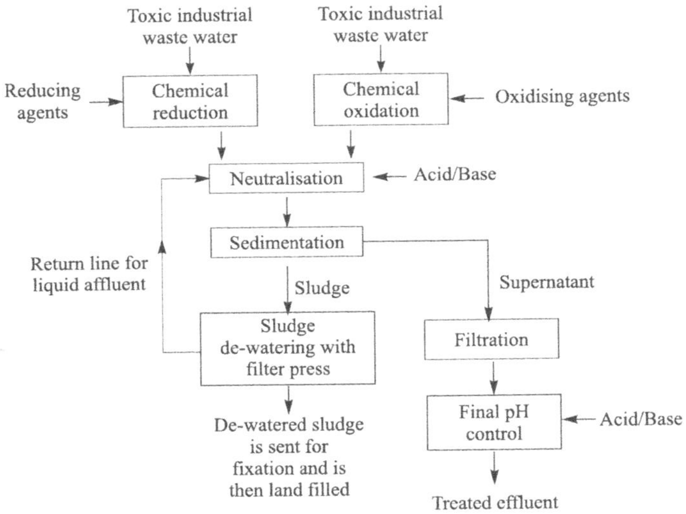


**Fig. 5.23** *Various methods used for the treatment of toxic industrial waste water*

Volatile Organic Compounds (VOCs) such as benzene, toluene, $CHCl_3$, $CCl_4$, etc.

(iv) After the removal of volatile substances, waste water is subjected to *adsorption*. In this process, hazardous chemicals are adsorbed onto Granular Activated Carbon (GAC) adsorbent. The adsorption process uses a series of large vessels partially filled with GAC. Contaminated water enters the top of each vessel, trickles down the GAC and is released at the bottom. GAC has large surface area of about 1000 $m^2/g$. It helps in removal of dissolved hazardous substance that cannot be removed by sedimentation.

(v) *Reverse osmosis*, *electrodialysis* and *ion-exchange* processes are other physical methods for treatment of hazardous waste.

## (B) Chemical Treatment Methods

**(i) Ion-exchange Process** Anion exchangers are used for the removal of anionic nickel cyanide complex and chromate ions from hazardous waste solution.

$$2\ Res^+\ OH^- + [Ni(CN)_4]^{2-} \rightarrow (Res^+)_2[Ni(CN)_4]^{2-} + 2\ OH^-$$

$$2\ Res^+\ OH^- + CrO_4^{2-} \rightarrow (Res^+)_2[(CrO_4^{2-}] + 2\ OH^-$$

Ion-exchange resins have also been used for the removal of radionuclides from radioactive wastes.

**(ii) Hydrolysis Treatment** It is given to those hazardous waste constituents (like halides, carbide, hydride, alkoxide and active metal) which are very reactive with water.

$$SiCl_4 + 2\ H_2O \rightarrow SiO_2 + 4\ HCl$$

$$CaC_2 + 2\ H_2O \rightarrow Ca(OH)_2 + C_2H_2$$

$$NaAlH_4 + 4\ H_2O \rightarrow 4\ H_2 + NaOH + Al(OH)_3$$

$$NaOC_2H_5 + H_2O \rightarrow NaOH + C_2H_5OH$$
$$Ca + 2\,H_2O \rightarrow Ca(OH)_2 + H_2$$

**(iii) Chemical Precipitation Technique** In this technique, the pH is properly adjusted for decreasing the solubility of toxic metals. A precipitate is subsequently formed that is then removed by settling and filtration.

(a) Chromium is precipitated as hydroxide using lime for precipitation.
$$Cr^{3+} + 3\,OH^- \rightarrow Cr(OH_3)\downarrow$$
(b) Sodium carbonate ($Na_2CO_3$) is used to precipitate metals as hydroxides ($Fe(OH)_3\ x\ H_2O$), carbonates ($CdCO_3$), basic carbonate salts ($2PbCO_3$, $Pb(OH)_2$).
(c) Sodium borohydride ($NaBH_4$) is a reducing agent. It is used to precipitate metal ions from solution in the elemental from.
$$4\,Cu^{2+} + NaBH_4 + 2H_2O \rightarrow 4\,Cu + NaBO_2 + 8H^+$$
(d) Metals can be precipitated as sulphides when they are present in very low concentrations. Metal sulphides have lower solubilities than metal hydroxides and hence can be precipitated out. Ferrous sulphide is used as a safe source of sulphide ions to produce sulphide precipitates with other metals that are less soluble than ferrous sulphide.
(e) Lime is used for precipitation of metal ions as metal hydroxides.
$$M^{2+} + Ca(OH)_2 \rightarrow Ca^{2+} + M(OH)_2\downarrow$$

**(iv) Neutralisation Process** Hazardous wastes are termed corrosive when their solution pH is less than 2 or more than 12.5. Neutralisation process converts these materials to less hazardous ones by changing their pH.

(a) Lime is the least expensive. It is used for neutralisation of acidic wastes.
(b) Relatively inexpensive sulphuric acid is used for neutralisation of alkaline wastes.
(c) Nontoxic and biodegradable acetic acid is preferable for some applications where alkaline wastes need neutralisation.
(d) Alkaline wastes can also be neutralised by bubbling gaseous carbon dioxide.
$$CO_2 + H_2O \rightarrow H_2CO_3$$
$$H_2CO_3 \rightarrow 2H^+ + CO_3^{2-}$$
$$H^+ + OH^- \rightarrow H_2O$$

The advantages of $CO_2$ is that it is often easily available in the exhaust gas from any combustion process at the treatment site.

In the same vessel, simultaneous neutralisation of waste acid and waste base can also be done (Fig. 5.24).

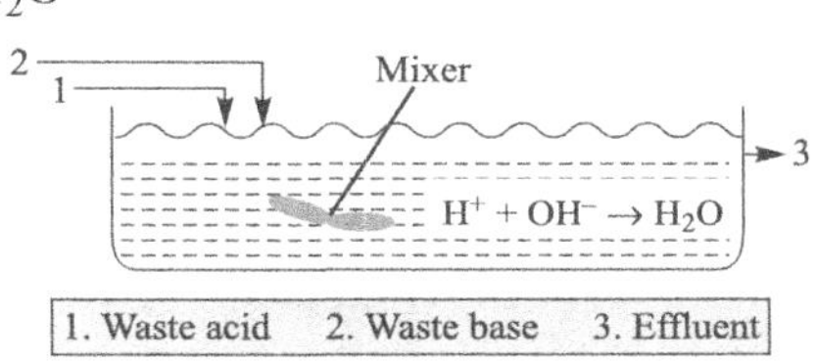


**Fig. 5.24** *Simultaneous neutralisation of acid and base*

**(v) Oxidation–Reduction Process** By redox reactions, toxic substances are converted to those oxidation states which are less toxic.

(a) Hexavalent chromium ($Cr^{6+}$) is reduced to trivalent chromium ($Cr^{3+}$) by using sulpher dioxide as a reducing agent.

$$\underset{\text{Carcinogen}}{Cr^{6+}} \xrightarrow{SO_2} \underset{\text{Less toxic}}{Cr^{3+}}$$

$$3\ SO_2 + 3H_2O \rightarrow 3H_2SO_3$$

$$2CrO_3 + 3H_2SO_3 \rightarrow Cr_2(SO_4)_3 + 3H_2O$$

(b) From metal finishing industry, toxic cyanide waste is oxidised using chlorine in alkali solution. First, the less toxic cyanate is formed. Finally, $CO_2$ and $N_2$ gases are evolved.

$$NaCN + Cl_2 + 2\ NaOH \rightarrow NaCNO + 2NaCl + H_2O$$

$$2NaCNO + 3Cl_2 + 4NaOH \rightarrow 2CO_2 + N_2 + 6NaCl + 2H_2O$$

## *(C) Thermal Treatment Process*

**(i) Incineration** It is a thermal treatment process that uses high-temperature oxidation to convert a waste to a less bulky, less toxic or less noxious material.

Many of the component elements of organic compounds (including C and H) are converted to gaseous form:

$$C(\text{Organic}) + O_2 \rightarrow CO_2$$

$$4\ H(\text{Organic}) + O_2 \rightarrow 2H_2O$$

Only noncombustible inorganic ash is left behind. Thus, incineration is a volume-reduction process.

The hazardous products of incineration are compounds of sulphur, nitrogen, halogen and heavy metals (lead, mercury, arsenic, cadmium, etc). Thus, air-pollution-control equipment is required if the gaseous combustion products of incineration contain undesirable compounds. Prior to ultimate disposal or discharge, the solid and liquid effluents also require treatment.

***(a) Performance of Incinerators*** It is measured in terms of **Destruction** and **Removal Efficiency** (DRE). It accounts for both the destruction in the combustion chamber (s) and the removal of organics in any air-pollution-control equipment. DRE can be calculated as the percentage of mass difference of input (feed) and output (stack emission) waste constituents through the incinerator.

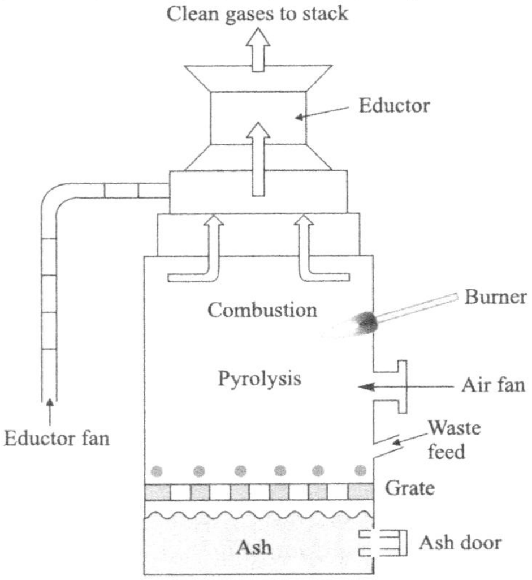


Fig. 5.25 *Incinerator*

**Example 1** *Suppose one gram of an organic compound is released in air for every 10,000 grams entering the incinerator. Find DRE.*

**Solution**

| | *Input (feed)* | *Output (stack emission)* |
|---|---|---|
| *Mass (g)* | 10,000 | 1 |
| *Per cent* | $\frac{10,000}{10,000}\times 100 = 100$ | $\frac{1}{10,000}\times 100 = \frac{1}{100} = 0.01$ |

$\therefore$ DRE = Percentage of mass difference of input and output waste constituents

$= 100 - 0.01\ \%$

$= 99.99\ \%$

***Alternative solution for Example 1:***

Let DRE = Destruction and Removed Efficiency

$A$ = Mass feed rate = 10,000 grams

$B$ = Mass emission rate = 1 gram

$$\text{DRE} = \left(\frac{A-B}{A}\right)\times 100\ \% = \left(\frac{10,000-1}{10,000}\right)\times 100\ \%$$

$$= \frac{9,999}{100}\ \% = 99.99\ \%$$

**Example 2** *A mixture described below is being incinerated at 200°F with 50% excess air and a residence time of 2.1 seconds. Calculate DRE for all organic compounds.*

| *Compound* | *A = Inlet mass flow rate (lb/h)* | *B = Outlet mass flow rate (lb/h)* |
|---|---|---|
| Benzene | 2015 | 0.537 |
| Toluene | 637 | 0.022 |
| xylene | 3040 | 1.25 |

**Solution**

$$\text{DRE} = \left(\frac{A-B}{A}\right)\times 100\ \%$$

For Benzene, $\text{DRE} = \left[\frac{2015-0.537}{2015}\right]\times 100\% = 99.9733\%$

For Toluene, $\text{DRE} = \left[\frac{637-0.022}{637}\right]\times 100\% = 99.9965\%$

For xylene, $\text{DRE} = \left[\frac{3040-1.25}{3040}\right]\times 100\% = 99.9589\%$

***Notes:***

- DRE is defined on a compound-specific basis. It is calculated and specified for each constituent of interest separately.
- According to **Resource Conservation** and **Recovery Act** (RCRA) requirements:

| *Compound* | *Minimum DRE* |
|---|---|
| (a) Organic Compounds | 99.99 |
| (b) Dioxins, Dibenzo-furans | 99.9999 |

***(b) Advantages of Incineration***

- Reasonably well-developed technology is available.
- Land requirement is not large.
- Tetragens (causing birth defects), mutagens (causing mutation in genes), and carcinogens (cancer causing) can all be completely detoxified in a properly operated incinerator.
- It is the best-known method for the disposal of mixed wastes.
- It is an excellent disposal method for biological hazardous wastes.
- It can be scaled up easily to handle large volumes of liquid waste.

***(c) Disadvantages of Incineration***

- Substances which volatilise at temperatures below 2000 °F are added in the air and cause air pollution.
- Incinerator is costly.
- The gaseous and particulate products of combustion may be hazardous and should be controlled by air-pollution-control technology.
- The ash may or may not be toxic, but it must be disposed of properly.

**(ii) Wet-air Oxidation** It is the aqueous phase oxidation of dissolved or suspended organic substances at elevated temperatures and pressures.

Water makes up the bulk of the aqueous phase and serves to catalyse the oxidation reactions which proceed at temperatures of 150–325°C. The high pressures (2000 kPa to 20,000 kPa) allow high concentration of oxygen to be dissolved in water.

This process is used for the removal of cyanide from electroplating waste solutions.

$$2Na^+ + 2CN^- + O_2 + 4\,H_2O \rightarrow 2\,Na^+ + 2\,HCO_3^- + 2NH_3$$

***(D) Biological Treatment Methods*** Microorganisms degrade certain organic hazardous wastes using them as nutrients (*bioremediation*).

Certain plants take up and concentrate metals from soil or water (*phytoremediation*).

***(E) Hazardous Waste Landfill*** It is a type of facility that is reserved for the disposal of toxic waste after suitable treatment.

In hazardous waste landfills, the dumping area is protected by using a double composite liner, for preventing the seepage of waste into the ground.

In addition, all liquids must be properly contained. The leachate (the leakage of liquids from the hazardous waste) that collects over each double liner is collected in a series of perforated drainage pipes and pumped to the surface for further treatment (Fig. 5.26).

Monitoring facilities must also be included to check the possibility of any contamination of groundwater from the landfill site. Landfills also have gas vent system with sufficient vent points so that if methane is generated, it may be burned off continuously.

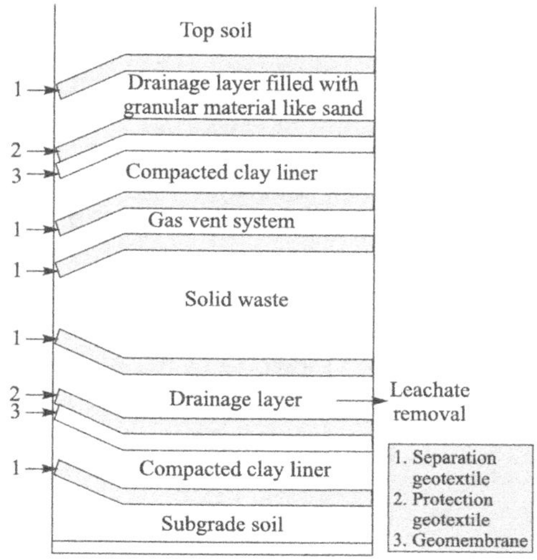


**Fig. 5.26** *Hazardous waste landfill section*

The landfill is sloped to permit adequate run-off so that infiltration is minimised and thus preventing the rainwater or snow melt from entering the soil. The stress on the liner material is reduced by providing slopes to the landfill at a maximum of 3:1.

## 5.11 NUCLEAR HAZARDS (RADIATION POLLUTION)

*Radiation* is a form of energy that can travel through any medium including vacuum. It is of two types:

*Ionising radiation* carries greater energy and can ionise atoms and molecules to create ions. It can travel both as particles like $\alpha$ and $\beta$ rays, or as a waves like X-rays, gamma rays, etc.

*Non-ionising radiation* carries lower energy than ionising radiation and so it cannot ionise atoms and molecules. Radio waves, heat and light are examples of non-ionising radiation.

*Nuclear hazard, or radiation pollution, is the danger or risk to human health or the environment posed by radiation emanating from the atomic nuclei of a radioactive substance or the possibility of an uncontrolled explosion originating from a fission or fusion reaction of atomic nuclei.*

Nuclear hazards following an accidental release of radioactivity from a nuclear reactor are illustrated in Fig. 5.27.

Nuclear hazards are caused by the addition of more ionising radiation to the environment and people are exposed to more radiation than they normally experience.

Huge clouds of fine radioactive particles and gases are thrown up in the environment, due to testing of nuclear weapons or disposal of radioactive wastes. These clouds of fine radioactive particles are carried away to distant areas by winds.

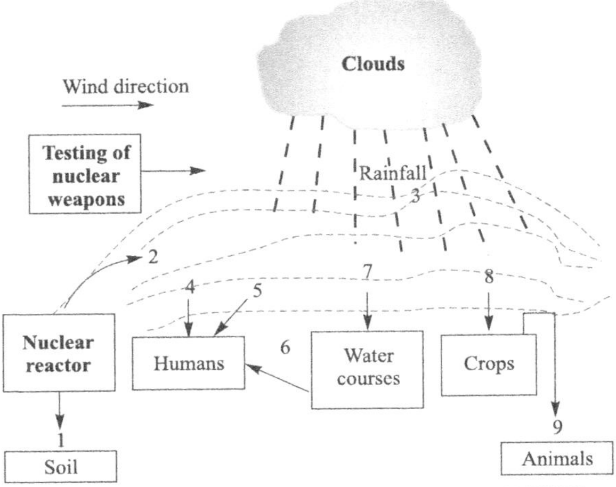


1. Direct radiation from reactor containment to soil
2. Accidental release of radioactivity from reactor
3. Increased deposition rate of radioactivity due to rainfall
4. Direct irradiation from the plume
5. Inhalation of radioactivity
6. Irradiation from deposited radioactivity
7. Deposition on water courses
8. Deposition on crops
9. Uptake by grazing animals and concentration in milk

**Fig. 5.27** *Nuclear (or radiological) hazards following an accidental release of radioactivity*

Gradually, they settle down on the earth as fall-out or are brought down by rainwater. When raindrops containing these radioactive particles fall on the earth, radioactivity is transferred to soil, water, etc., causing soil pollution, water pollution, etc. From the soil and water, radiation reaches crops, animals and aquatic organisms which absorb and accumulate them through food chains and may pass them to human beings.

## 5.11.1 Sources/Causes of Nuclear Hazards:

Natural and man-made sources of nuclear hazards are briefly described below:

### *(A) Natural Radioactive Sources*

**(i) Cosmic Radiation** It is a stream of ionising radiation that enters the earth's atmosphere from outer space. It consists mainly of protons, alpha particles, and other atomic nuclei including some high-energy electrons. The intensity of cosmic rays in the biosphere is low. Therefore, they are not a health hazard. However, cosmic rays are a major hazard in space.

**(ii) Terrestrial Radiation** It is long-wave electromagnetic radiation emitted by naturally radioactive materials on the earth including radon, uranium and thorium.

Humans have been exposed to low levels of radiation from these natural sources for thousands of years. But it is the man-made sources which are posing a threat to mankind.

### *(B) Artificial (or Man-Made) Radioactive Sources*

These sources of radioactivity are waste materials that contain radioactive nuclei produced during the

(i) mining and processing of radioactive ores,
(ii) use of radioactive materials in nuclear weapons,
(iii) use of radioactive isotopes in medical, research and industrial applications, and
(iv) use of radioactive materials in nuclear power plants.

Radioactive materials are composed of unstable atoms. *Radioactivity* is a process by which an unstable atom emits radiation until it becomes stable. Radiation cannot be detected by sight, smell, etc., but it has harmful effects on humans. The longer a person is exposed to radiation, the greater the risk.

## 5.11.2 Effects of Nuclear Hazards

The effects of nuclear hazards may be somatic or genetic.

**(i) Somatic Effects** Somatic Effects of nuclear radiation appear in the exposed person. The quantity of radiation that leads to the absorption of 100 erg per gram of the absorbing material is known as **Radiation Absorbed Dose** (RAD).

When an individual receives an acute dose (typically ≥ 10 RAD) in a short period of time, prompt somatic effects occurs.

For example, a dose of 400 RAD to the scalp results in temporary hair loss which occurs about three weeks after exposure. New hair is expected to grow within two months after the dose although the colour and texture may be different.

When an individual receives a small dose, *delayed somatic effects* are observed years after irradiation, for example, development of cataracts and cancer.

**(ii) Genetic (or Heritable) Effects** These effects appear as abnormalities in the future generations of the exposed person as a result of radiation damage to the reproductive cells.

### *Types of Radiation Doses*

Potential biological effects of radiation depend on how much and how fast a radiation is received. Radiation doses can be grouped into the following two types:

**(i) Acute Dose** A large dose (≥ 10 RAD) to the complete body delivered during a short period of time (~few days) is known as an acute radiation dose. It may result in effects which are observable within a period of days to weeks.

A pattern of clearly identifiable symptoms (syndromes) are caused by acute dose. These are known as *acute radiation syndromes*. The effects caused by acute doses are deterministic. It means the dose has some threshold level below which the effect will probably not occur, but above which the effect is expected. When the dose is above the threshold level, the severity of the effect increases as the dose increases.

*Effects from an acute dose* include the following:

(a) 50 RAD to the thyroid gland can result in benign (noncancerous) tumors.

(b) At > 100 RAD, damage to cells that divide at a fast rate (such as bone marrow, the spleen and lymphatic tissue) happens. It results in internal bleeding, fatigue, bacterial infections and fever.

(c) At 200 to 300 RAD, irradiation to the skin can result in the reddening of skin, damage to hair folicles and hair loss.

(d) At 125 to 200 RAD, the ovaries can result in prolonged or permanent suppression of menstruation in about 50% of women.

(e) At 600 RAD, the ovaries or testicles can result in permanent sterilisation.

(f) At > 1000 RAD, cells that divide less rapidly are damaged. These are cells in the linings of the stomach and intestines. It causes nausea, vomiting, diarrhoea, dehydration, electrolytic imbalance, loss of digestion ability, bleeding ulcers, etc.

(g) At > 5000 RAD, nerve cells (and other cells that do not reproduce) are damaged. It results in loss of coordination, confusion, coma, shock, etc. As a consequence, complications caused by internal bleeding, and fluid and pressure build-up on the brain result in death.

**(ii) Chronic Dose** It is a relatively small amount of radiation received over a long period of time. As smaller percentage of the cells need repair at any given time, the body has time to repair damage. Thus, the body is better equipped to tolerate a chronic dose than an acute dose.

## 5.11.3 Control Measures of Nuclear Hazards

Nuclear hazards can be controlled by practicising the following measures:

(i) Nuclear power plants should be located far from populated areas and should be provided with a suitable radiation-absorption zone around them to minimise the escape of radiation.

(ii) Safety measures should be enforced strictly to avoid nuclear accidents and occupational exposure.

(iii) Waste disposal must be effective, careful and efficient,

(iv) The following should be totally stopped:
Leakages from nuclear reactors, careless handling, transport and use of radioactive fuels and/or radioactive isotopes.

(v) Nuclear wastes have to be properly disposed off.
*High-Level Wastes (HLW)* like spent nuclear fuel have a very high radioactivity per unit volume. These are very dangerous. These wastes must be contained either by converting them into inert solids (ceramics) and then burying deep into earth or storing in deep salt mines.

Filters, reactor components, etc., are *Medium-Level Wastes* (MLW). These are solidified and mixed with concrete in steel drums before being buried in deep mines or below the sea bed in concentrate chambers.

Solids or liquids contaminated with traces of radioactivity are *Low-Level Wastes* (LLW). They are disposed of in steel drums in concrete-lined trenches in designated sites.

(vi) After the disposal of nuclear waste, drilling activity must be prevented in and around the disposal site, and radioactivity must be monitored periodically around the disposal sites.

## 5.12 SOLID WASTE AND ITS MANAGEMENT

### 5.12.1 Solid Waste

*The waste materials which have been rejected for further use and which can neither readily escape into the atmosphere nor can be transported by water into streams are called solid waste.*

All the discarded solid materials from municipal, agricultural and industrial activities are included in solid wastes.

### 5.12.2 Types and Sources of Solid Wastes

The various types of solid wastes are briefly described below:

***(A) Municipal Wastes*** These include *garbage* (i.e. biodegradable food waste), *rubbish* (i.e. nonbiodegradable solid waste from homes, offices, markets, hotels, etc.).

*Construction and Demolition Wastes*: *Sludges* from septic tanks; *wires*; conduits pipes; ashes; *abandoned vehicles*, etc.

***(B) Special Wastes*** These include hazardous wastes like toxic substances (pesticides, heavy metal sludges); radioactive wastes; biological waste; explosives, inflammable substances, corrosive materials, etc.

***(C) Domestic Wastes*** These include wastes generated from domestic cooking and serving of food.

*Examples* Garbage, waste paper, plastic, cloth, etc.

***(D) Agricultural Wastes*** These wastes result from farms, feed lots and livestock yards.

*Examples* Corn residues, baggasse from sugarcane manures, paddy husk, etc.

***(E) Industrial Wastes*** These include the following:

**(i) Process Wastes** Here, waste depends on the products being manufactured.

*Examples* Plastic wastes, rubber wastes, metal scraps, food-processing wastes, etc.

**(ii) Non-process Wastes** Here, waste is common to all industries.

*Examples* Office and cafeteria wastes, packing wastes, etc.

***(F) E-Waste*** It is a new form of waste from discarded mobile phones, mobile chargers, remotes, CDs, headphones, batteries, computers/TVs, monitors, printers, CPUs, LCD/Plasma TVs, etc. It is also known as electronic waste.

About 11,000 metric tonnes of E-waste is generated in Delhi annually (Hindustan Times, March 12, 2010). A recent report has labelled India as the second biggest E-waste contributor in Asia.

### 5.12.3 Causes of Generation of Solid Wastes

The main causes for the rapid growth in the quantity of solid-waste generation are described below:

**(i) Overpopulation** Solid waste generated per person multiplied by total population results in increased generation of solid waste every day.

**(ii) Urbanisation** Urbanisation requires various construction activities like construction of buildings, markets, shopping malls, roads, railways, airports, bridges, dams, water supply and sewage disposal systems. Each construction activity also generates solid wastes.

**(iii) Affluence** Consumers with high purchasing capacity discard 'obsolete goods'. This leads to solid waste generation.

**(iv) Advances in Technology** These lead to large-scale production of goods for *consumption-based* society preferring disposable items and almost every item 'packaged'. All these results in generation of huge quantities of solid wastes.

### 5.12.4 Effects of Solid Wastes

The accumulation and improper handling of solid wastes results in various health and environmental hazards. Some of these effects are described below:

(i) Flies and mosquitoes breed on choked drains and gully pits through solid wastes. These flies and mosquitoes then contaminate food and water. In turn, diseases like diarrhoea, amoebic dysentery, bacillary dysentery, malaria, dengue, etc. result.

(ii) Stray animals and scavengers invade the roadside garbage dumps. It results in harming the aesthetic beauty of the surroundings.

(iii) Bad odours pollute the air as a result of decomposition of organic solid wastes.

(iv) Percolation of decomposed garbage cause pollution of underground water and land. The crops and water supply get contaminated and result in occurrence of cholera, hepatitis, jaundice, gastro-intestinal diseases.

(v) Rats living in solid waste dumping sites rapidly multiply in numbers and may cause plague and other diseases.

(vi) E-waste is either burnt or buried, So it can have harmful effects on the environment. This is because E-waste contains many hazardous materials like lead, mercury, cadmium, flame retardants, etc.

### 5.12.5 Solid-Waste Management

Important solid-waste management practices are briefly described below:

**(i) Source Reduction** It involves changing the design manufacture or use of products and materials to reduce the amounts of solid-waste generation.
***Examples*** Two-sided copying of paper, backyard composting, etc.

**(ii) Recycling** From the wastestream; paper/glass/plastic/metal, etc., are sorted, collected, processed and then manufactured, sold and purchased as new products.

***Advantages*** Energy saving, prevention of emission of many greenhouse gases/water pollutants, job creation, resource conservation for future and reduced need of new landfills and incinerators.

**(iii) Treatment** Suitable treatment is given depending on the nature of solid wastes.

**(iv) Disposal** Solid wastes can be disposed in combustion facilities and land fills.

The most preferred method for solid waste management is source reduction (including reuse). It is followed by recycling and composting. Lastly, disposal of solid waste is done.

A hierarchy of waste management is illustrated in Fig 5.28.

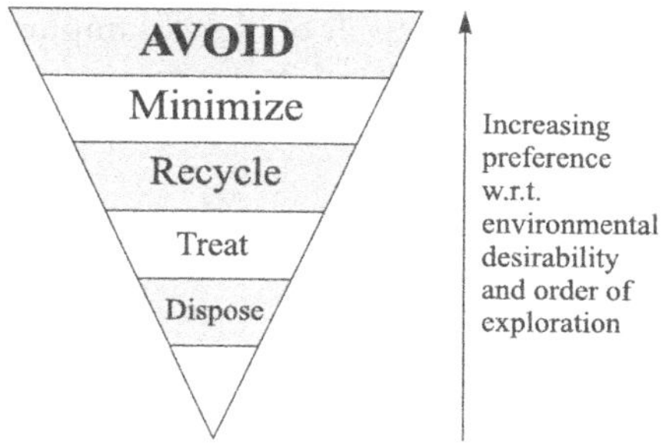


**Fig. 5.28** *Waste-management hierarchy*

## 5.12.6 Solid-Waste Disposal

The various methods commonly employed for disposal of solid waste are explained below.

***(A) Composting*** *Composting is the thermophilic and aerobic decomposition of organic matter present in solid waste by microorganisms, mainly bacteria and fungi.* As a result of this composting process, the organic matter is transformed into a stable humuslike substance, which is a valuable manure for crops.

### (i) Classification of Composting Techniques Based on Oxygen Use

***(a) Aerobic Composting*** It requires high temperature and results in rapid decomposition of organic matter. Odours are also absent.

***(b) Anaerobic Composting*** It requires low temperatures. Decomposition of organic matter of solid waste is slow. It needs minimum attention.

***(ii) Vermicomposting*** It uses a special kind of earthworm and a container of food scraps. After some time, the food is replaced with worm droppings, a rich brown matter that serves as excellent natural plant food (manure).

***Advantages of Vermicomposting Over Conventional Composting***

- Vermicomposting needs less space than normal composting.
- Vermicomposting is ideal for apartments in high-density urban areas.

- Vermicomposting provides excellent natural plant food.

### *(B) Illegal Dumping/Open Dumping/Fly Dumping/Midnight Dumping*

It is the disposal of solid waste by dumping in open areas, dumped from vehicles along roadsides, and/or dumped late at night.

**(i) Advantages** It is done to avoid either the time and effort required for proper disposal or to avoid disposal fees.

**(ii) Disadvantages**

(a) Illegal dumping of nonhazardous wastes often attract more waste, even the hazardous wastes.
(b) Illegal dump sites divert land from more productive uses.
(c) Property values decrease as a result of illegal dumping.
(d) Public nuisance is created by illegal dump sites.

### *(C) Land Dumping*

Solid wastes are dumped in low-lying areas outside the city/town limits. These areas have no provision of leachate collection and treatment. Moreover, landfill gas is neither collected nor used.

**(i) Advantages**

(a) It requires no planning.
(b) It is cheaper.

**(ii) Disadvantages**

(a) The waste is untreated, uncovered and not segregated. It is the breeding ground for flies, other insects, rats, etc., that spread diseases.
(b) Rainwater run-off from these dump sites contaminates nearby land and water thereby spreading diseases.

### *(D) Landfills*

A *landfill site is a pit that is dug in the ground.* The solid waste is dumped and the pit is covered with a layer of soil to form a cell. The process is repeated every day so that many cells completely fill the landfill site. Finally, about 1 m of earth-layer covering is done.

**(i) Advantages**

(a) Breeding of insects is prevented.
(b) Landfill sites can be developed as parks or parking spaces.

**(ii) Disadvantages**

All types of wastes are dumped in landfill sites without segregation. When rainwater seeps through them, it gets contaminated and in turn pollutes the surrounding area and groundwater.

## (E) Sanitary Landfills

*Sanitary landfill sites have liner systems and other safeguards to prevent groundwater contamination.* These sites are consistent with the economic considerations, hydrogeological requirements, climatic conditions and topography.

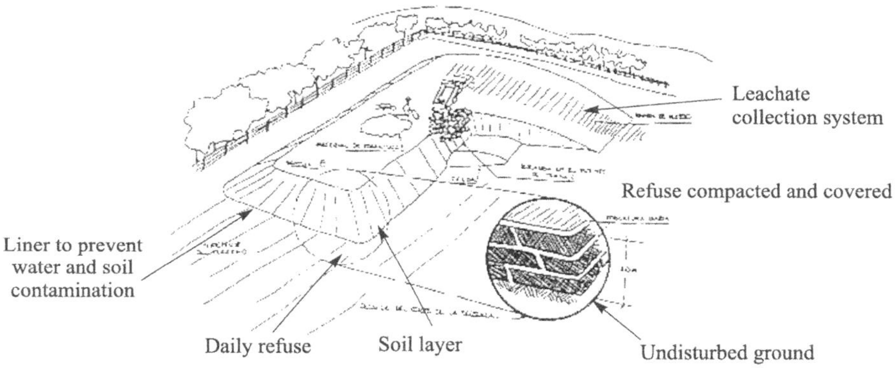


(a)

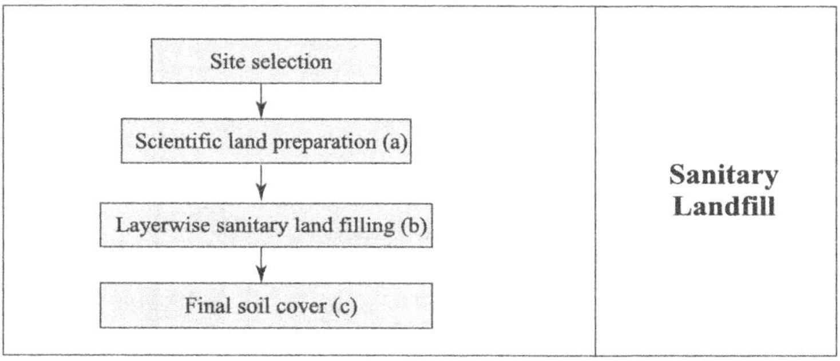


(a) A clay barrier and plastic liner is installed at the base of the ground to prevent water and soil contamination.

(b) Solid waste is compacted, spread and covered with a layer of soil to form a cell.

Each cell is equipped with a gas vent to collect methane gas formed for further use.

***Leachate*** is the liquid (say rainwater) that seeps through solid waste and has extracts of dissolved or suspended material from it.

***Leachate collection system*** consists of perforated pipes in a layer of sand. It helps in collection of leachate.

Many cells are made one above the another in a scientific way.

(c) The top cell is covered with about 1m of earth layer to prevent breeding of pests and disease vectors.

(b)

**Fig. 5.29** *Sanitary landfill*

### (i) Advantages

(a) The site is well above the groundwater table; so underground water pollution is avoided.
(b) The site is easily accessible; so the process is low in cost.
(c) The site is at least 1.5 km downwind from the commercial and residential areas; so it is not offensive to the surrounding environment.
(d) The finished sanitary landfill can be used for the development of regions of recreation like parks, golf courses, etc.

### (ii) Disadvantages

(a) Leachate from sanitary landfill site can contaminate the groundwater.
(b) The sites cannot be used in future as productive farmland.
(c) In a sanitary landfill, about 60% of methane gas (odourless) is generated. When its concentration in air reaches about 5%, it is explosive and so very hazardous.
(d) Aesthetic problems may arise as a result of poorly operated landfill operations.

## *(F) Combustion*

Solid waste is burned at high temperature in combustion facilities.

### (i) Advantages

(a) Energy is generated.
(b) Amount of waste is reduced by up to 90% in volume and 75% in weight.

### (ii) Disadvantages

(a) Cost increases with rise in the moisture content of solid waste. This is because energy is required for preheating the solid waste.
(b) Ash formed after combustion has high concentrations of dangerous toxins such as dioxins and heavy metals. It results in air and water pollution.

## *(G) Incineration*

*It is the controlled combustion of organic solid wastes so as to convert them into incombustible residue and gaseous products.* The weight and volume of solid waste is reduced and often energy is also produced.

### *(i) Advantages*

(a) As the volume of the waste is reduced, in taking the waste to the ultimate disposal site, less transportation cost is required.
(b) Larger wastes can be accommodated in a given landfill area because incineration reduces the land requirement to one-third.

### (ii) Disadvantages

(a) Not applicable for radioactive wastes
(b) High capital and operational costs
(c) Air pollution chances if incineration is not properly done
(d) Highly trained manpower is needed

## 5.13 ROLE OF INDIVIDUALS IN POLLUTION PREVENTION

Instead of complaining about the deteriorating environmental situation, individuals can play a very important role in pollution prevention.

Individuals can adopt the following activities in their daily routines for prevention of pollution:

(i) Whenever feasible, preferably use *public transport*, car pools and cycles instead of your own automobiles.

(ii) Purchase and use energy-efficient and pollution-free
- (a) vehicles
- (b) appliances (like refrigerators, ACs, etc.)
- (c) rechargeable batteries, etc.

(iii) Plant trees and help in *afforestation*.

(iv) *Conserve* natural resources; save water/electricity.

(v) *Reduce* consumption, waste generation, water leakages, etc.

(vi) *Reuse* paper and various products.

(vii) *Recycle* paper, metal, plastic, etc.

(viii) *Refuse* to buy and use toxic pesticides, fertilisers, lead-based paints, products without recycling symbol, products with unnecessary packaging, etc.

(ix) *Don't pollute* air, water, soil, etc.

(x) Advocate and *participate* in environment-friendly activities.

## 5.14 DISASTER MANAGEMENT

*Disaster can be defined as a man-made or natural event (like floods, earthquake, cyclone or landslides) which results in great damage or loss of life.*

A disaster is a consequence of inappropriately managed risk. The risk is the product of hazard and vulnerability.

$$\text{Disaster} = \text{Risk} - \text{Capacity}$$

or

$$\text{Disaster} = [\text{Hazard} \times \text{Vulnerability}] - [\text{Capacity of the community}]$$

*Hazard* is a situation which poses a level of threat to life, health, property or a dangerous condition or event that may deleteriously affect society or an environment.

*Vulnerability* is the extent to which damage will likely happen by the impact of a particular hazard.

*Capacity* means resources and strengths which exist in households and communities and enable them to cope with, withstand, prepare for, prevent or quickly recover from a disaster.

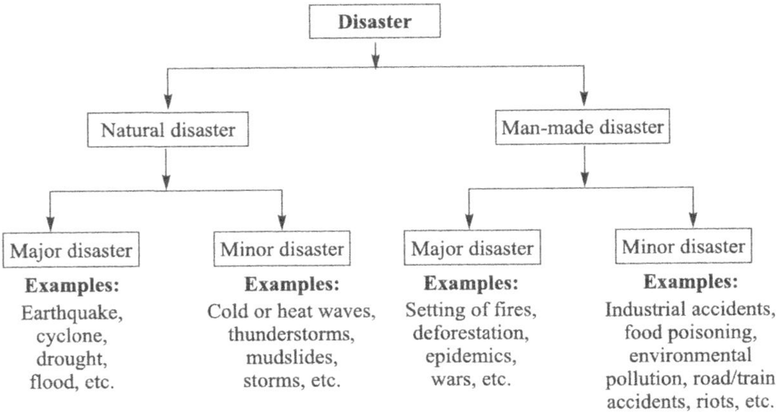


**Fig. 5.30** *Classification of Disasters*

**Risk = (Probability of the accident occurring) × (Expected loss in case of accident)**

Thus, risk is a measure of the expected losses due to a hazard event occurring in a given area over a specified time period.

## *Realisation of Risk is Disaster*

***Note:*** The term *disaster* is derived from a French word meaning bad or evil star.

**Disaster management** is the practice of successful management of natural and man-made disasters.

*The major objective of disaster management* is to reduce the adverse effects of a disaster on the affected community and to help them return to normal life within the shortest possible time.

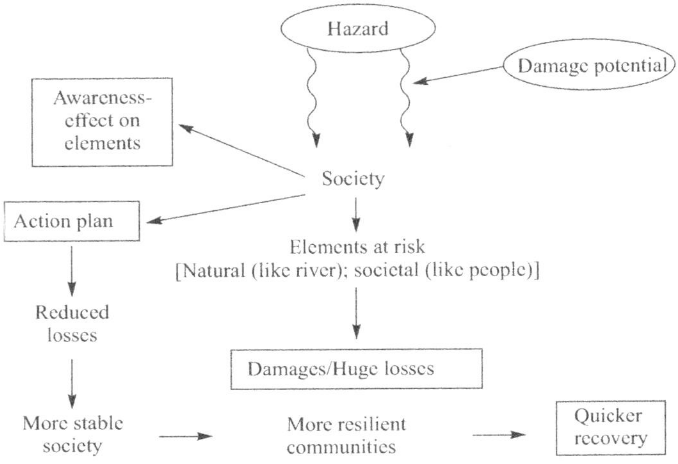


**Fig. 5.31** *Disaster and its management*

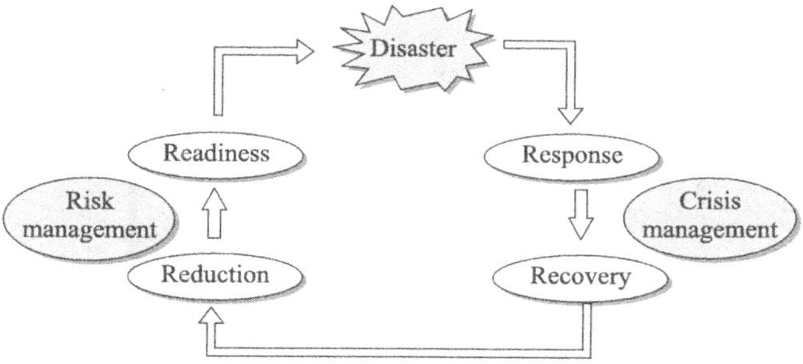


**Fig. 5.32** *Disaster management cycle (the 4 R's)*

**(i) Response** It includes activities during a disaster such as public warning systems, emergency operations, search, rescue (i.e. save life) and relief (i.e., food aid).

**(ii) Recovery** It includes activities following a disaster like rehabilitation and reconstruction which includes temporary housing; processing of insurance claims; distribution of grants; provisions for long-term medical care and counselling.

**(iii) Mitigation or Reduction** It includes activities that reduce the effects of disasters like building codes and zoning, vulnerability analyses; public education.

**(iv) Preparedness or Readiness** It includes activities prior to a disaster like preparation of emergency plans for disasters, emergency training through workshops; warning systems, etc.

*To sum up, disaster management means the organisation and management of resources and responsibilities for dealing with all humanitarian aspects of emergencies, in particular **response, recovery, reduction** and **readiness,** for reducing the impact of disasters, i.e. the 4R's.*

**Example 3** *Comment on the following statement:*

*(a) Flood is a hazard, but it may or may not become a disaster.*
*(b) Humans are all vulnerable in different degrees to various hazards.*

**Solution** (a) The given statement is TRUE.

Flood is a hazard, but it becomes a disaster when people are not ready to cope with it. Then it may result in loss of livelihood or property.

However, when houses are built to make them flood resistant, people are evacuated along with valuables to a safe shelter and cattle are moved to hilly areas then flood remains a hazard and does not become a disaster.

(b) The given statement is TRUE.

- People living in the Ganga–Brahmaputra plain are vulnerable to floods.
- People living in dry areas of Rajasthan and Western Orissa are vulnerable to drought.
- People living in the Himalayan regions are vulnerable to earthquakes.

Thus, humans are easily exposed to danger or attack and so they are all vulnerable in different degrees to various hazards.

## 5.14.1 Flood

*Flood is an overflowing of water onto land that is normally dry.* As a result of flood, the land is filled with an excess of water (Fig. 5.33).

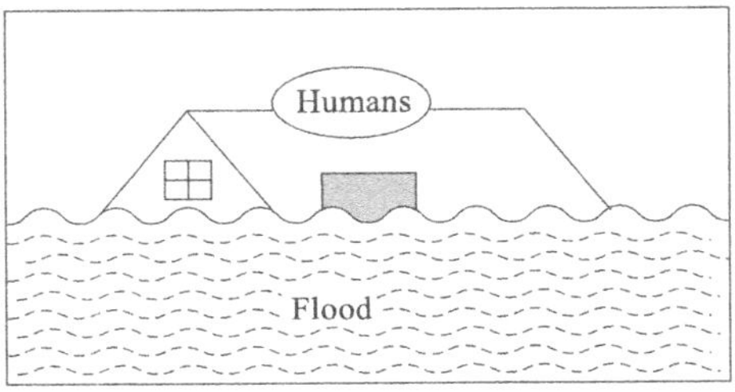


**Fig. 5.33** *Flood situation: Relatively high stream flow overtopping the natural or artificial banks in any reach of a stream*

### *(A) Causes of Floods*

A flood is caused by continuous heavy rain, bad drainage facility, poor design in the construction of dams/embankments, etc., blocking of river channels by landslides, silting of river bed, tsunami, cyclones, and melting of glaciers and sea tides.

### *(B) Types of Floods*

Depending on the cause, floods are of the following types:

(i) When a flood occurs within a short period of time, usually less than 6 hours, because of dam burst, melting of snow, heavy rains, etc., it is known as a *flash flood.*

(ii) When a flood takes place in urban areas because of lack of proper drainage facilities, it is known as an *urban flood.*

(iii) When a flood is caused by tides, storms at sea, cyclone and tsunami, it is known as a *coastal flood.*

(iv) When a flood is caused by an overflowing river, it is known as a *riverine flood.*

**Example 4** *Explain how flood is caused by (a) heavy rain, (b) deforestation, and (c) soil erosion.*

**Solution**

(a) The monsoon climate brings very heavy rain, which results in overflowing rivers. Excess water is not readily absorbed or stored within the drainage basin; so flood is caused.

(b) Trees are cleared for fuel and grazing land in deforestation. As less trees are available on land, evapotranspiration is less, and run-off is more. Thus, flood is caused.

(c) Rivers silt up due to increased soil erosion. This raises the river bed and reduces the capacity of the channel. As a result of this, the likelihood of flooding is increased.

### *(C) Effects of Floods*

**(i) Unavailability of Clean Water** Water in wells, groundwater and piped water supply gets contaminated as a result of flood, resulting in shortages of clean water.

**(ii) Damage to Crops and Food Shortages** Standing crops are damaged by flood. Flood can erode the top soil layer causing land to become infertile. If sea water floods the area, the land turns saline.

As a result of floods, godown and storage facilities get submerged in water resulting in spoilage of grains by fungus. Even entire harvests are lost as a result of flood resulting in sudden food shortages.

**(iii) Diseases and Deaths** Floods result in outbreaks of epidemics, diarrhoea, malaria and viral infections. Animals and humans die either due to these diseases or due to drowning.

**(iv) Physical Damage** In coastal areas, boats or fishing equipments may be lost or damaged. Property gets damaged or collapsed by flooding.

***(D) Flood Management*** Management of flood requires a cyclic pattern linking

(i) Planning: ideas, proposals, consultations, adopting proposals, preparing guidelines;

(ii) Design: design of flood control structure;

(iii) Implementation: construction; and

(iv) Operation: operating and maintaining finalised schemes.

A flood-management cycle is illustrated in Fig. 5.34.

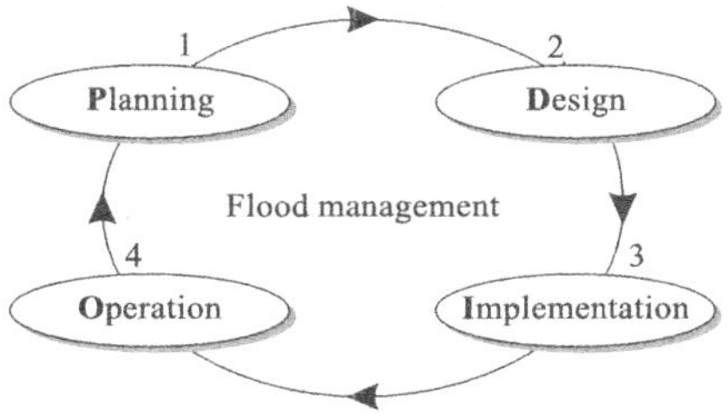


**Fig. 5.34** *Flood Management Cycle (PDIO)*

- *At the planning level*, proposed flood mitigation options are assessed and environmental impact assessments are prepared.
- *At the design stage*, civil works design criteria is determined, Flood Control and Drainage (FCD) structure is designed. Even inputs to flood preparedness programmes are provided at the design stage.
- *At the implemention stage*, flood prone-construction work is started; help is also given for flood preparedness training.
- *At the operation stage*, flood control and drainage structures are built by taking helps of flood forecasting. Emergency relief operations are also assisted in the operations stage (Fig. 5.35).

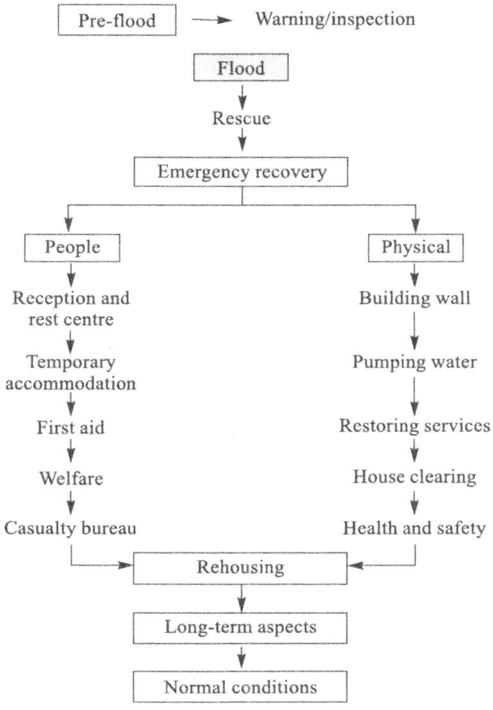


**Fig. 5.35** *Operation stage of flood management*

***(E) Flood-Control Measures*** Technological skills, better warning systems, positive human response to flood warnings and various control measures adopted by the government helps in minimising the impacts of floods.

> ***Notes:*** In India, the Central Water Commission (CWC) is equipped with 157 flood forecasting centres covering 62 interstate river basins. In collaboration with the Indian Meterological Department (IMD), these flood forecasting centres monitor rainfall situations and water levels in reservoirs. This information helps the CWC to issue forecast and warning about floods.

Some of the important flood-control measures are briefly described below:

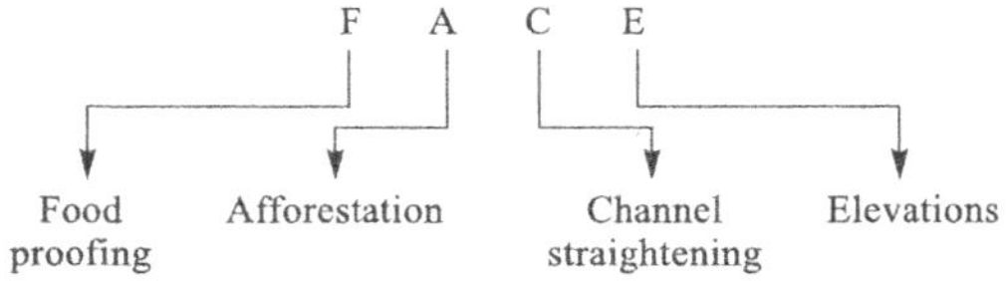


**Fig. 5.36** *Food control measures*

**(i) Flood Proofing** In flood-prone areas, artificial reservoirs can be built to divert flood water into a certain direction.

**(ii) Afforestation** In the hilly source catchment areas of rivers, afforestation helps in infiltration of rainwater, reduction of the soil erosion and reduction in the sediment load of the river.

**(iii) Artificial Straightening of Channels** It is a very useful strategy for flood control because water flows speedily during floods in straight channels.

**(iv) Elevations and Platforms** These structures help in preventing the flood water from entering residential areas (Fig. 5.37).

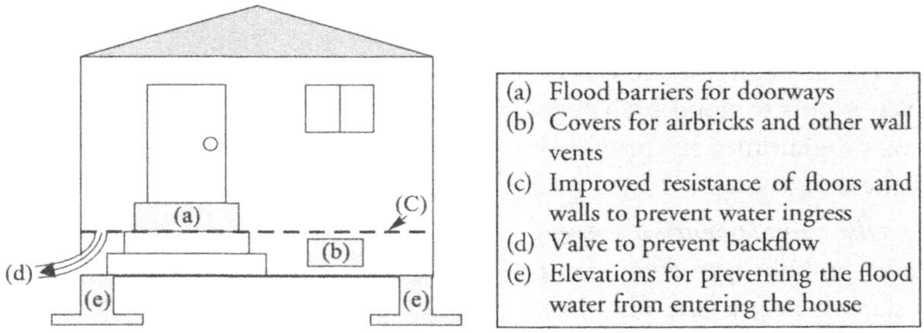


**Fig. 5.37** *Elevations and barriers for flood-risk minimisation*

Flow of water can be blocked by using sand bags.

***(F) Flood Disaster Management Maps*** To mitigate flood disasters, a flood disaster management map contains following informations:

(i) Information on flood inundation areas predicted by possible overflows in rivers due to heavy rain, etc. The prediction can be based on historical flood damage records and by scientific predictions.

This information provides flood hazard and vulnerable areas against floods in the community [refer Fig. 5.38(a) Flood Hazard Map (FHM)], A = 1$^{st}$ Day; B = 2$^{nd}$ Day; C = 3$^{rd}$; D = 4$^{th}$ Day; Inundation Areas

(ii) Information to provide flood disaster mitigation and information for protecting lives such as initiation time of inundation, inundation area, inundation depth, velocity of river water flow.

(iii) Evacuation sites, flood shelters, evacuation routes, etc. [refer Fig-5.38(b) Flood Disaster Management Map (FDMM)].

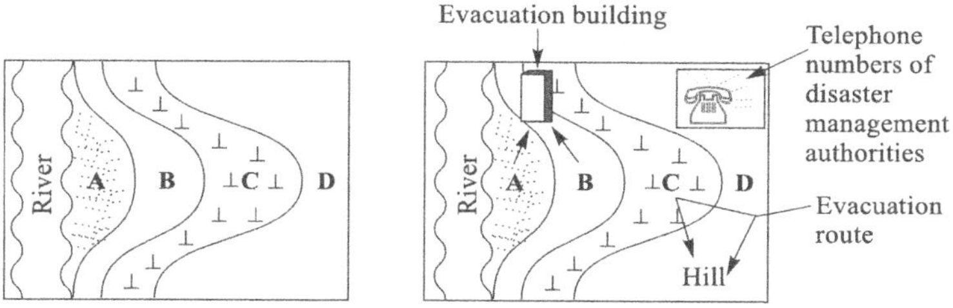


(a) FHM (b) FDMM

**Fig. 5.38** *(a) Flood Hazard Map (b) Flood Disaster Management Map*

In general, an FDMM contains information about

(a) hazard, (b) evacuation, (c) disaster awareness raising, and (d) disaster management.

## *(G) Guidelines for Preparing Flood Disaster Management Map*

(i) Investigate historical flood damage records, seek scientific advice and guidance as scientists can do numerical simulation of possible floods.

(ii) Estimate inundation prediction area.

(iii) Display inundation prediction area as a hazard risk area in the Flood Hazard Map (FHM)

(iv) Add disaster mitigation information such as

(a) basic knowledge of flood

(b) disaster prevention facilities

(c) evacuation tips

(d) evacuation sites and routes

By adding the above information in FHM, flood disaster management maps (FDMM) are constructed.

Information on the map is presented in a comprehensive but easy-to-understand manner. The following information could be included in an FDMM:

(a) Telephone number of authorities related to disaster management (police, hospital, fire department, etc.)

(b) Disaster prevention centers

(c) Land use

(d) Power, gas and water supply facilities, and sewerage facilities

(e) Elementary schools
(f) Others

(v) The flood disaster management map must be prepared in the workshops together with local residents, schools and scientific representatives:
   (a) To reflect local information
   (b) To promote the understanding of inundation areas and evacuation sites
   (c) To facilitate well-prepared persons for flood rescue teams
   (d) To establish the importance of the map within the community from childhood
   (e) To provide a chance for family members to talk about disaster prevention and mitigation
   (f) To promote more precise flood disaster management maps in flood-prone villages.

### *(H) Usefulness of Flood Disaster Management Map*

These maps can be used for

(i) *planning* of preventive measures,
(ii) *integration* of land use planning,
(iii) *evacuation*,
(iv) *construction* of required protection facilities,
(v) *evaluating* the effects of structural measures, and
(vi) increasing the people's *self-protecting* capability against disaster.

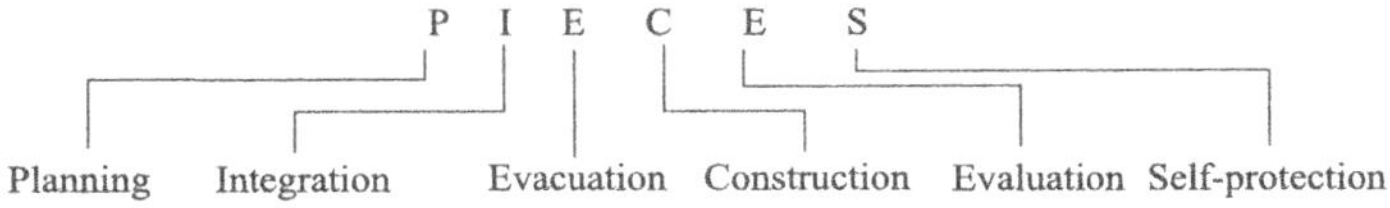


**To sum up**

A Flood disaster management map provides necessary graphical information about regional flood hazards and vulnerability

- to manage flood disasters
- to minimise damage in the region

A Flood disaster management map is a crucial tool to investigate and establish comprehensive disaster mitigation systems.

**Example 5** *Enumerate economic, environmental, political, social and cultural factors for reducing flood risk.*

**Solution** ***(a) Economic***

- Insurance
- Risk management as a normal part of business
- Sustainable flood-plan development

***(b) Environmental***

- Working with natural processes and systems
- Climate change

- Protecting life-supporting capacities and ecological values
- Integrated catchment management

***(c) Political***

- Adaptive and responsive to change
- Agreed roles and responsibilities
- Common long-term goals
- Long-term outcomes

***(d) Social and Cultural***

- People understand and accept the level of flood risk.
- Non-structural, structural and emergency management measures
- Risk management

**Example 6** *What are the requirements for good flood-risk management?*

**Solution** The requirements for good flood-risk management are summarized below:

(a) Political will
(b) Approved goal of risk reduction
(c) Resources
(d) Guidance
(e) Government policies and legislative framework to support risk management
(f) Availability of required information
(g) Monitoring
(h) Evaluation
(i) National capacity in allied fields
(j) Cooperation from society

**Example 7** *What are the drivers against good flood-risk management?*

**Solution** The drivers against good flood-risk management are summarised below:

(a) Unwillingness to pay
(b) Lack of resources, capacity and required information
(c) Vested interests
(d) Inertia and inability to change
(e) Not bearing the full cost of management

**Example 8** *Enumerate economic environmental, political, social and cultural factors which cause increased flood risk.*

**Solution** ***(a) Economic***

- Under-insurance
- Underestimated financial cost of flooding
- Urban and rural development at risk

***(b) Environmental***

- Ignorance of natural processes.
- Ignorance about probability of large floods
- Ignorance of climate change and variability

***(c) Political***

- Focus only on short-term outcomes
- Desire for short-term visible actions and achievable goals
- Enabling legislation with no explicit goals: social and cultural

***(d) Social and Cultural***

- Flood management is not considered as community need
- Non-sufficient protection of people and assets
- Past decisions not reviewed
- Over-reliance on engineering and hazard control

**Example 9** *Illustrate prevention, response and recovery activities of flood management using a simple diagram.*

**Solution**

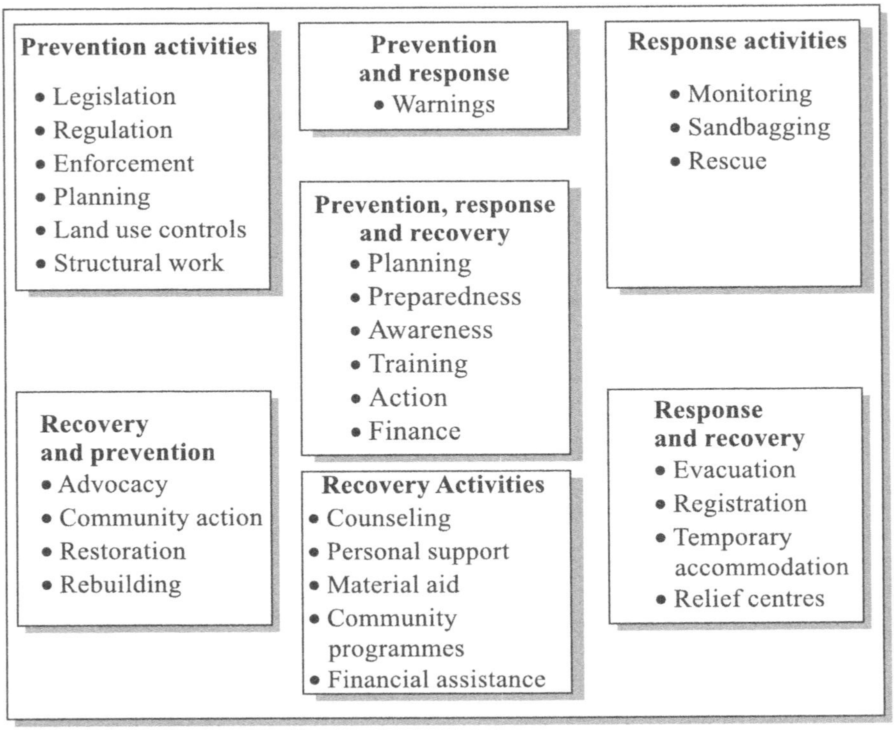

## 5.14.2 Earthquake

*An earthquake is the vibration (sometimes violent) of the earth's surface that follows a release of energy in the earth's crust.*

*Tectonic earthquake* is caused by the sudden dislocation of large rock masses along the faults within the earth's crust. A *fault* is a fracture within some particular rocky mass within the earth's crust. Faults along which the two sides of the fracture move with respect to each other are known as *active faults.* Earthquakes are caused by active faults. Depending on the movement of two sides of fracture, faults can be of following three types:

(i) In response to pulling or tension, *normal faults* occur. The overlying block moves down the dip of the fault plane.

(ii) In response to squeezing or compression, *thrust (reverse) faults* occur. The overlying block moves up the dip of the fault plane.

(iii) In response to either type of stress, *strike-slip (lateral) faults* occur.

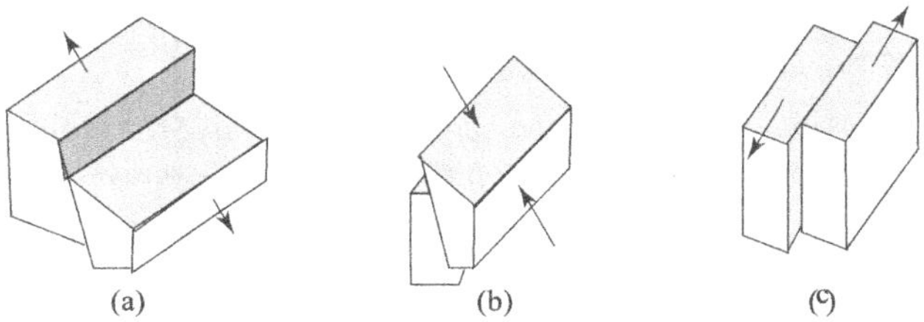


**Fig. 5.39** *(a) Normal fault (b) Thrust fault (c) Strike slip fault*

The *focus* is the point or centre where the energy release starts. The *epicentre* is the point on the earth's surface directly above the focus of the earthquake.

The slow and continuous movement of two sides of an active fault relative to one another (called fault slip), results in gradual build-up of elastic strain energy within the rock along the fault. The rock stores this strain like a giant spring being slowly tightened. At some stage, the build of strain will be too much for the rock to bear. The fault then suddenly moves a comparatively large distance in a short duration. The rocky masses which form the two sides of the fault then snap into a new position along with release of the strain energy. This suddenly released energy takes the form of *seismic waves.*

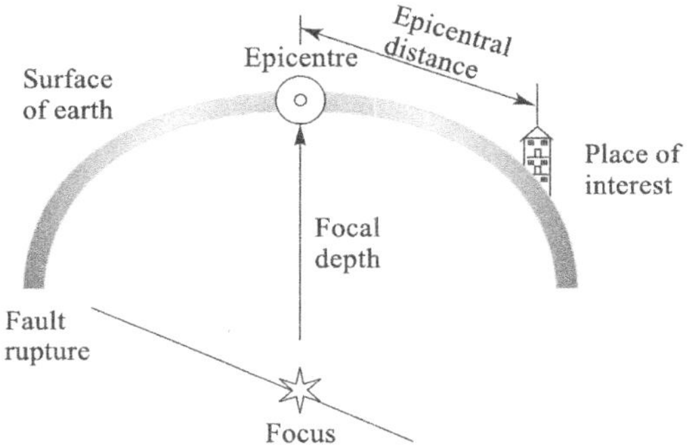


**Fig. 5.40** *Earthquake*

Thus, *seismic waves* are the waves of energy caused by the sudden breaking of rock within the earth or an explosion.

An earthquake radiates body waves in all directions. These are compressional or primary (P) and transverse or secondary (S) waves. The interaction of the P and S waves with the earth's surface and shallow structure produces surface waves.

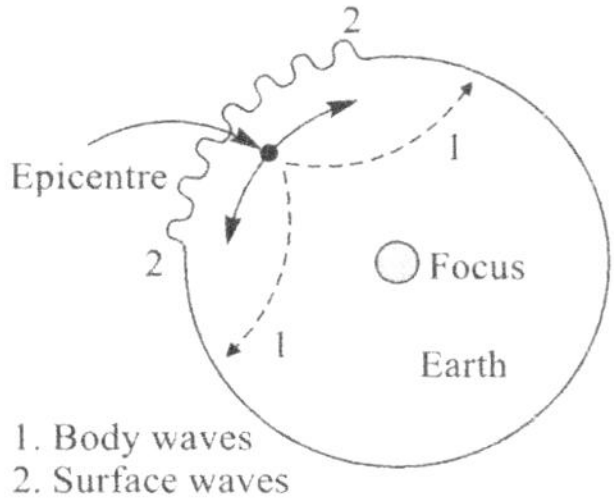


**Fig.5.41** *Seismic waves*

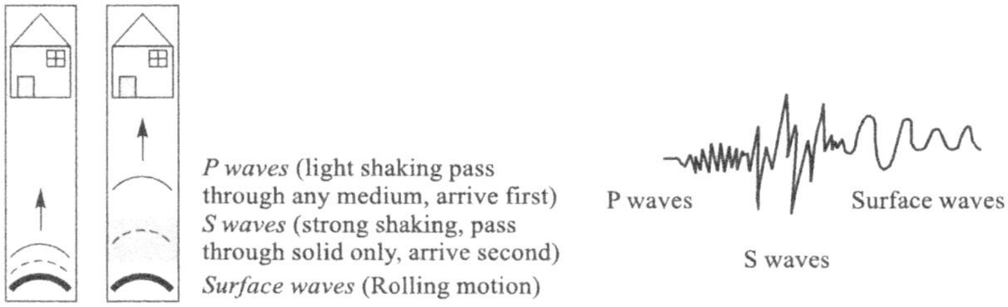


**Fig. 5.42** *The P waves, S waves and surface waves shake the ground in different ways and also travel through the earth at different velocities*

The seismometer must be able to move with the vibrations of seismic waves, yet part of it must remain nearly stationary. This is achieved by isolating the recording device (say pen) from the rest of the earth using the principle of inertia. For this, a pen is attached to a heavy mass which is suspended by a spring. Paper is attached to the earth. During an earthquake, the paper moves more than the pen and heavy mass, and so vibrations get recorded on the paper.

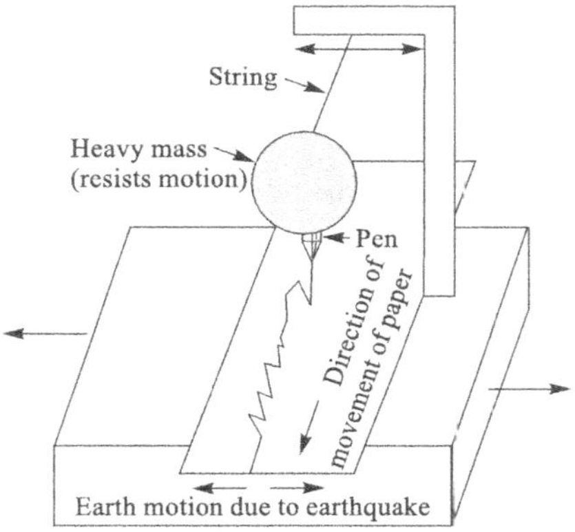


**Fig. 5.43** *The machine used to measure earthquakes is known as Seismometer*

The size of an earthquake is generally given in terms of the *Richter magnitude*. It is a scale of earthquake size developed by the seismologist Conrad Richter.

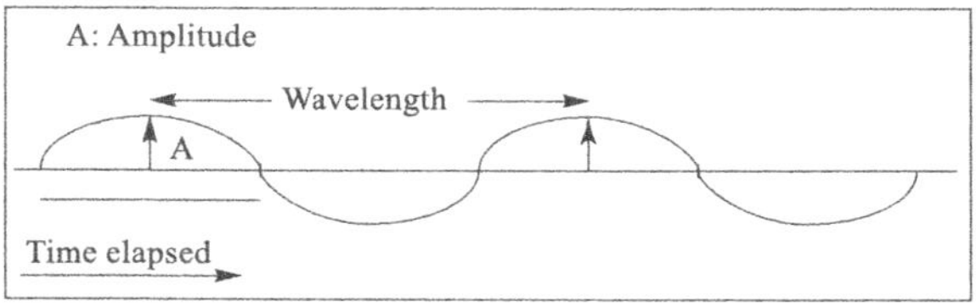


**Fig. 5.44** *The Richter magnitude involves measuring the amplitude (height) of the largest recorded wave at a specific distance from the earthquake. It is logarithm scale. For each increase of 1 in the Richter magnitude, there is a tenfold increase in amplitude of the wave*

The amount of energy released ($E$) in ergs by the earthquake is related to the Richter magnitude ($M$) by the following equation:

$$\log E = 11.8 + 1.5M \tag{5.3}$$

**Example 10** *Find the amount of energy released in ergs when the earthquake has magnitudes of 8 and 9 on the Richter scale.*

**Solution** (a) Given $M = 8$

As $\mathbf{\log E = 11.8 + 1.5M}$

$\Rightarrow \log E = 11.8 + 1.5 \times 8$

$\Rightarrow \log E = 23.8$

$\Rightarrow E = \text{antilog}\ (23.8)$

$\Rightarrow E = 6.3 \times 10^{23}$ ergs (1)

(b) When $M = 9$

$\Rightarrow \log E = 11.8 + 1.5 \times 9$

$\Rightarrow \log E = 25.3$

$\Rightarrow E = \text{antilog}\ (25.3)$

$\Rightarrow E = 1.995 \times 10^{25}$ ergs (2)

Dividing Eqn. (2) with Eqn. (1), we get

$$\frac{1.995 \times 10^{25}}{6.3 \times 10^{23}} = 31.6$$

Thus, *an earthquake of magnitude 9 releases 31 times more energy than an earthquake of magnitude 8.*

***Notes:***

(i) An amount of energy released by Hiroshima atomic bomb was equivalent to a magnitude of 5.5 earthquake.

(ii) In 1964, the Alaska earthquake had a Richter magnitude of 8.6.

(iii) Sensitive seismographs can greatly magnify earthquake-caused ground motions. They can detect strong earthquakes from sources anywhere in the world.

(iv) The magnitude of an earthquake is determined from the logarithm of the amplitude of waves recorded by seismographs.

### 5.14.3 Tsunami

In Japanese, *tsu* means harbour and *nami* means waves. *A tsunami is a series of waves in the ocean that can be hundreds of miles long and have been known to reach heights of 10.5 m.* The massive December 26, 2004 tsunami travelled at a speed of 480 km per hour.

***Origin of Tsunami*** The top layer of the earth (i.e, the **lithosphere**) is made up of a series of huge plates. They rest on an underlying viscous layer called the **asthenosphere**. On the earth, these plates are constantly in motion, moving along each other at a speed of 2.5 to 5 cm per year. When two plates come into contact at a plate boundary region, a heavier plate can slip under a lighter one, resulting in subduction. Part of the sea floor connected to the lighter plate may 'snap up' in some cases. It sends tons of rock shooting upward with tremendous force. When the energy of the force in transferred to the water, water is pushed upwards above normal sea level. This is the birth of a **tsunami**.

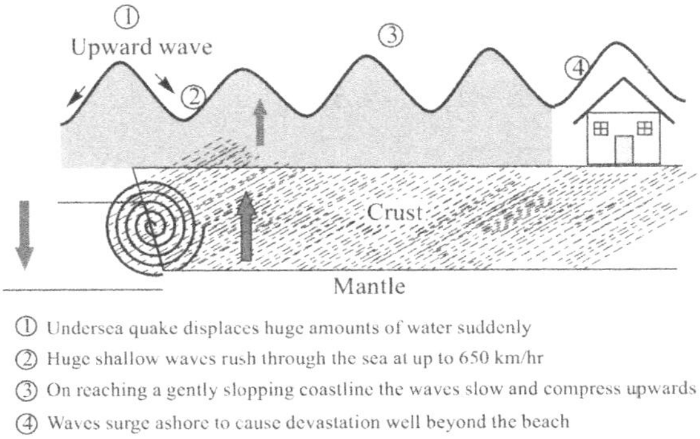


**Fig. 5.45** *Cause and effects of undersea earthquake and tsunami*

## 5.14.4 Cyclone

*Cyclones are swirling atmospheric disturbance in the form of huge revolving storms caused by powerful winds moving with very high velocities (sometimes exceeding 300 km/h).* Cyclones are accompanied by rain and generate enormous waves in the ocean.

***Notes:***

- The word cyclone is derived from the Greek word '*cyclos*' meaning coils of a snake.
- Cyclones are known as **willie-willie** in Australia; **typhoons** in the Pacific ocean; and **hurricanes** in the Atlantic ocean.
- Huge storms of cyclones revolve in clockwise direction in the northern hemisphere and in anti-clockwise direction in the southern hemisphere.

### *(A) Cause(s) of Cyclones*

(1) Near the Equator, over warm seas, air heated by the sun rises upwards quickly and creates areas of very low pressures. (2) To fill the void that is left, cool air rushes in. (3) We know that the earth is constantly revolving around its axis. (4) Thus, the air is bent inwards and spirals outwards with great force. (5) As the warm air rises, it becomes loaded with moisture which condenses into massive thunderclouds. (6) Due to the faster and faster rotation of the swirling winds, a huge circle of clouds get formed. The edge of a cyclone is called 'wall of the eye' which has a radius of 20 km–30 km. At the centre of the cyclone, wind velocity is less. This calm, cloudless area is called the 'eye' of the cyclone.

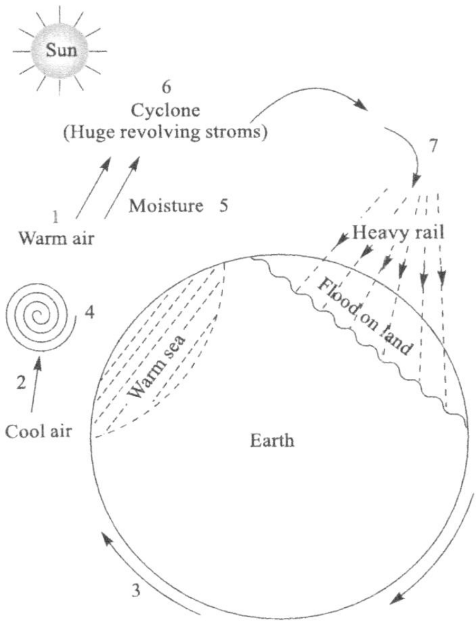


**Fig. 5.46** *Cause of cyclones*

When cyclones move over the ocean, they drag clouds and moisture. They can also pick up energy when they travel across warm water. (7) When cyclones move over land, they result in heavy rains leading to floods.

***(B) Effects of a Cyclone*** Cyclones can cause the following damages:

(i) The standing *crop* and food stock lying in low-lying areas will be *ruined* due to powerful winds and heavy rain. Banana, coconut and other plantation crops are extremely vulnerable.

(ii) Sea water dragged through cyclones result in *inundation of land.* This increases salinity as a consequences of which soil becomes unfit for cultivation.

(iii) Heavy rain can cause *flooding*. This can lead to contamination of groundwater and surface water. Viral outbreaks, diarrhoea and malaria are consequences of contamination of water.

(iv) Gable-ended roofs made from cement, asbestos or tin sheets get high uplift as a result of powerful winds of cyclones. As these sheets are blown away, these then strike against nearby buildings, animals and humans causing damage and deaths.

(v) Asymmetric buildings with empty pockets collapse due to the impacts of powerful winds.

(vi) Trees get uprooted and carried away along with powerful winds. These, then, destroy telephone lines, electricity poles, transmission line towers, etc. Thus, power supply and communication networks get disturbed.

(vii) Cyclones are powerful enough to damage loose or weak parts of buildings like doors, windows, etc.

***(C) Control Measures of Cyclones*** Control (or protection) measures and main mitigation strategies are briefly discussed below:

(i) On the coasts of India, 10 cyclone detection radars are installed by India Metrological Department (IMD). A geo-stationary satellite (INSAT-IB) also monitors cyclone movements. A special Disaster Warning System (DWS) provides a cyclone alert (48 hours in advance) and a cyclone warning (24 hours in advance). Such early warnings help in controlling the damages of cyclones.

(ii) Symmetrical designs of buildings, no empty pockets, hip or pyramidal roofs, no loose ACC sheets or tin sheets, reinforced cemented foundations, less wide doors and windows, etc., are engineered structures better in withstanding impacts of powerful winds and heavy rains of cyclones.

(iii) Strong rooted trees with needlelike leaves can be planted in the directions facing the wind along coastline. These shelter-belt plantations lessen the impact of strong cyclonic winds and thus stop soil erosion. As a result, houses, cultivable fields, etc., are protected (Fig. 5.47).

***(D) Cyclones in India***

(i) About two-thirds of the cyclones that occur in the Indian coastline occur in the Bay of Bengal.

(ii) West Bengal, Orissa, Andhra Pradesh and Tamil Nadu are the Indian states which are generally affected by cyclones in the east coast.

(iii) Gujarat, Maharashtra, Goa, Karnataka and Kerala are the Indian states which are generally affected by cyclones in the west coast.

(iv) About 8913 people died as a result of tropical cyclone-hit in Orissa in 1999. The death toll was 5000 for the 1985 cyclone in Andhra Pradesh. In 1977, about 14,204 people died as a result of cyclone which hit Chennai, Kerala and Andhra Pradesh.

(v) India is extremely vulnerable to cyclones and its associated hazards like high winds, heavy rainfall and storm surge. This is because India has a long coastline of about 7516 km, flat coastal terrain, high population density and the vulnerable geographical location.

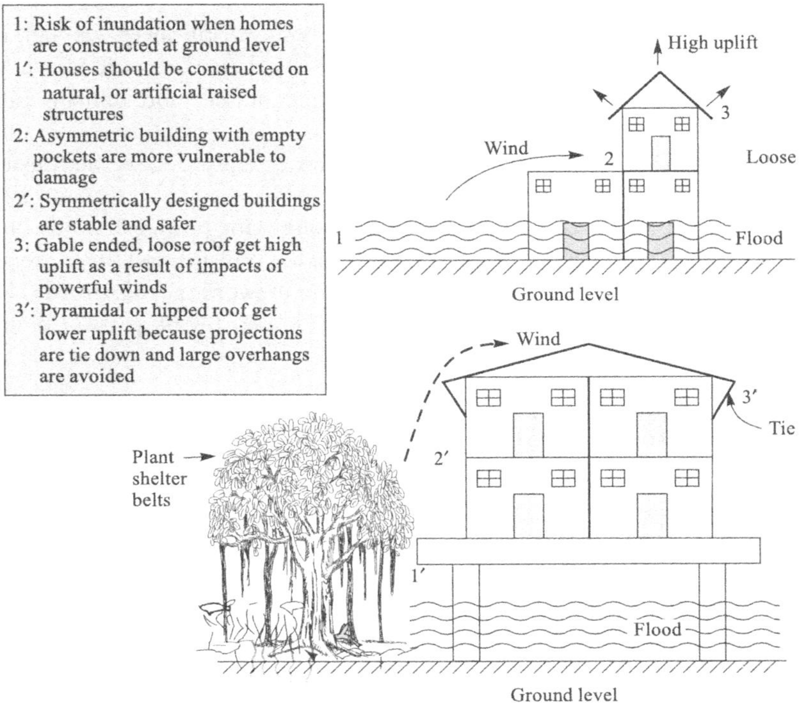


**Fig.5.47** *Main mitigation strategies w.r.t. engineered structures for protection against cyclones*

## 5.14.5 Landslides

*Landslide means downward sliding of a relatively dry mass of land and rock. It is also known as landslip.*

### (A) Effects of Landslides

(i) Landslides block or bury roads, lines of communication, railways lines, etc.

(ii) They destroy anything that comes in their path. They destroy settlements.

(iii) They destroy agricultural areas leading to loss to food production.
(iv) They block river flow; flooding may also occur.
(v) The flow of debris in landslides causes heavy casualties.

## *(B) Landslides (Disaster) Management*

Main mitigation strategies and preventive measures of landslides are illustrated in Fig. 5.48 and important solutions are described below:

**(i) Grow Trees** The most effective way of arresting landslides is *afforestation*. Shelter belts with several rows of trees, both on slopes and the bottom, should be maintained. This helps to bind the top layer of soil with layers below, while preventing excessive run-off and soil erosion.

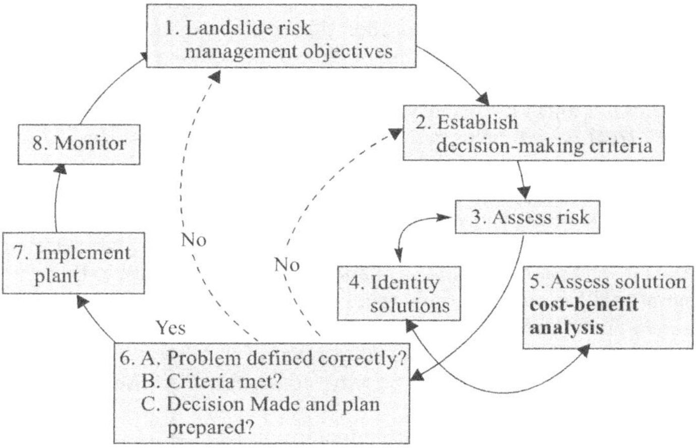


**Fig. 5.48** *Important mitigation strategies for landslide risk management*

**(ii) Stabilise Slopes** Build retaining walls along roads in hill stations to stop land from slipping. Property owners at the toes of steep slopes should construct reinforced walls as barriers to trap landslides.

**(iii) Avoid Blockage of Natural Drainage** This should be done in construction of roads, irrigation canals, etc., when landslides are accompanied by infiltration of rainwater and spring flows, redirect surface run-off towards safe places through straight channels.

**(iv) Install Flexible Pipes and Cables Underground** Flexible pipes and cables should be installed underground in order to withstand forces caused by the landslides.

**(v) Build Strong Foundations** To withstand ground movement forces of landslides, structures must be engineered with strong foundations.

**(vi) Encourage Proper Land Use** Vulnerable areas must be protected. Land-use practice must be improved.

## Important Definitions

- *Pollutant* is a material which is present in excess of the natural concentration and produces a bad effect upon the environment.
- *Pollution or environmental pollution* can be defined as an undesirable change in the physical, chemical or biological characteristics of our environment by the introduction of substances or energy by humans into the environment.
- *Air pollution* is the presence of substances in the air (which generally originate from human activities) in sufficient concentrations and sufficient time, to interfere with the comfort, health, safety or full use and enjoyment of property.
- *Air pollutants* are gaseous, liquid or solid substances present in such concentration as may be or tend to be injurious to human beings or other living creatures or plants or the environment.
- *Water pollution* is defined as presence of any foreign substance or energy in water in such concentration and for such duration that tends to degrade the quality of water so that humans, animals or any other organism cannot enjoy the beneficial qualities of water but the use constitutes a hazard.
- *Waste-water treatment* is carried out in the following steps:
  (a) *Primary Treatment*: It helps in the removal of suspended particles.
  (b) *Secondary Treatment*: It helps in the aerobic decomposition of organic matter.
  (c) *Tertiary Treatment*: It helps in the production of safe water, free from harmful chemicals and pathogenic bacteria.
- *Soil pollution* can be defined as the introduction of substances, biological organisms, or energy into the soil that lead to a change in the quality of soil so that plant growth and animal health is adversely affected.
- *Marine pollution* is defined as the direct or indirect discharge of matter or energy by humans into marine water bodies that is harmful to living organisms, hazardous to human health, hinders marine activity, adversely affects-sea-water quality and reduces its amenities.
- *Noise pollution* is defined an environmental noise or an unwanted sound that is annoying, distracting, or physically harmful. Harms include hearing loss, stress, sleeplessness, etc. Noise pollution is also known as *sound pollution*.
- *Thermal pollution* may be defined as the degradation of water quality by any process that changes ambient water temperature.
- *Hazardous wastes* can be defined as useless, unwanted and discarded material that may pose a threat to human, plant or animal life.
- *Radiation* is a form of energy that can travel through any medium including vacuum.
- *Nuclear hazard, or radiation pollution*, is the danger or risk to human health or the environment posed by radiation emanating from the atomic nuclei of a radioactive substance or the possibility of an uncontrolled explosion originating from a fission or fusion reaction of atomic nuclei.
- The waste materials which have been rejected for further use and which can neither readily escape into the atmosphere nor can be transported by water into streams are called *solid waste*.
- *Composting* is the thermophilic and aerobic decomposition of organic matter present in solid waste by microorganisms, mainly bacteria and fungi.
- *Sanitary landfill* sites have liner systems and other safeguards to prevent groundwater contamination. These sites are consistent with the economic considerations, hydrogeological requirements, climatic conditions and topography.
- *Incineration* is the controlled combustion of organic solid wastes so as to convert them into incombustible residue and gaseous products.

- *Disaster* can be defined as a man-made or natural event (like floods, earthquake, cyclone or landslides) which results in great damage or loss of life.
- *Hazard* is a situation which poses a level of threat to life, health, property or a dangerous condition or event that may deleteriously affect society or an environment.
- *Vulnerability* is the extent to which damage will likely happen by the impact of a particular hazard.
- *Capacity* means resources and strengths which exist in households and communities and enable them to cope with, withstand, prepare for, prevent or quickly recover from a disaster.
- *Disaster management* is the practice of successful management of natural and man-made disasters.
- The major *objective of disaster management* is to reduce the adverse effects of a disaster on the affected community and to help them return to normal life within the shortest possible time.
- *Flood* is an overflowing of water onto land that is normally dry.
- An *earthquake* is the vibration (sometimes violent) of the earth's surface that follows a release of energy in the earth's crust.
- *Seismic waves* are the waves of energy caused by the sudden breaking of rock within the earth or an explosion.
- A *tsunami* is a series of waves in the ocean that can be hundreds of miles long and have been known to reach heights of 10.5 m. The massive December 26, 2004 tsunami travelled at a speed of 480 km per hour.
- *Cyclones* are swirling atmospheric disturbance in the form of huge revolving storms caused by powerful winds moving with very high velocities (sometimes exceeding 300 km/h).
- *Landslide* means downward sliding of a relatively dry mass of land and rock. It is also known as *landslip*.

# EXERCISES

**Based on Environmental Pollution**

1. What is meant by environmental pollution? Discuss the requirement of a nonpolluted environment.
2. How can you, as an individual, prevent environmental pollution? Why is such an effort at individual level important?
3. What is pollution?
4. Enlist global environmental problems and discuss any one. (b) What are the common toxic chemicals in the environment?
5. What are the causes of environment pollution? List all environment pollutions and their sources in detail.

**Based on Water Pollution**

1. What do you mean by water pollution?
2. Discuss water treatment with layout of a water-treatment plant.
3. Suggest various remedial and control measures to minimise water pollution.
4. What are the common sources of water pollution? Discuss the various types of water pollutants.
5. How do synthetic detergents, pesticides and plant nutrients present as water pollutants create problems?
6. Write a short note on groundwater pollution.
7. Describe sources of water pollution.

8. What is potable water? Describe the different water pollutants which make water unfit for drinking.
9. Define water pollution and enlist various sources of water pollution. Explain any one in detail.
10. What are effects of water pollutants on environment and humans?
11. Discuss important characteristics of waste water. Give the outline of the waste-water treatment methods.
12. Write classification of water pollutants with suitable examples.

**Based on Land Pollution**

1. Define land pollution. Discuss the causes of land pollution and their control.
2. Enumerate various effects of soil pollution. How do industrial, agroproducts and pesticides deteriorate the soil?
3. (a) Discus the effect of the following in soil pollution:
 (i) Agro technology, (ii) Urban waste
(b) Write short notes on
 (i) Soil profile, (ii) Soil pollution control
4. What is land degradation? What are its causes.
5. What are the effects of soil pollution?
6. What are the sources of soil pollution?
7. How is the monitoring of soil pollutions made?
8. Mention various/major types of soil pollution.
9. Describe uses and overuses of land. What are causes and effects of land pollution?
10. What are the major sources of land pollution? How does land pollution affect soil productivity?
11. What are causes of land degradation? Narrate common pollutants responsible for causing land pollution.

**Based on Air Pollution and Air Pollutants**

1. What is air pollution? What are its effects on human health?
2. Explain four major air pollutants and their consequences.
3. Enumerate various air pollutants for which National Ambient Air Quality Standards have been given in India.
4. Explain the various nonpoint sources contributing to air pollution.
5. Discuss the measures used for controlling air pollution.
6. Discuss the natural and man-made (synthetic) pollutants that cause air pollution.
7. What are the sources of (a) $NO_x$ and (b) CO pollution in the atmosphere? What measures can be adopted to control these emissions?
8. Enlist natural and human sources of air pollution. Explain any one briefly and list out the common atmospheric pollutants.
9. Write the characteristics and biological effects of the following in air:
 (a) $SO_x$ (b) $NO_x$ (c) CO
10. Discuss the sources and effects of particulates in the atmosphere.
11. Discuss hazardous effects of air pollutants.

**Based on Solid Waste and its Management**

1. What are 'solid wastes'? Discuss their types, effects, and name the various methods used to dispose solid wastes. Explain any one of them with merits and demerits.
2. Explain the various methods commonly employed for disposal of solid waste, with their advantages and disadvantages.
3. Briefly describe the idea of "solid-waste management". Distinguish between the incineration and combusting methods of solid-waste disposal.

4. Classify the composting techniques based on oxygen use. What are the advantages of 'vermicomposting' over conventional composting? What are the advantages of solid-waste incineration?

5. What do you understand by the term *solid waste*? Discuss in brief the various types of solid wastes.

6. What is solid waste? Discuss its sources and effects.

7. What does the term *incineration* mean? What are its disadvantages?

8. What are the advantages and problems of a sanitary landfill?

9. How is solid waste segregated?

10. Discuss the treatment and disposal methods for solid waste.

11. What are the different types and sources of waste? Explain the properties and effects of solid wastes.

12. Give an informative appraisal on recycling of solid wastes.

**Based on Marine Pollution**

1. What are the sources and effects of marine pollution?

2. (a) Oceans are ultimate sinks for most of the waste we produce. Explain.
(b) List offshore sources of marine pollution.
(c) Explain the effects of oil pollution on the ocean.

**Based on Noise Pollution**

1. Discuss effects of noise pollution and its control.

2. Define noise pollution. Describe all effects of noise pollution.

3. Define noise. Enlist major sources of noise pollution and explain any one briefly.

4. Write an account of decibel scale for the measurement of sound.

5. Write short notes on (a) noise control, and (b) speech interference level.

6. What are physical, physiological and psychological effects of noise?

7. Discuss the harmful effects of noise pollution. State the measures to control noise pollution.

**Based on Radioactive Pollution**

1. What are the sources of radioactive pollution and how can we control the radioactive pollution?

2. What are the different methods for treatment of hazardous waste and how can it be disposed safely?

3. Write a note on hazardous wastes and the environmental problems and health risks associated with these.

4. Describe the causes of the Chernobyl disaster and what are the lessons to be learnt from this incident?

# OBJECTIVE TYPE QUESTIONS

**I. Fill in the Blanks**

1. Automobiles are the largest sources of _________ pollution in cities.

2. PAN is formed by the interaction of oxides of nitrogen and _________ in presence of sunlight.

3. Increase in the concentration of soluble salts in _________ is called salinisation.

4. _________ is a combination of smoke and fog.

5. Taj Mahal at Agra may be damaged by _________.

6. _________ measures dissolved oxygen needed by microbes to decompose organic waste.

7. Air pollution from automobiles can be controlled by fitting _________.

**8.** __________ (air pollutant) gets combined with haemoglobin of the blood more rapidly than oxygen.

**9.** Electrostatic precipitators are used to remove __________.

**10.** The largest range of sound frequency measured in vibration/second which humans can hear is __________.

**11.** __________ is the most effective and cheap method of noise control.

**12.** World Health Day is recalled on __________.

## II. Match the following terms.

Match the terms of column I with appropriate terms of column II.

**A.**

| Column I | Column II |
|---|---|
| 1. Soil pollution | (a) Cancer |
| 2. Noise pollution | (b) Cooling tower |
| 3. Nuclear pollution | (c) Fish kills |
| 4. Thermal pollution | (d) Hearing loss |
| 5. Oil pollution | (e) Sewage |

**B.**

| Column I | Column II |
|---|---|
| 1. Asthma and bronchitis | (a) Ozone-layer depletion |
| 2. Absorbtion and removal of UV light | (b) Inhalation of pollen |
| 3. Peroxyacetyl nitrate (PAN) | (c) Ozone |
| 4. Skin cancer | (d) Secondary pollutant |

**C.**

| Column I | Column II |
|---|---|
| 1. Grit chamber | (a) Treatment of domestic liquid wastes |
| 2. 1° sedimentation tank | (b) Removes gravel and sand |
| 3. 3° treatment | (c) Remove (45–60)% suspended solids and (30–45)% BOD |
| 4. Septic tanks | (d) Removes residual, volatile and suspended solids |

## III. Multiple Choice Questions

**1.** The gas leaked in Bhopal tragedy was

(a) methyl isocyanate
(b) ethyl isocyanate
(c) butyl isocyanate
(d) sodium isocyanate

**2.** Noise is ________ sound.

(a) constant
(b) unwanted
(c) loud
(d) high frequency

**3.** The disease caused by eating fish inhabiting mercury-contaminated water is

(a) cancer
(b) fever
(c) minamata disease
(d) osteosclerosis

**4.** Increasing skin cancer and high mutation rate are the result of

(a) acid rain
(b) $CO_2$ pollution
(c) global warming
(d) ozone depletion

**5.** Which of the water samples from the following four sources have the excessive quantity of calcium, sulphate, magnesium, bicarbonate?

(a) River water (b) Lake water
(c) Sea water (d) Well water

**6.** One among the following is not essential for biomagnification to take place. Identify it.

(a) Phytoplankton
(b) Animals

(c) Zooplankton
(d) Fishes

7. Eutrophication of surface waters occur due to the presence of
(a) organic carbon
(b) pesticides
(c) nutrients
(d) inorganic carbon

**IV. Indicate True or False for the following statements**

1. Noise is measured in decibel scale. True/False
2. All pollutants are degradable. True/False
3. Underground water is nonpolluted and safe. True/False
4. PAN is a primary pollutant. True/False
5. Ozone causes cracking of rubber. True/False
6. The residue of fertilisers and pesticides ramains for a long time and delivers deleterious effects on soil. True/False
7. Run-off from land area contributes to marine pollution. True/False
8. Permissible noise level during daytime in a residential zone is 55 dB. True/False
9. Sludge from a water-treatment plant is a hazardous waste. True/False
10. The production, processing and distribution of food all changes the environment. True/False
11. Desertification is the primary environmental effect of agriculture. True/False
12. House is a health hazard when factors such as poor design, environmental contamination and poverty combine to cause or exacerbate disease. True/False
13. For the development of a country and the prosperity of its people, industry is essential. True/False
14. Industrial practices release enormous quantities of air and water pollutants. True/False
15. Transportation activities support increasing mobility demands for passengers and freight but at the cost of adverse environmental effects. True/False

## Answers to Objective Type Questions

**I. Fill in the Blanks**

1. Air
2. Hydrocarbons
3. Soil
4. Smog
5. $SO_2$
6. BOD
7. catalytic converter
8. CO
9. Particulates
10. 20 to 20,000
11. Control at source
12. April, $7^{th}$

**II. Matching the terms.**

**A.** 1. (e) 2. (d) 3. (a) 4. (b) 5. (c)
**B.** 1. (b) 2. (c) 3. (d) 4. (a)
**C.** 1. (b) 2. (c) 3. (d) 4. (a)

**III. Multiple Choice Questions**

1. (a) 2. (b) 3. (c) 4. (d)
5. (a) 6. (b) 7. (c)

**IV. True or False**

1. True 2. False 3. False 4. False
5. True 6. True 7. False 8. True
9. False 10. True 11. False 12. False
13. True 14. True 15. True

# 6

# SOCIAL ISSUES AND THE ENVIRONMENT

## Learning Objectives

After studying this chapter, you should be able to

- name and discuss the contribution of greenhouse gases to global warming
- explain the effects of global warming
- describe the remedial measures of greenhouse effect
- explain urbanisation, automobile pollution
- define acid rain and describe its causes and effects
- discuss the problems associated with ozone-layer depletion and its remedial measures
- explain the causes of ozone-hole formation
- name and discuss the types of rainwater harvesting
- define rainwater harvesting
- describe the role of government and legal aspects in environmental protection
- discuss the salient features of the
  - ❏ Environment (Protection) Act, 1986
  - ❏ Air (Prevention and Control of Pollution) Act, 1981
  - ❏ Forest (Conservation) Act, 1980
  - ❏ Wildlife Protection Act, 1972
  - ❏ Water (Prevention and Control of Pollution) Act, 1984
- explain power of the state government to declare our pollution-control areas and restrictions or use of certain industrial plants as given in air prevention and control of pollution acts
- describe the Government Organisation/Department responsible for the protection of the environment
- explain the role of NGOs and environment education in environmental protection
- explain the meaning of industrial symbiosis, carbon credits, animal husbandry and nuclear holocaust.

## 6.1 INTRODUCTION

The human population continues to increase, and with population growth comes more requirements of food, shelter and clothing. Thus natural and managed ecosystems are being exploited to provide increasing goods and services. Humans may have reached the limit in some resources, and most of the ways we use the other resources are presently unsustainable. In addition, humans extract and burn fossil fuels much faster than they could ever be replenished, bringing on change in global climate. Living beings are forced to tolerate the continued loss of biodiversity. Rich persons are unwilling to take the necessary steps to help those suffering from poverty, hunger and illnesses.

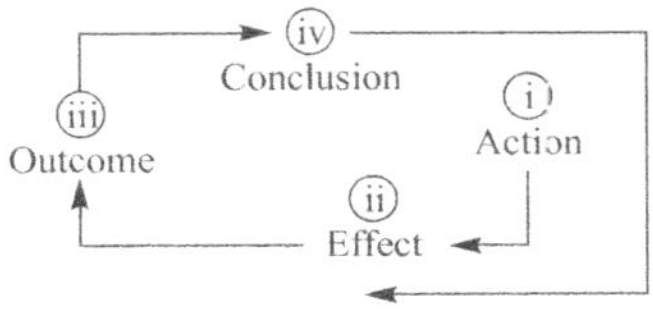


### 6.1.1 Developing Countries

**(i) Action** Escalating human activities to meet survival needs of food, clothes, shelter, etc.

**(ii) Effect** Resource depletion

**(iii) Outcome** Environmental degradation, poverty, ill-health, etc.

**(iv) Conclusion** The vicious cycle will continue with increasing human population, hunger, malnutrition, lack of safe/clean water, sanitation problems, inadequate shelter, etc.

### 6.1.2 Developed Countries

**(i) Action** Industrial revolution to meet desires of affluent persons

**(ii) Effect** Reliance on fossil fuels and nuclear power

**(iii) Outcome** Mass production and consumption, environmental pollution, climate change, global warming, acid rain, ozone layer depletion, nuclear accidents, etc.

**(iv) Conclusion** The vicious cycle will continue towards unsustainable society.

(a) Will it continue forever?

(b) Can you suggest some innovative way of economic development that would improve the well-being and the sustainable well-being of the entire planet?

***A summary of the current global and Indian crisis is presented below:***

**(i) Global Poverty** (as per the World Bank, August 2008)

- 1.4 billion people live at or below poverty line of $ 1.25 a day.

**(ii) State of the World's Children** (as per UNICEF, Feb. 2009)

- 2.5 billion people lack access to improved sanitation.
- 1 billion children are deprived of one or more services essential to survival and development.

- Over 24,000 children die every day around the world. This is equivalent to 16–17 children dying every minute.

The silent killers are poverty, hunger, easily preventable diseases and illnesses.

**(iii) Climate Change** (as per NASA, October 2009)
For 6,50,000 years, atmospheric $CO_2$ has never been above 300 parts per million. But current levels are very high due to the Industrial Revolution. Increased greenhouse gases and the greenhouse effect is expected to contribute to an overall warming of the earth's climate, leading to a global warming. 1900s was the warmest century during the last 1000 years. Due to global warming, agricultural yield will decline and biodiversity will be lost.

**(iv) Scarcity** (as per UNDP estimates, June 2009)
Due to climate change effects, an additional 600 million people will face food shortages and malnourishment in the coming years, and 1.8 billion will struggle to find water.

Many environmental problems arise from the abuse, misuse and overuse of natural resources by humans. The world is heading for an environmental disaster, and humans do not want to see, hear, or talk about it. If it continues, humanity will no longer be able to co-exist with nature.

Development activities are necessary for humans in order to enhance the quality of life and fulfill the needs of a fast-growing population. However, it should be done in such a manner that natural resources are not exploited. Developments must meet present needs without compromising the ability of future generations to meet their own needs. Thus, it is essential to learn about the causes and effects of current environmental issues of importance so that innovative solutions can be implemented.

## 6.2 SUSTAINABLE DEVELOPMENT

*Sustainable development means meeting the needs of the present without compromising the ability of future generations to meet their own needs.*

Sustainable development is a continuous process. To be sustainable is a constant challenge for humanity.

The pillars of sustainable development are the following:

(i) Human and social capital, their culture and knowledge constitute *social pillar.*

(ii) Nature, biodiversity or natural capital constitute *environment pillar.*

(iii) Money, goods or man-made capital constitute *economic pillar.*

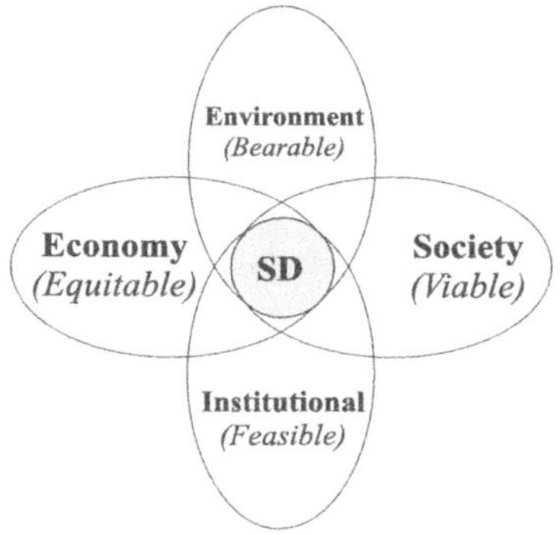


Fig. 6.1 *Pillars of Sustainable Development (SD)*

(iv) The actions of governments to implement sustainable development in their policies are considered the *institutional pillar*.

Social development, economic development and environmental protection are equivalent *objectives of sustainable development*. These are illustrated in Fig. 6.2.

Sustainable development involves the simultaneous pursuit of economic prosperity, environmental quality and social equity.

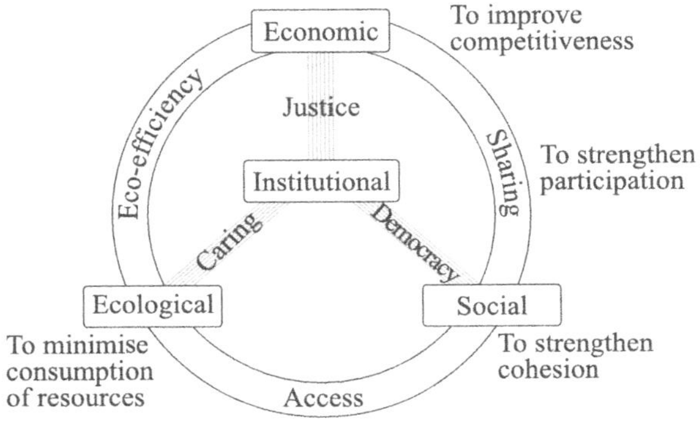


**Fig. 6.2** *Objectives of sustainable development*

***(A) Frameworks to Measure Sustainable Development*** (i) Agricultural, industrial or other production and consumption activities are *driving forces*. (ii) They impose stresses or *pressures* on the environment by releasing polluting substances.

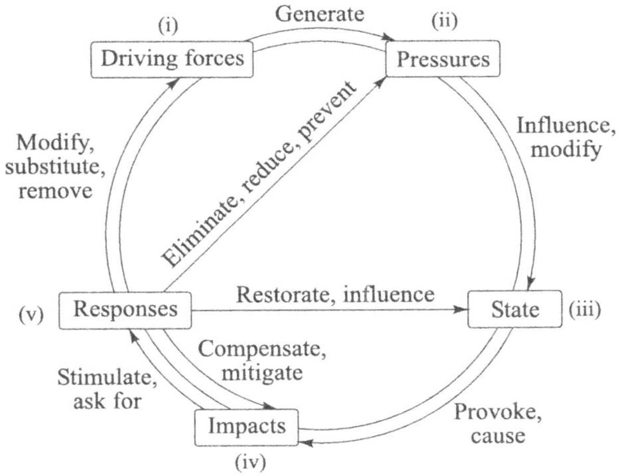


**Fig. 6.3** *DPSIR framework: Driving force (D) – Pressure (P) – State (S) – Impact (I) – Response (R)*

(iii) These in turn, affect the *state* of the environment like global warming. (iv) As a result, *impacts* (negative consequences on human health, economic loss in production activities, floods, etc.) are caused or provoked. (v) For pollution prevention, actions like sustainable use of resources are generated as *responses*.

***(B) Environmental Sustainability*** When the sum total of nature's resources (natural capital) is used up faster than it can be replenished, *degradation* of the environment occurs. However, if human activity only uses nature's resources at a rate at which they can be replenished naturally, *sustainability* occurs. These situations are summarised below in Table 6.1.

**Table 6.1** *Environmental sustainability*

| *Consumption of Renewable Resources* | *State of Environment* | *Sustainability* |
|---|---|---|
| (i) Less than nature's ability to replenish | Environmental renewal | Environmentally sustainable |
| (ii) Equal to nature's ability to replenish | Environmental equilibrium | Steady-state economy |
| (iii) More than nature's ability to replenish | Environmental degradation | Not sustainable |

***(C) Major Obstacles in the Path of Sustainable Development in India*** The major obstacles in the path of sustainable development are

(i) *Population explosion*
(ii) Absence of adequate political and *industrial willingness* for sustainable future
(iii) Non-availability of eco-friendly and resource efficient *technology*
(iv) Non-availability of sufficient *funds*
(v) Insufficiency of environmental *awareness*, non-conservation of resources
(vi) Absence of appropriate *land-use* planning
(vii) Absence of strict environmental *laws* and practices; absence of practice of effective methods of pollution control

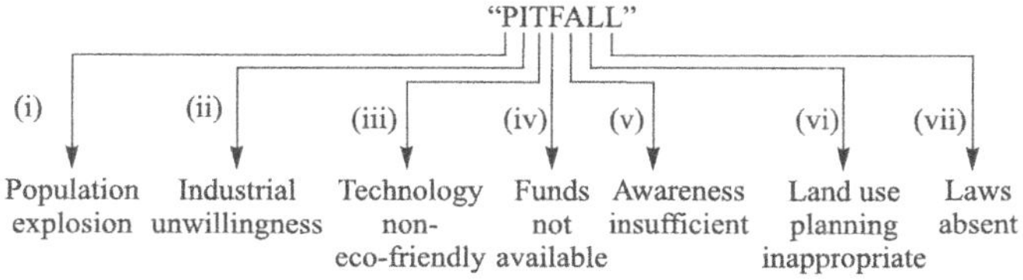


**Fig. 6.4** *Obstacles for sustainable development (India)*

***(D) Sustainable Lifestyle*** A lifestyle that attempts to reduce an individual's or society's use of the earth's natural resource and his/her own resources is known as a sustainable lifestyle.

***(E) Equitable Use of Resources for Sustainable Lifestyle*** Equitable use of resources for sustainable lifestyle requires that the rate of use of renewable resources do not exceed regeneration rates and rates of use of nonrenewable resources do not exceed rates of development of renewal substitutes.

Some examples of lifestyle changes which we can adopt to assist in environmental sustainable development are

(i) installing a rainwater harvesting tank,
(ii) washing our car on the lawn,
(iii) growing our own vegetables in the back garden,

(iv) using cloth bags for shopping and their re-use, instead of use of plastic bags and discarding them in environment, and

(v) installing energy-saving light bulbs in our homes, schools, colleges, offices, etc.

# Case Studies

### (i) Sustainable Development and Green Innovations

In today's globalised world, an essential precondition is environmentally sustainable industrial development, which can preserve the long-term interests of communities who depend on the industry as well as societies whose livelihood are affected by the operations of the industry. Societal expectations, competition and the health of business environment influence decisions, actions and their ultimate impact of businesses. The above aspects can be brought together in the framework of responsible competitiveness that can help in achieving environmental and social improvement, and economics of scale.

[Raja Rajeswari et al., Global Conference on Innovations in Management, London, UK, 2011]

### (ii) Constructed Wetland: A Cost-effective Alternative for Wastewater Treatment

The city of Arcata, California, was faced with a dilemma in 1974: buy into an expensive regional sewage processing plant to bring their wastewater discharge into Homboidt Bay up to an acceptable quality or use some other alternative. They used wastewater to create and nourish a wetland to provide prime wildlife habitat and recreation for the community. Simultaneously, the wetland helped in the purification of waste water.

[Amanda Suvtari http//www.ecotipping points.org]

### (iii) Sustainable City

Portland is the 25th largest metropolitan area in the US. Portland was experiencing the typical problems of urban growth, such as urban sprawl, traffic congestion, and pressure on natural areas within the city. The process in which the spread of development across the landscape far outpaces population growth is termed *sprawl*. It is characterised by four dimensions: (i) population widely dispersed in low-density development, (ii) widely separated homes, stores, and workplaces, (iii) a network of roads marked by poor

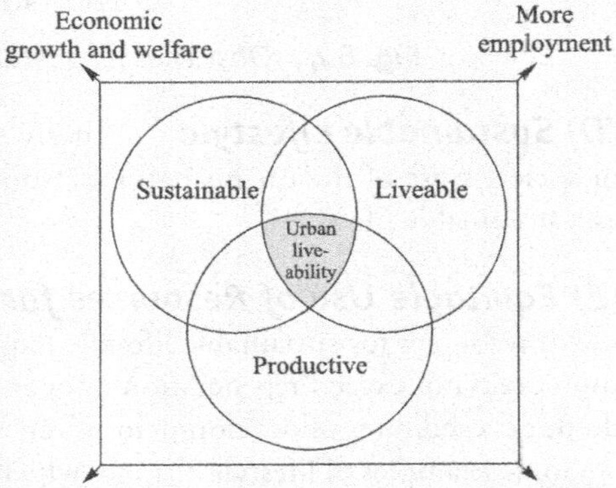


**Fig. 6.5** *Determinants and advantages of urban liveability*

connections and huge superblocks that concentrate traffic onto a few routes, and (iv) absence of activity centres.

Now Portland provides an excellent example of *intelligent urban planning.* They do so through an involving process of collaboration between citizens, city officials, planners and other professionals to understand and work with fundamental connections between transportation, economics, employment, population and land use.

A movement in portland towards *urban liveability* consisted of

(a) orienting growth clustered along transportation corridors,
(b) creating various options for public transport,
(c) encouraging compact urban design, and
(d) proactively anticipating long-term growth.

They have been able to reduce pollution, traffic congestion, generate employment, raise economy and build a sense of community.

[Amanda Suutari; http: //www.ecotipping points.org]

## 6.3 URBANISATION

From country to country, the definition of *urban* varies widely. Some countries distinguish between urbal and rural based on

(i) Size or density of localities
(ii) Percentage of persons who are not dependent on agriculture
(iii) Administrative considerations (only major cities are classed as urban)

An *urban area* means a town or a city plus its adjacent suburbs with a population of more than 2500 people. In contrast, a rural area means an area with less than 2500 people and less buildings.

Urban areas grow in two ways:

(i) Natural increase of its population by births
(ii) Immigration, mostly from rural areas [this is the biggest cause of urban growth]

*Urbanisation is defined as movement of people from rural to urban areas with population growth equating to urban migration* or it can also be defined as *the physical growth of urban areas as a result of global change.*

Cities with populations greater than 10 million people are known as *megacities.*

For the first time in history, more than half of the world's population is living in towns and cities in 2008. This number will swell to almost 5 billion by 2030, with urban growth concentrated in Asia and Africa.

### 6.3.1 Causes of Urbanisation

People move from rural areas to urban areas because cities offer more favourable conditions for the resolution of environmental and social problems than rural areas. A few specific reasons for urbanisation are summarised below:

(i) People move into cities to seek jobs and income.
(ii) With good governance, cities can deliver education, health care and other services more efficienty than rural areas.

(iii) Cities provide opportunities for women's empowerment and social mobilisation.
(iv) Density of urban life relieves pressure on areas of biodiversity and natural habitats.
(v) It is through cities that foreign money flows into a country (whether the source is tourism or trade).
(vi) Restaurants, movie theaters, theme parks and other varieties of entertainment are available in cities.

### 6.3.2 Drawbacks of Urbanisation

Often people who leave rural areas to find better jobs in the city have no choice but to settle in slums, where they lack access to decent housing, drinking water, sanitation, health care and education.

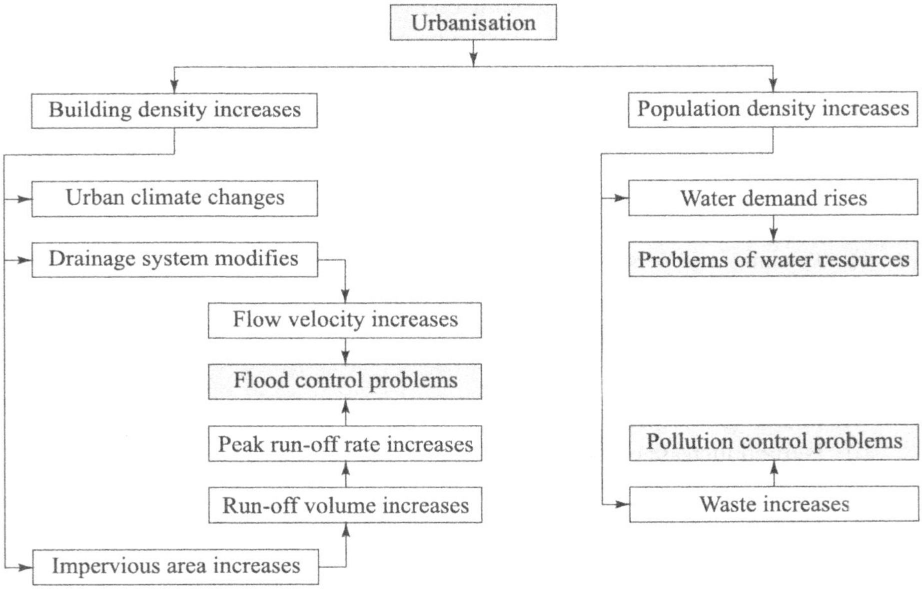


**Fig. 6.6** *Drawbacks of urbanisation*

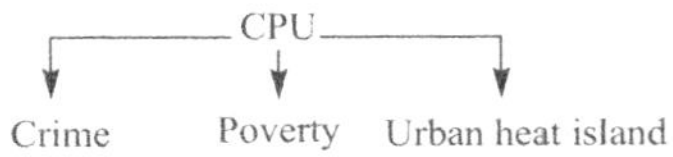


**Fig. 6.7** *Consequences of urbanisation*

**(i) Crimes** Chances of robbery, murder, assault, etc., increases with unplanned urbanisation.

**(ii) Poverty** Poverty is growing faster in urban than in rural areas. One billion people live in urban slums, which are typically overcrowded, polluted and dangerous. Urban areas are not self-sustaining. They survive only by importing food, water, energy, etc. However, they also produce large quantities of waste.

**(iii) Urban Heat Island** As urban and industrial areas are developed, the majority of the sun's energy is absorbed by urban structures and asphalt. Thus, during warm daylight hours, less evaporative cooling in cities allows surface temperatures to rise higher than in rural areas. Additional city heat is given off by vehicles and factories as well as by domestic and industrial cooling and heating units. This effect causes the city to become 1 to 6°C warmer than surrounding landscapes. Impacts also include intensification of carbon dioxide emissions and reducing soil moisture. The *urban heat island* has become a growing concern and is increasing over the years.

Urbanisation is also responsible for pollution (air, water, noise, etc.) and traffic congestion.

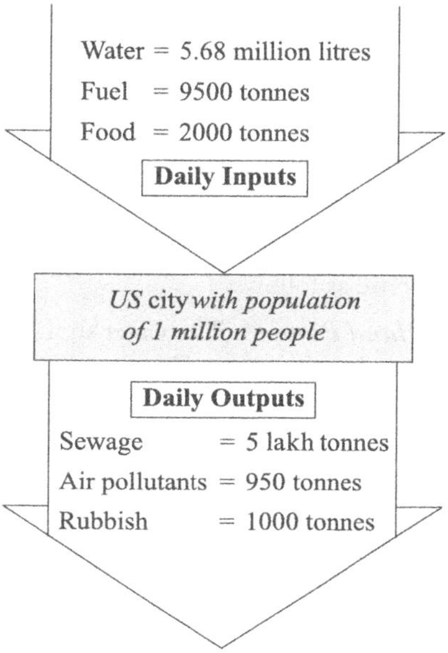


**Fig. 6.8** *Not self-sustaining urban city*

### 6.3.3 Positive Environmental Effects of Urbanisation

Some positive effects of urbanisation are the following:

(i) Prevention of overpopulation in future is possible because the birth rate of new urban dwellers falls immediately to the replacement rate, and keeps falling.

(ii) Spending per person on *environmental* protection is higher in urban areas.

(iii) Urbanisation puts a stop to slash and burn *agriculture* and other destructive subsistence farming techniques.

(iv) *Recycling* is more feasible economically because of large concentration of materials.

(v) It minimises *land* use by humans, leaving more for nature.

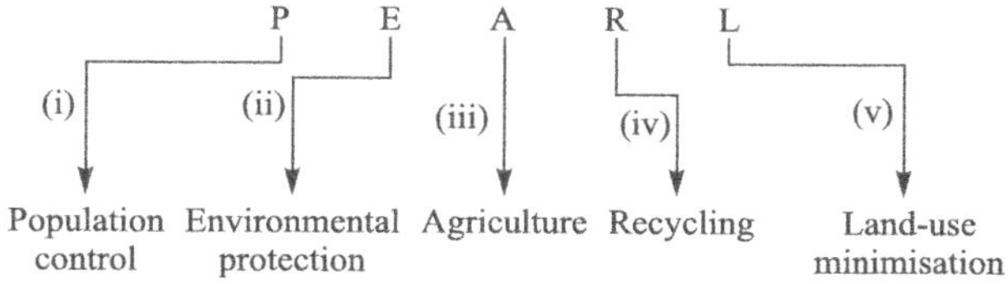


**Fig. 6.9** *Positive environment effects of urbanisation*

## 6.4 WATER CONSERVATION

### 6.4.1 Rainwater Harvesting

*Rainwater harvesting means collecting rainwater and storing/conserving it for a later use.*

## (A) Various Methods of Rainwater Harvesting

There are two main methods of rainwater harvesting:

**(i) Storage of Rainwater on the Surface for Future Use** In this method, traditional water-harvesting structures like artificial lakes, ponds, etc., are used as such or after proper revival.

**(ii) Recharge of Ground Water** The structures used for recharge of ground water are as follows:

***(a) Hand Pumps*** The water should pass through a filter bed before percolation in existing hand pumps. They are used for recharging aquifers.

***(b) Pits*** They are 1–2 m wide and 3 m deep. They are also back filled with gravel and coarse sand to aid filtration before percolation to the ground. They are used for recharging a shallow aquifer.

***(c) Dug Wells*** The rain water, after filtration, is put into existing dug wells for storage.

***(d) Roof-Top and Road-Rop Collection of Rainwater*** In urban areas, these methods are very useful to recharge aquifers.

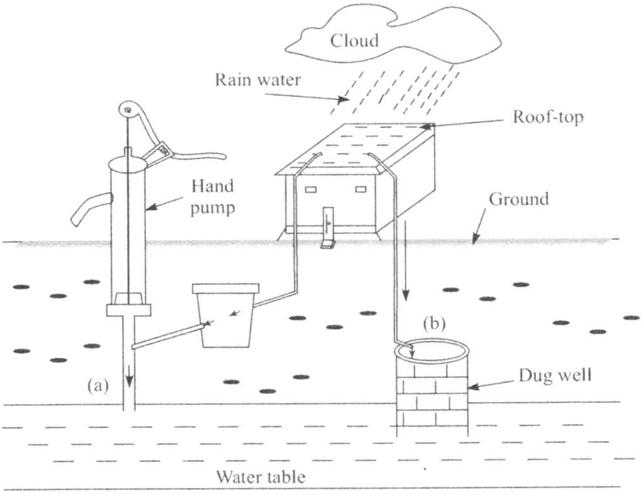


**Fig. 6.10** *Roof-top rainwater harvesting by recharging (a) through hand pump, or (b) through an abandoned dug well*

# Case Studies

### (i) Rainwater Harvesting to Replenish Underground Water (Rajasthan, India)

As a consequence of logging the forest, the wells in Rajasthan's Alwar District had dried up, thrusting the people into hopeless and unavoidable poverty. The revival of *johad* (a crescent half-moon shapped dam of earth and rocks) to capture rainwater for recharging the underground water supply provided a turning point that brought the wells back to life. And with the water came a better life for the people. It started in the Gopalpura village. Now, nearly a thousand villages are following this practice.

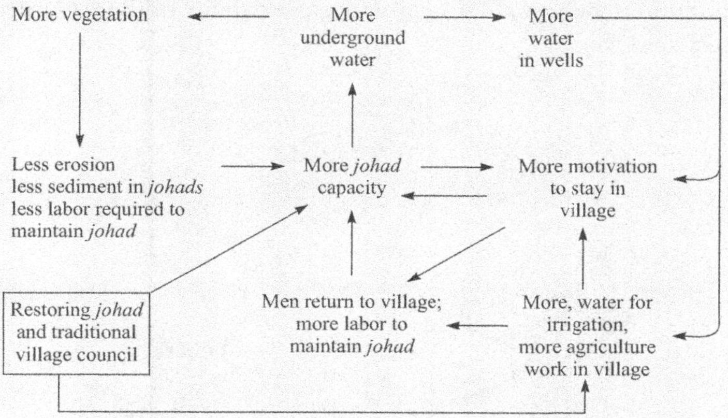


**Fig. 6.11** *Rainwater harvesting in Rajasthan [http: //www.ipptaonline.org/circular]*

**(ii) Water Conservation in India in the Pulp and Paper Industry**

The annual paper production will be double from the existing 10–11 million tones by 2020. However, the availability of water will be the same or lower. Thus, for sustainable development, it is essential to work towards stringent water management. The available options are use, reuse, recycle, treatment of waste water, reduction of consumption of fresh water, etc.

## 6.4.2 Watershed Management

*Watershed is a geographic area of land that collects, stores, and releases water.*

The area collects water from rain, snow, etc. This collected water is stored in lakes and ponds. The stored water is then released through streams, rivers, etc. Thus, a *watershed* means a land from where water drains into a particular stream, lake, river, estuary and even the ocean. *Drainage basin or catchment basins* are other terms used interchangeably with watershed.

A watershed is a natural system. All lands, humans, wildlife and their activities are part of one or the other watershed. When a watershed is kept in good condition, it provides sustainable benefits to humans, wildlife, etc.

*Watershed management refers to the conservation, protection and restoration of a watershed to secure water—both in quantity and quality for drinking, sanitation and agriculture in a sustained manner.*

### *(A) Strategies for Watershed Management: CUBS*

***C:*** ***Cost-benefit*** (i) Do the *cost-benefit analysis* of ecological effects of alternative actions.

*U:* ***Usage goals*** (ii) Develop goals for use of water and land resources.

*B:* ***Background*** of people and functions of watershed (iii) Find out present and historical structure and functions of the watershed system among people of diverse social backgrounds and values.

*S:* Solutions for ***sustainability*** (iv) Suggest innovative cost-effective *solutions* of optimum resource use for long-term sustainability of the watershed and the ecosystem.

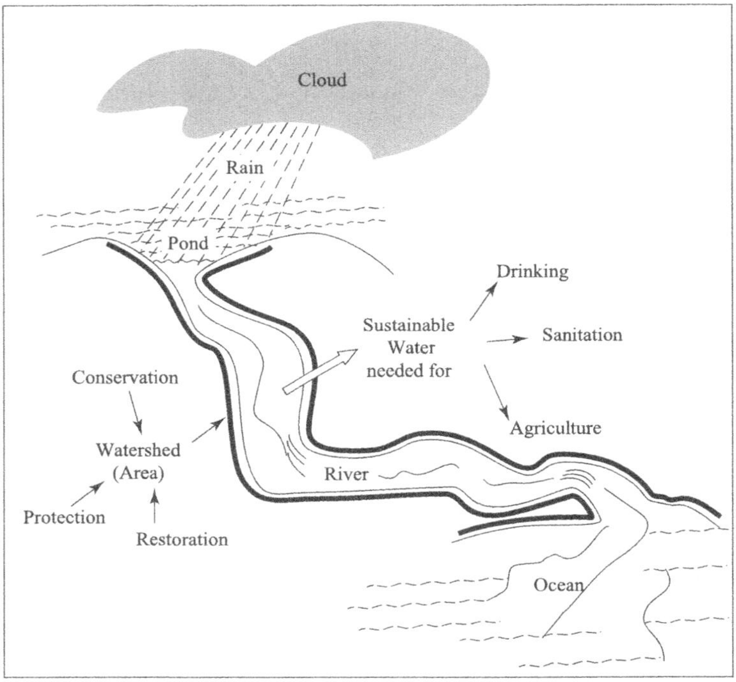


**Fig. 6.12** *Watershed and its management*

***(B) Action Plan for Watershed Management*** Watershed management can be done through MCD:

(i) ***Mapping*** of the watershed area, planting trees and grass for enhancing seeping of water to the ground and for preventing water contamination, torrents and landslides. Thus, plantation helps in the recharging of groundwater.

(ii) ***Constructing*** a series of long *trenches* and mounds along hill contours to hold rainwater and allowing it to percolate into the ground.

(iii) Making ***dams*** for preventing large amounts of water from rushing down the hillside. This helps in recharging of an underground aquifer. Moreover, streams and rivers flow for the whole year.

### (C) Need of Watershed Management: MUD

***M:*** ***Misuse*** (i) The *misuse* of Himalayan slopes are increasing. Our water regimes are threatened resulting in the depletion of water resources.

***U:*** ***Unsustainable*** (ii) The society is becoming *unsustainable.*

***D:*** ***Damage*** (iii) The *damage* to irrigation systems and reservoirs are increasing. Every year, during the 'rainy season', costs for controlling floods is increasing.

### (D) Advantages of Watershed Management

The main benefits of watershed management are summarised below:

(i) Watershed management reduces water shortage during summers by facilitating recharge of groundwater.

(ii) It provides wildlife and fish, food, habitat, and resting areas, etc., for sustainable survival.

(iii) It protects stream and river banks from erosion.

(iv) By retaining water and releasing it during summers, watershed management reduces chances and associated damages of floods.

(v) It provides good quality water and food for human use.

## 6.5 RESETTLEMENT AND REHABILITATION OF PEOPLE; ITS PROBLEMS AND CONCERNS

The displacement or the involuntary and forced relocation of people is the most significant negative impact of many development projects such as power plants, oil refineries, fertiliser and chemical industries, river-valley projects, dams, reservoirs and mining.

This involuntary movement of people from one place to another, for *resettlement*, gives rise to the following problems:

### (A) Economical Problems

(i) *Compensation* is awarded only to real owners of property taken for developmental projects. Tenants, wage labourers, artisans and encroachers are not considered eligible for compensation but ironically they are the most vulnerable and in need of support.

(ii) Families are forced to face long-term *hardships*.

(iii) Under the *acquisition process*, community assets and common resources like grazing grounds and forests are not compensated. But these areas were critical for the livelihood of the poorest.

(iv) Sources of *income* are lost or ruined or jeopardised,

(v) The *resettlement cost* is generally underestimated and under-financed. Whenever the development project runs into financial problems, mostly the resettlement and rehabilitation budget is reduced.

(vi) Most programmes have failed to facilitate self-employment in critical areas of employment, skills and capacity building.

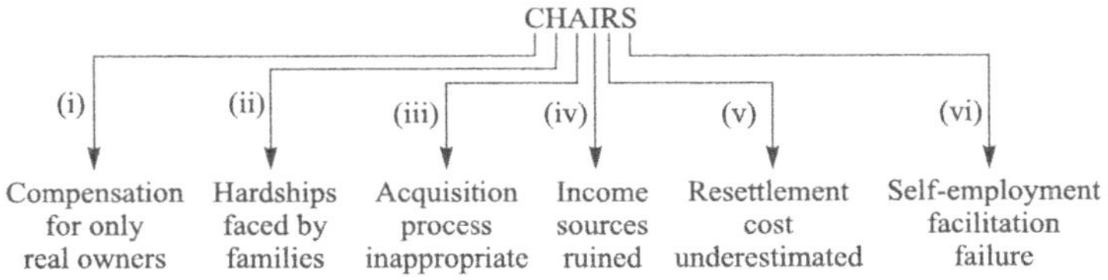


**Fig. 6.13** *Economical problems of resettlement and rehabilitation*

## (B) Social Problems

(i) *Social networks* are disturbed.

(ii) *Large families and communities* are broken up and resettled over a wide area.

(iii) Poor people are transplanted from a social ecology in which they were primary actors to one in which they are *aliens*. Such people are very vulnerable and forced to become underclass members of a new socio-cultural milieu. Traditions are weakened. Cultural identity is lost.

(iv) Generally, *participation* of the victims has been superficial or treated as unimportant by those responsible for the development project.

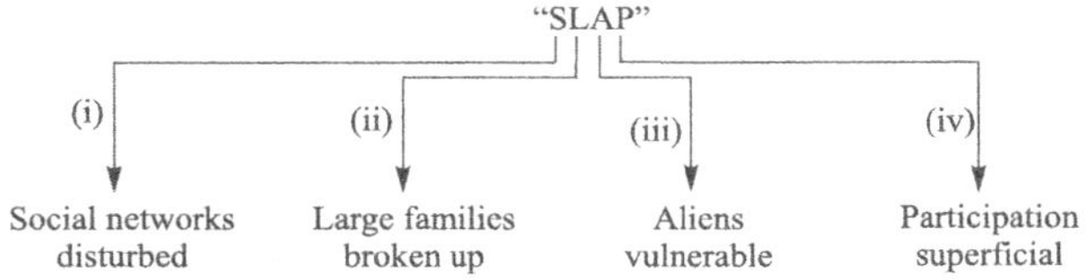


**Fig. 6.14** *Social problems of resettlement and rehabilitation*

## (C) Educational, Psychological and Environmental Problems

(i) Effective resettlement is never ensured nor implemented because of institutional weaknesses, confusions between various departments, absence of policy, absence of legal instruments and absence of an effective mechanism to monitor compliance. All these result in

 (a) Interruption of *education of children*

 (b) Separation of joint families resulting in *psychological tensions* among members of the family

 (c) *Unemployment*, *debt bondage*, hunger and *assetlessness*

(ii) Resettlement sites are not selected with respect to availability of livelihood opportunities. Even proper residential houses are not provided. The People are forced to live in small, temporary structures made of tin or other inappropriate material and design. *Basic amenities* and essential infrastructure such as health, schooling and credit are not available in these resettlement sites.

(iii) Planning time for most of the projects is long. There is a huge time gap between initial notifications and the actual physical relocation. The interim period is full of *uncertainties* and enormous psycho-social anxieties for the to-be-relocated communities. People in such areas thus become poorer even before they are relocated for settlement.

## *(D) Rehabilitation and Resettlement Concerns*

Rehabilitation should be envisioned as a process that would reverse the risks of resettlement.

The objective of the Ministry of Rehabilitation and Resettlement (Government of India) is to transfer the benefits, in lieu of the losses occurred to displaced people due to involuntarily displacement.

This objective may be ensured by implementing the following policies:

(i) To keep them intact in a *family* or community in which they were settled prior to being displaced

(ii) To provide them essential *infrastructure* such as health, schooling and credit in resettlement sites

(iii) To govern the displacement process by *laws* for avoiding possible problems, to create new rights for them that will enable them to directly share the benefits of the development project and to provide adequate compensation

(iv) To relocate them to a *locality* of their preference

(v) To provide them increasing *incomes* through opportunities of employment and livelihood, to provide them opportunities for the enhancement of capabilities and to improve their standard of living

(vi) To give them proper *participation* and choice for their resettlement and rehabilitation

(vii) To provide them *social infrastructure* and *community services*

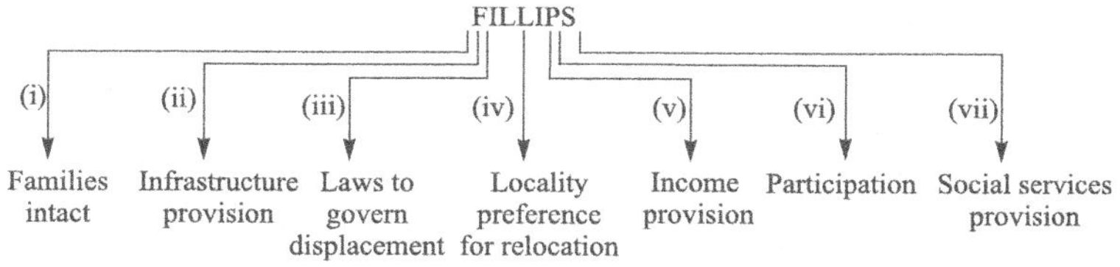


**Fig. 6.15** *Policies to ensure fulfillment of objectives of rehabilitation and resettlement*

### 6.5.1 Environmental Ethics: Issues and Possible Solutions

The work '***ethic***' is derived from the Greek work '***ethos***' which means the character of a person as described by his or her actions.

*Ethics* deals with moral duty and obligations. It gives rise to a set of values, which are used to judge whether one's behaviour or conduct is right or wrong.

Trust, honesty, justice, trustworthiness, competence and accountability are the basis of ethics.

***(A) Issues*** In early 1970s, the world started realising the dangers of human actions on the environment. The main reasons of such concerns were the following issues whose effects were changing the global environment: "ACNE"

**A** (i) Reduced purification of ***atmosphere*** because of increased deforestation

**C** (ii) Increased $CO_2$ ***concentration*** and presence of harmful elements in atmosphere due to *enhanced* burning of fossil fuels

N (iii) Depletion of ***natural resources*** at a rapid rate due to consumption by increasing population

E (iv) Deteriorating quality of ***environment*** because of urbanisation, industrialisation, deforestation, consumerism, etc.

These issues result in the following serious consequences:

Greenhouse effect, global warming, acid rain, ozone layer depletion, etc.

***(B) Possible Solutions*** Awareness of consequences like global warming, etc., has led the society to adopt environmental ethics. These environmental ethics give equal importance to growth and sustainability.

The Following ethical guidelines are very useful as possible solutions for most environment-related problems:

(i) Humans must keep themselves *informed* about ecological changes as a result of development.

(ii) Humans must *understand* that they cannot survive alone on the earth. The earth is the habitat of all living species. Living species depend on each other for survival.

(iii) All humans must be cooperative, honest, *affectionate* and polite to nature and society.

(iv) Humans must *respect* nature. They are part of it.

(v) Humans should *act* locally for protection of environment globally.

(vi) Humans must *protect* natural resources and energies as resources are depleting fast.

(vii) Humans must *reserve* scarce resources for the future and the future generations.

(viii) Humans must *involve* themselves in the care of the earth.

(ix) Everyone should *try* to bring about a change in the attitudes of other persons. They should collectively work towards conservation, protection and restoration of the environment.

(x) Each human should *plant* at least one tree on his or her birthday.

(xi) Humans should *purchase* only that much which is essential because excess turns into wastage.

## Case Study

### Resettling Project-Affected People in India

A concern for the fate of people who are forced to relocate, leaving behind forever their homes, land and communities is a contemporary issue of great importance. A variety of irrigation, hydropower, mining, and thermal-power projects are responsible for resettlement. The affected people face risks like insecurity of livelihood; problems of access to common property and public services; health and nutrition hazards; loss of community structures and social unrest, etc. The simple solution is *proactive planning* so that corrective measures can be taken.

[http: //www.usp.ac.fj]

# 6.6 SOCIAL ISSUES AND THE ENVIRONMENT

## 6.6.1 Climate Change

***Weather** is the reflection of atmospheric humidity, temperature and rainfall.*

***Climate** is the average weather pattern over longer duration in a place.*

*Climate change* is reflected from the following facts:

(i) Since the late 19th century, the earth has warmed by 0.3 to 0.6°C on an average.

(ii) By the year 2100,

  (a) Temperatures would rise by 1 to 3.5°C

  (b) Global mean sea levels would rise by 15 to 95 cm

*Causes of climate change* on a global scale are linked with changes in the amount of heat that is either let into the earth system or let out of the earth system.

Depending on proximity to oceans and altitude and amounts of sunlight received by different regions of the world, climate differs from place to place.

For example, hot climates are due to greater concentrations of greenhouse gases in the atmosphere which reduce the amount of heat that is let out of the atmosphere.

*Ill-effects of climate change* are linked to

(i) cyclones,

(ii) floods,

(iii) dry or wet spells of rainfall, and

(iv) cold and hot spells of temperature.

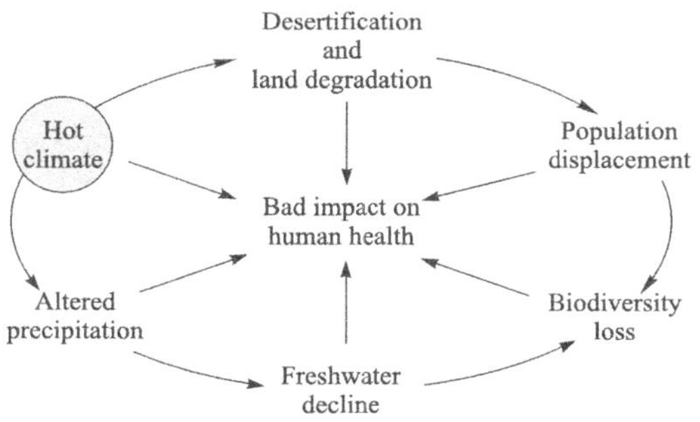


**Fig. 6.16** *Impacts of climate change*

**Table 6.2** *Ill-effects of climate change*

| *Phenomenon* | *Probability* | *Likelihood* |
|---|---|---|
| (i) More extreme sea levels | >66% | Likely |
| (ii) More intense tropical cyclones | " | " |
| (iii) More area hit by drought | " | " |
| (iv) More heavy rain events | >90% | Very likely |
| (v) More warm spells and heat waves | " | " |
| (vi) Warmer days/nights | >99% | Virtually certain |

### *(A) Impacts of Climate Change*

**(i) Positive Impacts** Reduced deaths from cold and higher agricultural output in northern regions (at least for a while) are few positive impacts of global warming.

### (ii) Negative Impacts

(a) Submergence of low-lying islands (especially in the Pacific), vast saline inundations in countries like Bangladesh, etc., are linked to rise in sea level. (Global warming is responsible for melting of ice caps and glaciers, which lead to rise in sea level.)

(b) The tourism industry will suffer in parts of southern Europe because of change of climate towards hotter than ever before (prediction by the European Acacia Project).

(c) Damaging changes in wildlife behaviour like failure of Scottish seabirds to raise young during the 2004 breeding season.

(d) The poorest of the poor are most likely to be hit by the impacts of climate change.

***(B) Solutions of Climate-change Problems*** Some of the simple solutions are

(i) Reduce the emission of greenhouse gases
(ii) Use renewable energy resources
(iii) Use energy efficient technologies

**Example 1** *What are the important mitigation technologies and commercially available latest technologies for tackling the climate change as per Fourth Assessment Report of Intergovernmental Panel on Climate Change (IPCC)?*

**Solution** As per the IPCC Fourth Assessment Report, the important mitigation technologies and practices presently commercially available are described below:

***(i) Industry***

- Recovery of heat and power
- Recycling of material
- Green technology
- Efficient equipment, etc.

***(ii) Buildings***

- Passive and active solar design for cooling and heating
- Efficient electrical appliances
- Improved cooking stoves
- Efficient lighting and day lighting, etc.

***(iii) Transport***

- Use of public transport systems
- Cycling, walking, etc., nonmotorised transport
- Use of best fuel-efficient vehicles
- Biofuels
- Proper transport planning, etc.

***(iv) Energy supply***

- Renewable energy
- Improved efficiency in supply and distribution, etc.

***(v) Agriculture***

- Improved nitrogen fertiliser-application techniques to reduce $N_2O$ emissions,
- Improved rice-cultivation techniques and livestock-and-manure management to reduce $CH_4$ emissions,
- Dedicated energy corps to replace fossil-fuel use,
- Improved crop and grazing-land management, etc., to increase soil carbon storage,

***(vi) Waste***

- Recycling and waste minimisation
- Compositing of organic waste
- Waste incineration with energy recovery
- Waste-water treatment, etc.
- Landfill methane recovery

***(vii) Forests***

- Reduced deforestation
- Forest management
- Afforestation
- Reforestation
- Harvested wood-product management
- Use of forestry products for bio-energy to replace use of fossil fuel, etc.

## Case Study

**Climate Change, Perspectives from India**

Climate change is the biggest development challenge for the planet. There is not much difference between managing a local forest and the global climate—we need a framework which encourages cooperation. Then a country can have both growth and less carbon emissions. To tackle the impact of climate change on food security in India, it has been suggested to practice soil and water conservation. Small-scale industries emit substantial greenhouse gases and have the potential for saving huge amounts of energy.

[http: //www.undp.org.in]

### *(C) Environment Security and Climate Change*

The security of the entire global community is increasingly exposed to risk by humans through air, water and/or land pollution. Economic activities cause environmental changes that lead to conflict. This can be understood through the illustration shown in Fig. 6.17.

It is important for all countries to cooperate in order to reduce the effects of environmental degradation. Everyone should contribute by limiting greenhouse gas emission, conserving natural resources, and developing and sharing energy-efficient technologies.

[Elizabeth b. Chaleck;, http: //www.pacinst.org/reporst]

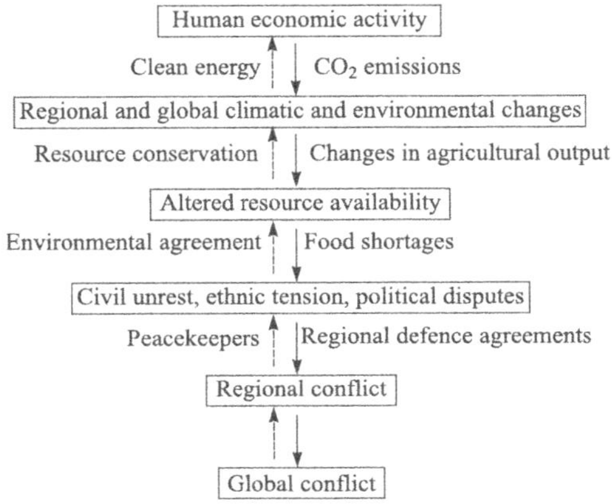


**Fig. 6.17** *Origin of conflict and methods to reduce them*

## 6.6.2 Global Warming—The Greenhouse Effect

*Greenhouse effect is a process by which infrared radiation leaving the earth's surface is trapped by some greenhouse gases; so the temperature is higher than it would be if direct heating by solar radiation were the only warming mechanism.*

Greenhouse effect is illustrated in Fig. 6.18.

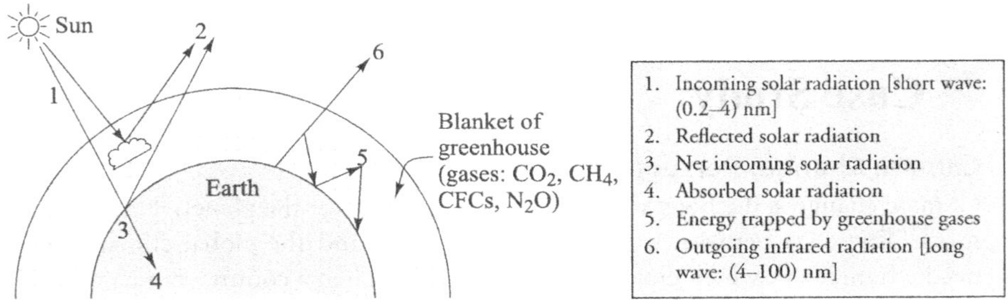


**Fig. 6.18** *Greenhouse effect*

1. Short-wave radiations from the sun penetrate the earth's atmosphere.
2. Some solar radiation is reflected by the atmosphere and the earth's surface.
3. Remaining solar radiation reaches the earth's surface.
4. Solar energy is absorbed by the earth's surface and warms it and is converted into heat causing the emission of long-wave (infrared) radiation back to the atmosphere.
5. Some of the infrared radiation is absorbed and re-emitted by the greenhouse gases. The direct effect is the warming of the earth's surface and the troposphere. The earth's surface gains more heat and infrared radiation is emitted again. By this process, some of the infrared radiation is trapped in the atmosphere by greenhouse gases causing a warming of the earth's climate.
6. Some of the infrared radiation passes through the atmosphere and is lost in space.

To sum up, the greenhouse effect is the rise in temperature that the earth experiences because certain gases in the atmosphere trap energy from the sun.

| Greenhouse effect | Natural | Enhanced |
|---|---|---|
| • Concentration of greenhouse gases: | Normal | Excessive |
| • Surface temperature of earth: | ≤14°C | (14.1 to 14.5)°C |

***Greenhouse gases*** are gases in an atmosphere that absorb and emit radiation within the thermal infrared range.

***Examples:*** Carbon dioxide, nitrous oxide, methane, water vapour, ozone

Greenhouse gases greatly affect the temperature of the earth. Without them, the earth's surface would be (on average) about 33°C colder than at present, and life would be impossible.

### (A) Contribution of Greenhouse Gases to Global Warming

Greenhouse gases vary in their ability to absorb and hold heat in the atmosphere. For example, nitrous oxide absorbs 270 times more heat per molecule than carbon dioxide, and methane absorbs 21 times more heat per molecule than carbon dioxide.

The Global Warming Potential (GWP) depends on both the efficiency of the molecule as a greenhouse gas and its atmospheric lifetime. GWP is measured relative to the same mass of $CO_2$ and evaluated for a specific time scale. For example, if a greenhouse gas has a high GWP on a short time scale (say 20 years) but has only a short lifetime, it will have a large GWP on a 20-year scale but a less GWP on a bigger time scale. If a molecule has a longer atmospheric lifetime than $CO_2$, its GWP will increase with the increase in the time scale.

**Table 6.3** *Global Warming Potential (GWP)*

| *Greenhouse Gas* | *Atmospheric Life-time (years)* | *Global Warming Potential (GWP)* | | |
|---|---|---|---|---|
| | | *20 years* | *100 years* | *500 years* |
| (i) $CO_2$ | $10^4$ | 1 | 1 | 1 |
| (ii) $CH_4$ | 12 ± 3 | 72 | 25 | 7.6 |
| (iii) $N_2O$ | 114 | 289 | 298 | 153 |
| (iv) CFC–12 | 100 | 11000 | 10900 | 5200 |
| (v) HCFC–22 | 12 | 5160 | 1810 | 549 |
| (vi) $SF_6$ | 3200 | 16300 | 22800 | 32600 |
| (vii) $NF_3$ | 740 | 12300 | 17200 | 20700 |

### (B) Global Warming

*Global warming* means a rise in temperature over the earth's surface.

It is primarily caused by $CO_2$ emission by man-made activities.

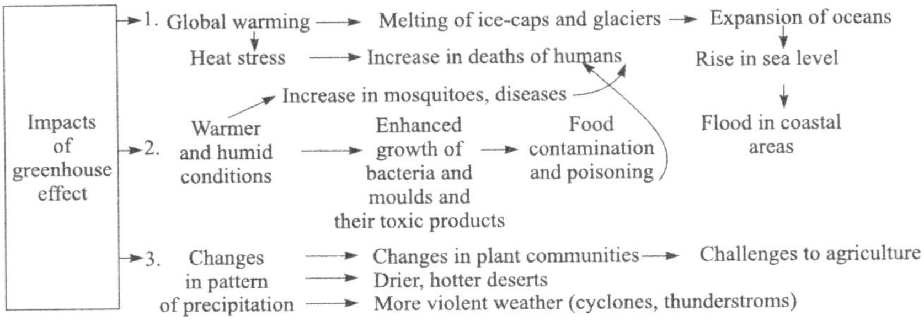


**Fig. 6.19** *Impact of global warming and greenhouse effect*

### (i) Consequences of Global Warming

(a) Melting of polar ice caps, and increase of sea/ocean levels
(b) Flooding of low-lying land
(c) Less water vapour in the atmosphere leading to more drought
(d) Causes extremes of weather hurricanes, flooding and droughts, difficulties in growing crops and survival problems

More industrialised countries are responsible for causing high levels of $CO_2$ in the atmosphere and less industrialised countries are also contributing by destruction of the rainforest.

Sustainable development is the only solution,
[http: //www.thw.coventry.sch.uk/]

### (ii) Remedial Measures

(a) Enhance energy efficiency during use by adding insulation to your walls, and by using CFL bulbs, etc.
(b) Reduce transport sector emissions by less and smart driving.
(c) Promote renewable energy (like solar energy) usage.
(d) Remove subsidies on fossil fuels.
(e) Favour sustainable agriculture.
(f) Recover methane emissions through waste management.
(g) Promote afforestation and reforestations—a single tree will absorb approximately one ton of $CO_2$ during its lifetime.
(h) Reduce energy consumption by using energy-efficient home appliances.
(i) Avoid methane production from biomass decay through controlled combustion.
(j) Enhance energy efficiency during generation, transmission and distribution.
(k) Reduce waste, prefer reusable products, recycle paper, plastic, metals, etc.
(l) Eat locally grown fruits and vegetables and not the imported ones. The latter requires the burning of fossil fuels for transport.

## 6.6.3 Acid Rain

*Acid rain is rain which is unusually acidic (pH of less than the natural range of 5 to 6); caused mainly by atmospheric pollution with sulphur dioxide and nitrogen compounds.*

**Table 6.4** *Environmental effects of acid rain*

| *pH Value* | *Examples* | *Environmental Effects* |
|---|---|---|
| 4.2–4.4 | Acid rain | All fishes die at pH = 4.2 |
| 4.5 | Acidic lake | Frog eggs, tadpoles, cray fish die at pH = 4.5 |
| 5.6 | Clean rain | |
| 6.5 | Healthy lake | |
| 6.5–6.8 | Milk | |
| 7 | Pure water | |
| 8 | Sea water | |

*Notes:* pH = – logarithm [hydrogen-ion concentration]

As the pH is log $[H^+]$, so each pH unit represents a tenfold change:

(i) A pH of 5 is ten times more acidic than a pH of 6.

(ii) A pH of 4 is hundred times more acidic than a pH of 6.

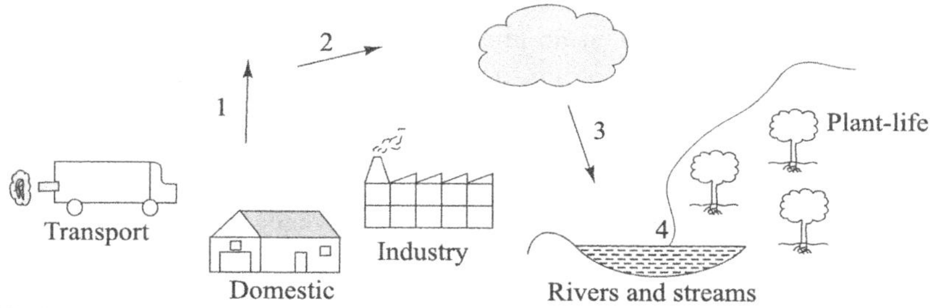


1. Acidic gases (sulphur dioxide and nitrogen oxides) released into atmosphere from industries, vehicle's exhaust, etc.
2. Acidic gases carried by the wind
3. Acidic gases dissolve in rainwater to from acid rain
4. Acid rain kills plant life, pollutes rivers and streams and erodes stonework.

**Fig. 6.20** *Causes and effects of acid rain*

# Case Study

### Acid Rain and Taj Mahal

Tourist traffic is not allowed near the Taj Mahal in an effort to control the deleterious effects of pollution. The degradation of the Taj Mahal's marble facades has still not slowed down. This is due to acid rain generated from local foundries and an oil refinery. The once brilliant white Taj Mahal has been losing its luster, dulling into a sickly pale shade. Two common air pollutants, $SO_2$ and $NO_X$, cause acid rain.

[John mink, http: //archive.cyark.org]

## 6.6.4 Ozone-Layer Depletion (Ozone Hole)

***(A) Ozone*** Ozone ($O_3$) is an allotropic form of oxygen ($O_2$). It is a pale blue gas. It helps in sustaining life on earth by filtering out the sun's harmful ultraviolet radiation.

***(B) Ozone Layer*** The total amount of ozone in an overhead column of the atmosphere is measured in dobson unit (after the atmospheric ozone pioneer GMB Dobson). One Dobson Unit (DU) indicates that a 0.01 mm thick ozone layer would be formed if ozone is compressed into one layer at 0°C and 1 atm pressure. Across the globe, in the stratosphere, the average thickness of the ozone layer is about 3 mm at 0°C and 1 atm pressure (or about 300 DU). The stratospheric pool of ozone is known as the *ozonosphere.*

***(C) Ozone Hole*** When the level of ozone in the stratosphere falls below 200 DU, it is considered to represent the beginnings of an ozone hole.

***(D) Causes of Ozone-Hole Formation*** The gradual thinning of the ozone layer and ozone-hole formation occurs by the destruction of ozone due to its reactions with nitric oxide, chlorine, hydroxyl radicals, etc., in the stratosphere.

Flying of supersonic aircrafts, nuclear explosions and various chemical/photochemical reactions in the atmosphere generate nitric oxide. Burning of biomass generates hydroxyl radicals. Volcanic activity releases chlorine in the atmosphere. Chlorofluorocarbons (CFCs), fluorochloro methane (freons), difluorodichloro methane ($CF_2Cl_2$) and fluorochloroform ($CFCl_3$) release chlorine by ultraviolet radiation induced homolytic cleavage in the atmosphere.

It is estimated that about 6.5% of the total ozone-layer depletion is due to chlorine radicals from various CFCs. The chemical reactions leading to the destruction of ozone layer by CFCs are given below:

Step (a): $CCl_2F_2 + h\nu \rightarrow Cl^\bullet + {}^\bullet CClF_2$

Step (b): $O_3 + Cl^\bullet \rightarrow O_2 + ClO^\bullet$

Step (c): $ClO^\bullet + O^\bullet \rightarrow Cl^\bullet + O_2$

Net reaction: $O_3 + O^\bullet \xrightarrow{Cl} 2O_2$

As $Cl^\bullet$ atoms are regenerated in step (c), a long-chain process is followed which keeps on consuming ozone. It is estimated that each atom of chlorine can destroy one lakh ozone molecules when they diffuse to the stratospheric level.

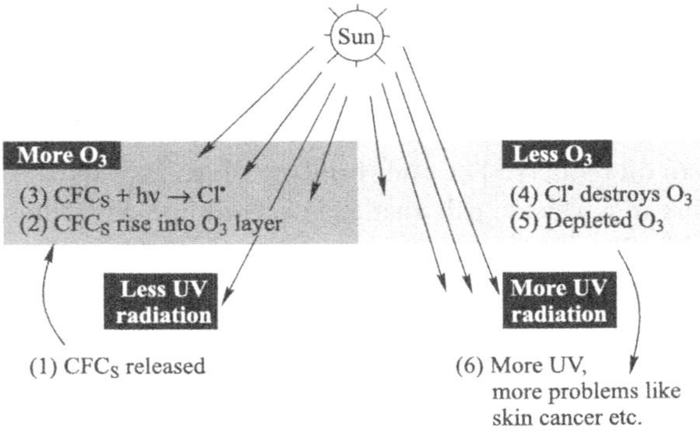


**Fig. 6.21** *Ozone depletion process*

***(E) Problems Associated with Ozone-Layer Depletion*** The ozone layer absorbs most of the harmful ultraviolet radiations coming from the sun in the region (220–330) nm.

In the absence of an ozone layer, these ultraviolet radiations could cause the following problems:

(i) Swelling of *skin* and skin cancer; skin aging, burning sensation
(ii) Death of *phytoplanktons* in marine environment (the sole producers) leading the entire ecosystem to collapse
(iii) Reduction in the body's ability to fight off disease, as UV suppresses the *immune system*; premature aging.
(iv) Inhibition and alteration of DNA replication and formation of DNA adduct; leukamia, breast *cancer*
(v) Visual impairment, dizziness, cataracts of *eyes*
(vi) Damage to plants; reduction in crop yields; faster deterioration of *paints, fabrics, plastics*

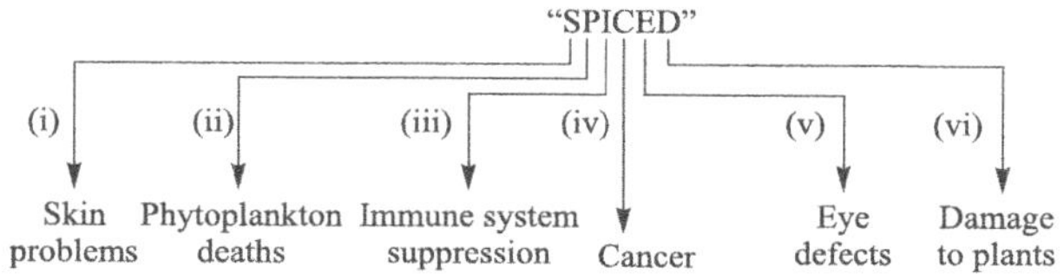


**Fig. 6.22** *Problems associated with ozone-layer depletion*

### *(F) Remedial Measures to Control the Depletion of Ozone Layer*

(i) Avoid any fire extinguisher that contains bromine-based halons. Preferably use water, carbon dioxide or dry chemical fire extinguishers.
(ii) Spread awareness about the restricted use of CFCs for the healthy survival of mankind.
(iii) Avoid purchasing and using refrigerators, air conditioners, etc., which use CFCs, freons, etc., as coolants.
(iv) Avoid purchasing and using pressurised aerosol cans which use CFCs, freons, etc., as propellants.
(v) Ban atmospheric nuclear explosions, as they emit $NO^{\bullet}$ and deplete the ozone layer.
(vi) Reduce the air traffic of supersonic aircrafts that fly at the ozonosphere altitude, as they release large amounts of $NO^{\bullet}$ and deplete the ozone layer.
(vii) Facilitate advanced research to plug the ozone holes that have already been formed.

## Case Study

### Ozone-Layer Depletion

Concern over the annual ozone hole over Antarctica led the Reagan administration to agree to the Montreal Protocol (1987). This was the landmark international agreement to phase out chlorofluorocarbons (CFCs) and other ozone-depleting compounds.

It has been recently demonstrated (2010) that the ozone hole is able to influence the tropical circulation and increase rainfall at low latitudes in the southern hemisphere. Thus, the **ozone hole is a big player in the climate system.**

This means that international agreements about reducing climate change should concern stopping ozone-layer depletion along with reducing $CO_2$ emissions.

[Anthony Watts, April 2011, http: //wattsupwiththat.com/2011]

## 6.6.5 Nuclear Accidents and Holocaust

### *(A) Nuclear Accidents*

Nuclear accidents can occur at any stage of the nuclear fuel cycle. They may also result from the failure of nuclear devices.

*Types of nuclear accidents* are discussed below:

**(i) Lost-Source Accidents** In such accidents, a radioactive source is lost, stolen or abandoned. The source then might cause harm to the environment or humans.

**(ii) Human-Error-Based Accidents** Humans have made errors while

(a) Calculating the activity of a teletherapy source for radiotherapy
(b) Attempting to service nuclear equipment and plants
(c) Performing nuclear experiments.

For instance, in 1946, Canadian Manhattan Project physicist Louis Slotin was performing "tickling the dragon's tail" experiment. This experiment involved two hemispheres of plutonium being brought together until separated only by a screwdriver. One day, the screwdriver slipped unfortunately. The plutonium hemispheres then touched and set off a chain reaction criticality accident. The lab was filled with harmful radiation and a flash of blue light. To stop the chain reaction, Louis Slotin bravely pushed the hemispheres apart with his bare hands. His efforts saved the life of several other co-workers present in the room. Unfortunately, Louis Slotin died within 10 days because his body absorbed a lethal dose of radiation.

**(iii) Equipment Failure Type of Accidents** A simple initial failure of a semiconductor diode followed by an equipment failure can result in malfunction of a particle accelerator, resulting in the overexposure of a patient undergoing treatment for cancer.

**(iv) Transport Accidents** A defective gamma radiography set was transported as cargo in a passenger bus in Cochabamba. The gamma source was outside the shielding, and it irradiated some bus passengers.

A court case had revealed that a radiotherapy source with defective shielding was transported from Leeds to Sellafield in the United Kingdom. On the underside, the shielding had a gap from which radiation was escaping and harming the passengers.

**(v) Decay Heat Accidents** The heat generated by the radioactive decay in a nuclear reactor (without a coolant) can result in a accident causing partial melting of the core and damaging the nuclear fuel.

**(vi) Criticality Accidents** Criticality accidents are smaller scale accidents in which

(a) only a few people can be harmed,
(b) no or small release of radioactivity occurs outside the experimental hall,
(c) limited off-site release of gamma and neutron radiation occurs, and
(d) the system remains critical for longer durations (even few days) before it could be stopped.

### *(B) World's Major Nuclear Accidents*

**(i) Three Mile Island Accident** On March 28, 1979, a nuclear accident occurred in US at the Three Mile Island nuclear power plant. One of two reactors lost its coolant, which caused overheating and partial meltdown of its uranium core. This resulted in release of intense radiation as well as radioisotopes. Fortunately, the plume emitted into the atmosphere was quite low for causing toxic effect.

**(ii) Chernobyl Nuclear Accident** On April 26, 1986, a nuclear accident occurred in Chernobyl, near Kiev, Ukraine. Explosion and fire in the graphite core of one of four reactors released radioactive material that spread over part of the Soviet Union, eastern Europe, western Europe and Scandinavia. This is one of the world's worst nuclear accidents involving 237 confirmed cases of chronic radiation illness and 37 deaths. Hundreds of thousands of Ukrainians, Russians and Belorussians had to abandon entire cities within a 30 km zone of extreme contamination. About 3 million people, more that 2 million in Belarus alone are still living in contaminated areas. Figures from the Ukraine Radiological Institute suggest that over 2500 deaths were caused by the Chernobyl nuclear disaster.

**(iii) Tokaimura Nuclear Accident** On September 30, 1999, a nuclear accident occurred in a nuclear plant in Tokaimura, Japan. In a uranium-processing nuclear fuel plant, a chain reaction went uncontrolled, resulting in emission of high levels of radioactive gas into the air. Because of this accident, two workers were seriously injured and one worker got killed.

Populations living in contaminated areas and persons who helped with the clean-up of the accident were found to have thyroid or other cancers. Among Ukrainian young children (up to 15 years of age), the average thyroid cancer rate was 4–6 incidents per million before the accident. After the incident, the cancer rate rose to 45 incidents per million.

People were not told the truth until several years after the accident. The lack of public information available after the accident, the stress and trauma of evacuation and concerns of the people affected and concerns about their children's health resulted in significant increases in psychological health disorders such as depression, anxiety, helplessness, social withdrawal, mental stress and lack of hope for the future.

## (C) Nuclear Holocaust

*Holocaust means large-scale destruction of human lives by intense heat and fire.*

### Holocaust: Great Destruction Resulting in the Extensive Loss of Life

*Hiroshima–Nagasaki disaster* is a nuclear holocaust. America was involved in the development of an atom bomb while World War II was at its peak. The bomb was made and was test fired on 16 July, 1945 in a desert in Mexico. The 30 m tower on which the bomb was placed completely melted. The blinding light that spread for a few minutes turned the sun into a pale ball. Frightened to the core by the resulting blast, scientists vehemently opposed using it on Japan or anywhere in the world. US President Harry Truman was determined to win the war against Japan. He ordered dropping of the bomb on Japan. The *uranium bomb* named the *Little Boy* was dropped on *Hiroshima* on *August 6, 1945*. Within minutes, one lakh persons were burnt to death like moths near a lamp. Just three days later, a *plutonium bomb* named *Fatman* was dropped over *Nagasaki*. The whole area was burnt and looked like a desert. Devastating shock waves, deadly gamma radiations and enormous amount of heat created conditions where any life cannot survive.

It is estimated that by December 1945, as many as 1,40,000 had died in Hiroshima by the uranium bomb and its associated effects. In Nagasaki, roughly 74,000 people died of the plutonium bomb and its after effects. In both cities, around 2,14,000 people in total, most of them were civilians, were killed. In Nagasaki alone, up to 60,000 people were injured. The radius of total destruction was about 1.6 km, followed by fire across the northern portion of the city to 3.2 km south of the bomb.

**Table 6.5** *Classification of accidents according to the International Nuclear and Radiological Events Scale (INES)*

| | *Accident* | *Description* | *Example* |
|---|---|---|---|
| 1. | Anomaly | At a nuclear facility any break of pre-scribed operating limits | — |
| 2. | Incident | Spread of contamination to an area of the facility which is not expected by design | Cadarache, France (1993) |
| 3. | Series incident | Release of large quantity of radioactive material, contained with the installation | Sellafield, Britain (2005) |
| 4. | Accident with local consequences | At a nuclear facility, fatal overexposure of workers after a criticality event | Tokaimura, Japan (1999) |
| 5. | Accident with wider consequences | Severe damage to reactor core | Three Mile Island, US (1979) |
| 6. | Serious accident | Explosion of waste tank and significant release of radioactive material | Kyshtyn, Russia (1957) |
| 7. | Major Accident | External release of significant fraction of core, widespread environmental and health effects | Chernobyl, Ukraine (1986) |

Radiation poisoning and necrosis caused illness and death after the bombing in about 1% of Hiroshima residents who survived the initial explosion. In the years between 1950 and 1990, It is estimated that hundreds of deaths are attributable to radiation exposure among atomic-bomb survivors from both Hiroshima and Nagasaki.

## 6.7 WASTELAND RECLAMATION

**Land** is a precious resource because it is used for agriculture, pastures and grazing fields, housing, agroforestry, roads, industrial areas, forestry, etc.

*Wasteland is a land which is*

(i) *abandoned,*
(ii) *degraded* and thus ecologically unstable,
(iii) *incapable* of producing material or service of value,
(iv) *eroded,*
(v) *unfit* for cultivation, unproductive, unfit for grazing as greenery cannot be sustained, and
(vi) *saline*, waterlogged, not being utilised to its potentials.

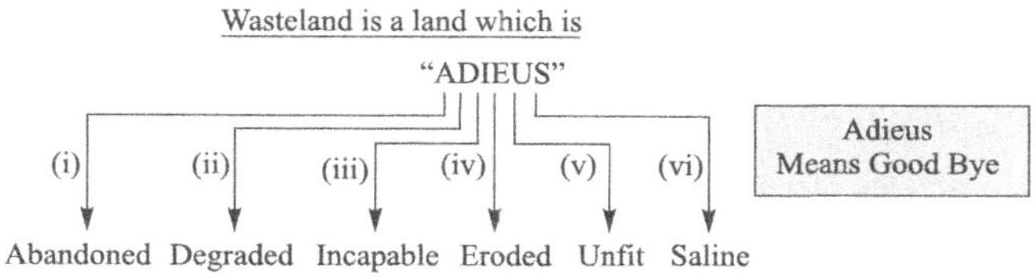


**Fig. 6.23** *Wasteland attributes*

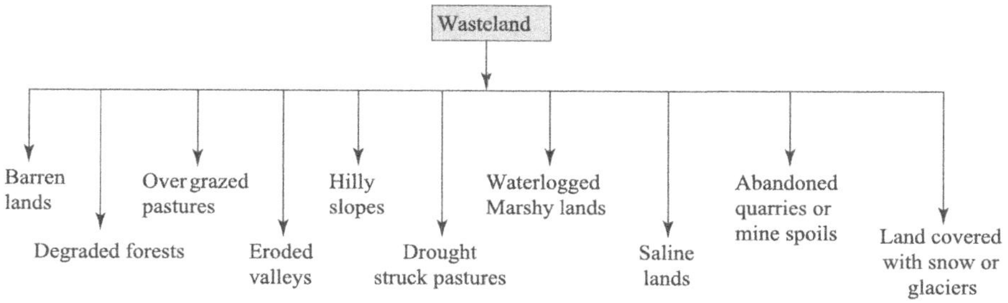


**Fig. 6.24** *Wasteland examples*

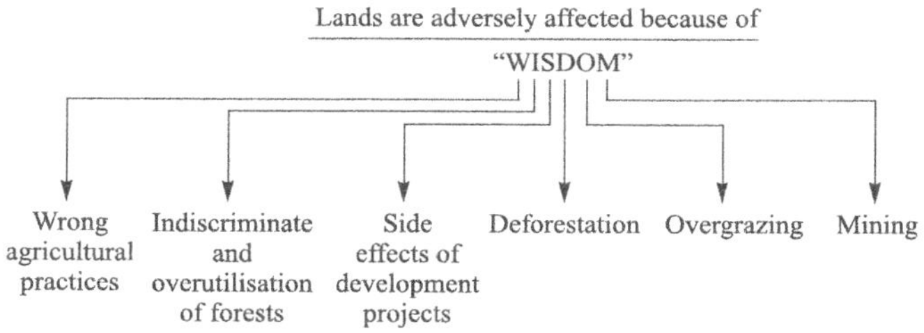


**Fig. 6.25** *Causes of formation of wastelands*

## (A) Classification of Wastelands

Wasteland is broadly classified into the following two types:

**(i) Cultivable Wastelands** The lands are cultivable but not cultivated for more than five years due to various reasons such as being declared as notified forest area or state or private occupation.

***Examples:*** Waterlogged marshy lands, saline lands, degraded forests, degraded pastures (or degraded grasslands), shifting cultivation land, gullied land, strip land, etc.

**(ii) Uncultivable or Barren Wastelands**

These wastelands cannot be brought under cultivation or economic use except at a very high cost.

***Examples:*** Barren rocky lands, areas covered by snow or glaciers, steep sloping areas.

***(B) Drawbacks of Wastelands Formation*** Formation of wastelands result in the deterioration of ecological balance. The various components of the ecosystem directly or indirectly dependent on that particular wasteland are adversely affected.

## (C) Wasteland Reclamation

*Wasteland reclamation is the process of converting sterile, barren wasteland into something that is fertile and suitable for habitation and cultivation.*

## (D) Wasteland-Reclamation Practices

Some of the important wasteland-reclamation practices are briefly described below:

**(i) Changing Agricultural Practices** Jhoom or shifting cultivation should be replaced by crop rotation, mixed cropping or developing plantation crops which would improve fertility of land and support a large population.

***(a) Mulching*** It means providing protective cover to stop the shifting of sand. A mulch is a protective layer formed by the stubble, i.e., the basal parts of herbaceous plants, especially cereals, attached to the soil after harvest. Dry stems of maize, tobacco, and cotton are used as multch. For mulching, artificial protective covering can also be used. Mulches act as wind barriers; so soil erosion due to wind is reduced. By addition of organic matter, mulches reduce evaporation and increase soil moisture. Even mulching is useful against water *erosion.*

***(b) Managing Topography*** Water running down the hill erodes soil. The faster it runs, the more soil it carries off the fields. *Soil* erosion and wasteland formation can be minimised by the following methods:

- *Strip Farming* In alternating strips along the contours, different kinds of crops are planted. When one crop is harvested, the other is still present to protect the soil and keep water from running straight downhill.
- *Contour Ploughing* The ploughing of land is done across the hill, and not in up and down style as in contour ploughing, The ridges created by cultivation make little dams that trap water and allow it to seep into the soil rather than running off.

- *Tied Ridges* This method involves a series of ridges running at right angles to each other, so that water run-off is blocked in every direction and is encouraged to soak into the soil. Tied ridges are very useful in very heavy rainfall areas.
- *Terracing* means shaping the earth in the form of levelled terraces to hold soil and water. The edges of the terraces are planted with soil-anchoring plant species. This method makes it possible to farm very steep hillsides. However, this method is costly, requires expensive machinery or much hand labour.

**(ii) Leaching** By providing adequate drainage to flood-prone and irrigated lands, salinity can be prevented.

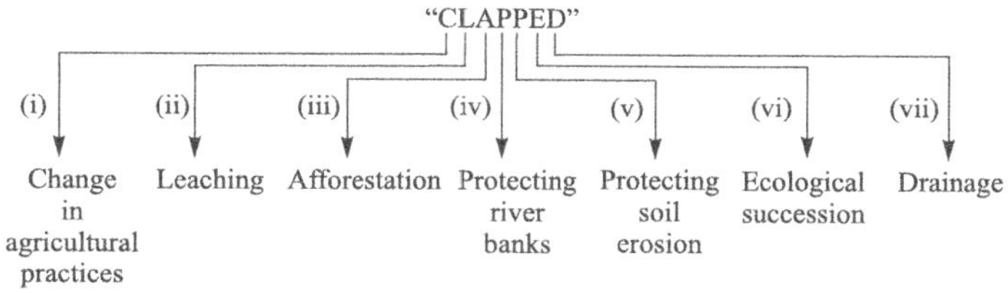


**Fig. 6.26** *Wasteland reclamation practices*

By leaching with more water, salt-affected lands can be recovered, especially in the areas where groundwater table is not high.

**(iii) Afforestation** It means growing forests over culturable wastelands for the first time. Previously there were no forests there due to lack of seeds or other adverse factors.

***Reforestation*** It means growing the forests over the lands where they were existing earlier; and had destroyed or degraded by forest fires, overgrazing, excessive felling, shifting cultivation, floods, waterlogging, soil erosion, etc.

**(iv) Protecting the River Banks** By providing stone, wooden or concrete pitching or by plantation of trees/vegetation along the river banks, it is possible to protect river banks against caving and cutting.

***Controlling Formation of Gullies*** The gullies get widened due to excess run-off water. This can be checked by constructing dams, diversion drains, bounds, etc.

**(v) Protecting Soil Erosion by Providing Ground Cover** After harvesting, the crop residues are left on the ground. They resist wind and water from creating erosion. The ground cover reduces soil temperature and evaporation in the hot season. It thus protects the ground organisms which are helpful in aerating and rebuilding the soil.

**(vi) Ecological Succession** This is a natural process of establishment or re-establishment of an ecosystem. In ecological succession, the slow-growing native grasses assist in reclaiming the minerally deficient soils in mining and industrial wastelands.

**(vii) Drainage** It is required for waterlogged soil reclamation where excess water is removed by artificial drainage.

In areas where waterlogging happens after heavy rains, surface drainage is facilitated to remove the excess water.

Subsurface drainage is better because chances of evaporation of water leading to accumulation of salt almost become nil in this method.

## 6.7.1 Consumerism and Waste Products

***Consumerism** is a process and habit of the chronic purchasing of new goods and services, with less attention to their true need, durability, origin of the product or the environmental impacts during manufacture and disposal.*

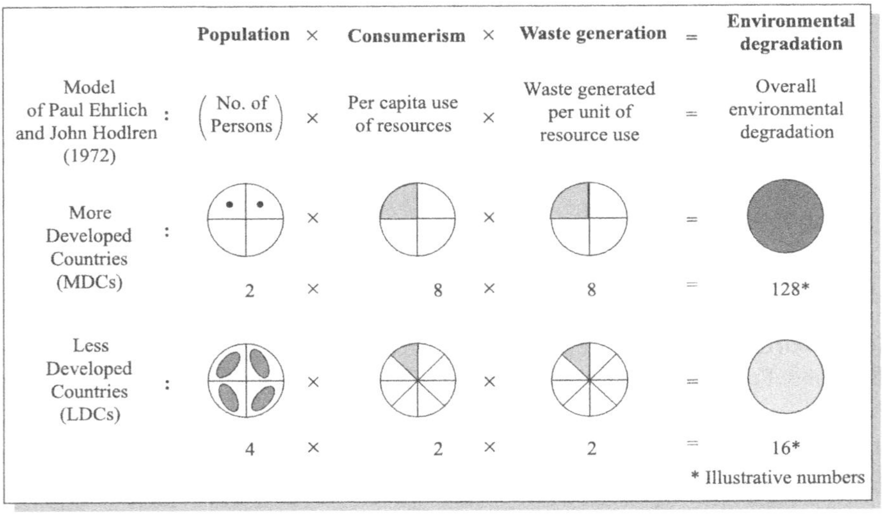


**Fig. 6.27** *Relationship of population, consumerism, waste generation and environmental degradation*

In More Developed Countries (MDCs), population is less and resources are in abundance. Lifestyle is luxurious and per capita consumption of resources is very high. More consumption of resources result in more waste generation and greater environmental degradation.

In Less Developed Countries (LDCs), population is large. Adequate resources are not available for all, so per capita consumption is less. However, overall consumption is high. Thus, environmental impact is same or slightly less in comparison to MDCs.

For example, the population of India is 3.4 times more than that of USA but its overall resource use and waste generation is less than $1/8^{th}$ that of USA.

### *(A) Creators of Consumerist Culture*

**(i) Artificial Beauty** Millions of people use soaps, detergents, hair dyes, skin-care creams and other cosmetic items to enhance their beauty artificially. Manufacturers have been able to create demands for these cosmetic items and the public is busy spending their money for this temporary beauty enhancement.

***Fashion*** Manufacturers of clothes, textiles, shoes and apparel keep changing fashions to accelerate the speed of consumerism through advertisements.

**(ii) Greed of Industry** To make more profits, industry and large businesses want to sell more products. Generally, products are made for a one-time use. Through regular advertising a "throw-away society" has been created. This society prefers disposable items discarding notions of inherent value, longevity and the environmental consequences of manufacture and disposal of the product. In the developed world, 200 billion paper cups, bottles, cans and plastic cartons are thrown away each year.

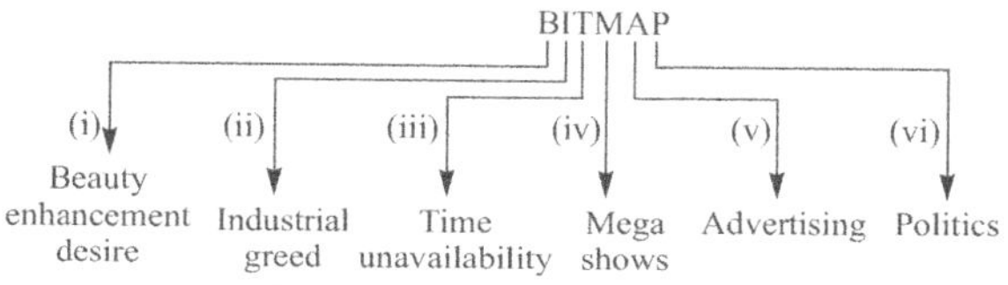


**Fig. 6.28** *Creators of consumerism*

**(iii) More Money, Less Time** Family ties, friendship, everything becomes mediated through the spending of money on goods and gifts and services. A generation is growing up without knowing what quality goods are. Relations are promoted only as a vehicle of giving and taking gifts.

**(iv) Mega Shows** Manufacturers of items of consumerism (like automobiles, televisions, radios, refrigerators, air conditioners, dishwashers, cosmetics) sponsor megashows. They kindle passion and unquenchable desire for latest items through prizes and other incentives.

**(v) Advertising** Advertising is designed to create both a desire to follow fashions, and the resultant personal self-reward system based on acquisition. Thus, a consumerist culture is not based on natural demand, but on a created demand.

**(vi) Politics** Consumerism is encouraged politically so that population remains satisfied by material needs and politicians can do whatever they wish for.

## *(B) Drawbacks of Consumerism*

### Consumerism

(i) causes more *pollution*, creates more waste products, causes wasteful use of material and energy,

(ii) helps lowest-wage, environmentally unregulated *overseas* manufacturers,

(iii) promotes social *unrest*,

(iv) *promotes riots*, and other criminalities, and

(v) leads to *societal suicide.*

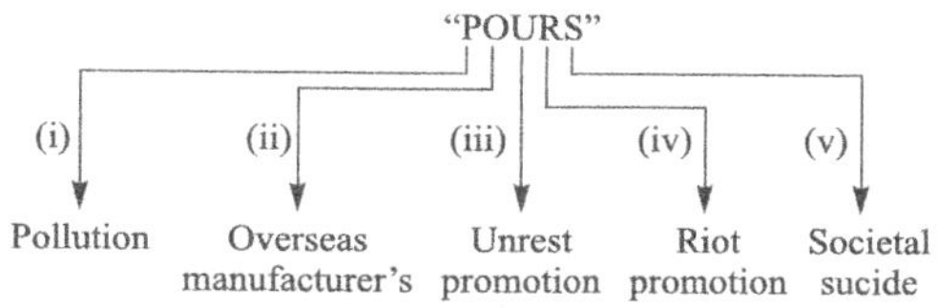


**Fig. 6.29** *Drawbacks of consumerism*

### (C) Measures to Prevent Excess Consumerism

Some of the measures to prevent excess consumerism include the following:

**(i) Pigourian Taxes** For encouraging industrial ecology and waste reduction, manufacturers are taxed for some or all the cost of recycling or waste disposal.

**(ii) Ecolabelling** It involves the marking of products to indicate that they are environmentally friendly. Ecolabelling assesses environmental impact and communicates this to the consumer. It also encourages manufacturers to reduce the impacts of their products. *Ecomark is used in India for ecolabelling.*

**(iii) Green Marketing** It involves communicating green image of the better, environmental friendly products to the consumer. It helps in conserving the environment and achieving better marketing niche. People also happily buy green products like refrigerators that do not leak ozone-layer-depleting chlorofluoro carbons (CFC's) and consumes less electricity,

**(iv) Self-awareness and Self-control**

Having fewer things means enjoying what you have more and actually getting to use it. It also helps in fewer distractions from the essentials such as food, family, nature, study, friends.

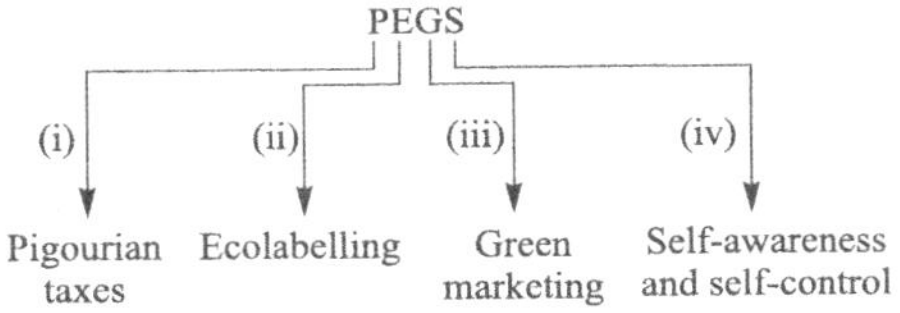


Fig. 6.30 *Measures to prevent excess consumerism*

## 6.8 ACTS FOR ENVIRONMENTAL PROTECTION

### 6.8.1 Environment Protection as a Common Goal for All

Paul Bigelow Sears once said, "*How far must suffering and misery go before we see that even in the day of vast cities and powerful machines, the Earth is our mother and that if we destroy her, we destroy ourselves.*"

So, we should act today for a better tomorrow. Adopt a strategy (like outlined below) for environmental protection. Government, industry, public and law must have only one goal, viz. environment protection.

**(i) Industry** Replace nonrenewable inputs in energy with renewable ones.

**(ii) Government** Educate and involve all in environmental protection drive. They must act together to fight corruption in government and ruthless exploitation by the industry.

**(iii) Public** Consume less, share more, control population and reduce pressure on natural resources.

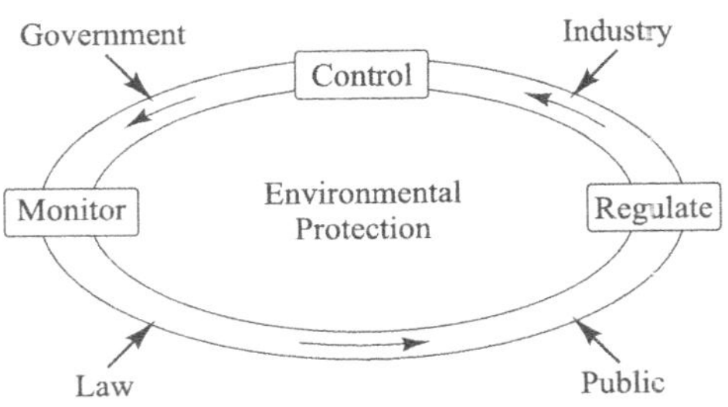


Fig. 6.31 *Environment protection as a common goal for all*

**(iv) Law** Take help from law, if needed, for protection of our environment.

### 6.8.2 Role of Government and Legal Aspects in Environmental Protection

***(A) Government*** Increased government intervention is a must for solving environmental problems because of the following reasons:

(i) The world is facing very serious environmental problems like loss of biodiversity, global warming, water pollution, air pollution, etc.
(ii) The world is facing increased probabilities of natural disasters due to global warming.
(iii) The health of millions of people is at risk if companies are left free to sell their products, vehicles, etc., just for profits. This is because we are exposed to thousands of chemicals a year, many of which interact in ways that are not yet fully understood.

***(B) Legal Aspects*** The Ministry of Environment and Forests (MoEF) in India is the apex administrative body for

(i) undertaking conservation and survey of fauna, flora, forests and wild life;
(ii) formulating the environmental policy framework in the country;
(iii) planning, promotion, co-ordination and overseeing the implementation of environmental and forestry programmes; and
(iv) regulating and ensuring environmental protection.

The responsibility for prevention and control of industrial pollution is primarily executed by the *Central Pollution Control Board (CPCB)* at the central level which is a statutory authority, attached to the MoEF. The State Pollution Control Boards and the State Departments of Environment are the designated agencies to perform the function at the state level. The administrative framework in India for protection of the environments is shown in Fig. 6.32.

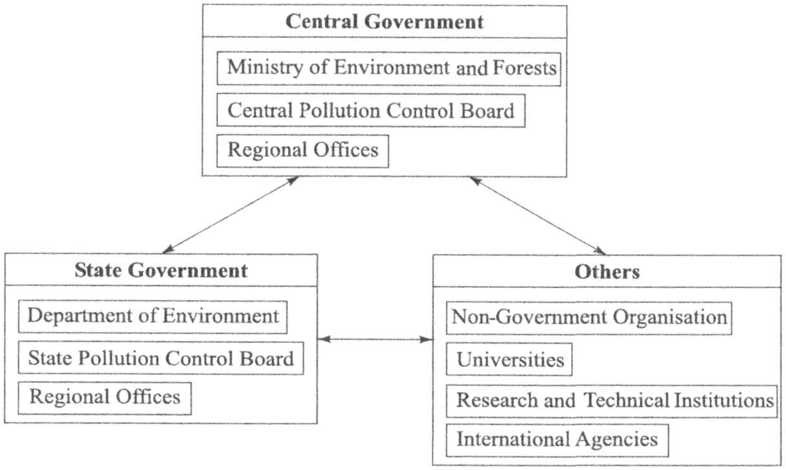


**Fig. 6.32** *Administrative framework for environment protection in India*

## 6.8.3 Environment (Protection) Act, 1986

It is the umbrella legislation which authorises the Central Government to

(i) protect and improve environmental quality,

(ii) control and reduce pollution from all sources, and

(iii) restrict or prohibit the selling and/or operation of any industrial facility on environmental grounds.

According to the Act, the term "*environment*" *includes water, air and land and the inter-relationship which exists among and between water, air, land, human beings, other living creatures, microorganisms, plants and property.*

The main provisions of the Act are given below:

(i) The Central Government shall have the power to take all such measures as it deems necessary or useful for the purpose of protecting and improving the quality of the environment and preventing, controlling and decreasing environmental pollution.

(ii) No person carrying on any industry, operation or process shall discharge or emit any environmental pollutants or permit to do so in excess of such standards as may be prescribed.

(iii) Where the discharge of any environmental pollutant in excess of the prescribed standards occurs or is bound to occur due to any accident or other unexpected act or event, the person responsible for such discharge and the person in charge of the place at which such discharge occurs or is expected to occur, shall be bound to prevent or reduce the environmental pollution caused as a result of such discharge and shall also immediately inform the fact of such occurrence or fear of such occurrence; and be bound, if called upon, to render all assistance to such authorities or agencies as may be advised.

(iv) No person shall handle or cause to be handled any hazardous substance except in accordance with such procedure and after complying with such safeguards as may be prescribed.

(v) The Central Government or any officer empowered by it in this behalf, shall have power to take, for the purpose of analysis, samples of air, water, soil or other substance from any premises, factory, etc., as may be prescribed.

(vi) The Central Government may, by notification in the Official Gazette, establish one or more environmental laboratories; and recognise one or more laboratories or institutes as environmental laboratories to carry out the functions assigned to an environmental laboratory under this Act.

(vii) Whoever fails to comply with or violate any of the provisions of this Act, or the rules made or orders or directions issued thereunder, shall, in respect of each such failure or violation, be punishable with imprisonment or with fine or with both.

## 6.8.4 Air (Prevention and Control of Pollution) Act, 1981

This is an act to provide for the prevention, control and reduction of air pollution in the country so as to preserve the quality of air.

The salient features of the Air (Prevention and Control of Pollution) Act 1981 are given below:

(i) Act is applicable to the whole of India.

(ii) Under Section 19 of the Act, the State Government in consultation with the State Pollution Control Board (SPCB) has the power to declare Air Pollution Control Area, in which provisions of the Act shall be applicable.

(iii) As per provisions in Section 21(1) and (2), no person can establish or operate any industrial plant without the previous consent of the State Pollution Control Board.

(iv) Every application for consent shall be made in Form I and shall be accompanied by a prescribed fee. Within a period of four months after the receipt of application, the Board shall complete the formalities to either refuse or grant consent. During the course of processing consent for the application, the Board may seek any information about the industry after giving notice in Form II.

(v) Under Section 22, 22(A) operating any industrial plant so as to cause emission of any air pollutant in excess of standard laid down by the State Board is liable for legal action by the Board.

(vi) Under Section 2(a), the term ***air pollutant*** is defined as *any solid, liquid or gaseous substance present in the atmosphere in such concentration as may be or tend to be injurious to human beings or other living creatures or plants or property or environment.*

### *Power of State Government*

Under Section 19, the State Government may, after consultation with the State Board, by notification in the official gazette, declare, in such manner as may be prescribed, any area or areas within the state as air pollution control area or areas for the purposes of this Act.

Under Section 21.3, no person shall without the previous consent of the State Pollution Control Board, establish or operate any industrial plant in an Air Pollution Control Area.

Under Section 22.4, no person operating any industrial plant in any Air Pollution Control Area shall discharge or cause or permit to be discharged, the emission of any air pollutant in excess of the laid-down standards by the State Pollution Control Boards.

Under Section 19.1, the State Government after consultation with the State Pollution Control Board, may

(i) Prohibit burning of any material causing or likely to cause air pollution in an Air Pollution Control Area

(ii) Prohibit the use of any appliance or fuel causing or likely to cause air pollution in an Air Pollution Control Area

## 6.8.5 Wildlife Protection Act, 1972

***(A) Objectives*** The objectives of the Wildlife Protection Act are

(i) to maintain essential ecological processes and life-supporting systems;

(ii) to preserve biodiversity; and
(iii) to ensure protection and conservation of wildlife.

### *(B) Salient Features*

(i) Under Section 3, the appointment of Director, Chief Wildlife Warden and other officers is done by the Central Government.
(ii) Under Section 6, Wildlife Advisory Board is constituted by the State Government or the Union Territory Administration. Under Section 7, the Wildlife Advisory Board shall meet at least twice a year.
(iii) The duties of the Wildlife Advisory Board, under Section 8, are to advise the State Government about
  (a) The selection of areas to be declared as.
    - National parks under Section 35,
    - Sanctuaries under Section 18, etc.
  (b) The formulation of the policy for protection and conservation of wildlife and specified plants
  (c) The measures to be taken for harmonising the protection and conservation of wildlife with the needs of the tribals and other forest dwellers
(iv) Under Section 44, the Act prohibits dealing in animal articles without licence.
(v) Under Section 38 A, the Central Government shall constitute the Central Zoo Authority which has various roles or functions as described in Section 38 C.

### *(C) Major Highlights of the Wildlife Protection Act, 1972*

(i) It provides for protection to listed species of flora and fauna and establishes a network of ecologically important protected areas.
(ii) The act consists of 60 sections and VI schedules divided into eight chapters.
(iii) It empowers the central and state governments to declare any area a wildlife sanctuary, national park or closed area. Industrial activities are banned in these protected areas.
(iv) It provides for authorities to
  (a) administer and implement the Act;
  (b) regulate the hunting of wild animals;
  (c) protect specified plants, sanctuaries, national parks, etc., and
  (d) restrict trade in wild animals or animals articles.
(v) The Act prohibits hunting of animals except with permission of an authorised officer when an animal has become dangerous to human life or property or as disabled or diseased as to be beyond recovery.

## 6.8.6 Forest (Conservation) Act, 1980

In 1980, the Forest (Conservation) Act was enacted for providing protection to forests and to regulate diversion of forestlands for nonforestry purposes.

### *Salient Features*

(i) Prior approval of the Central Government is essential for de-reservation of forest lands and/or diversion of forest lands for nonforestry purposes.
(ii) It is a regulatory act, not prohibitory.
(iii) The Forest (Conservation) Act is an interface between conservation and development.
(iv) It permits sensible and regulated use of forestland for nonforestry purposes.

During 1950–80, the rate of diversion of forestland for nonforestry purposes was 1.5 lakh hectares per annum. After enactment of the Forest (Conservation) Act, 1980, the rate came down to about 35 thousand hectares per annum.

At the time of granting approval under the Forest (Conservation) Act, following conditions are insist upon:

- Compensatory afforestation
- Treatment of catchment area
- Reclamation of mining area in phases
- Provisioning for safety zone area
- Rehabilitation of project affected families
- Plan for wildlife management, etc.

On the directions of the apex court in 2002, a new present value of the forestland being diverted is being charged from the user agencies.

## 6.8.7 Water (Prevention and Control of Pollution) Act, 1974

*"As defined in the Act,* **water pollution** *means such contamination of water or such alteration of the physical, chemical or biological properties of water or such discharge of any sewage or trade effluent or of any other liquid, gaseous or solid substance into water (whether directly or indirectly) as may, or is likely to, create a nuisance or render such water harmful or injurious to public health or safety, or to domestic, commercial, industrial, agricultural or other uses or to the life and health of animals or plants or of aquatic organism."*

(i) The water (Prevention and Control of Pollution) Act was enacted for prevention and control of water pollution and maintaining or restoring of wholesomeness of water.

The Central and State Pollution Control Boards have been constituted under sections 3 and 4 of the later Act respectively.

(ii) Obligations on the part of industries and local bodies are
  (a) To obtain prior consent to establish and operate industry for new discharge of domestic sewage or trade effluent under section 25 of the Act
  (b) Board within four months will either refuse or grant consent
(iii) Power of State Board:
  (a) To obtain information under Section 20
  (b) Carry out any related work under Section 30
  (c) Collect and analyse samples of streams/wells or trade effluent under sections 17(2) and 52

(d) To give direction for closure/prohibition or regulation under Section 33A
(e) Enter and inspect any place, examine any plants/records, etc., and seize if necessary under Section 23

## 6.9 CARBON CREDITS

(i) An industrial house needs to meet its pollutant emission limit.
(ii) The industrial house invests in carbon offsets (either directly or usually through an offset provider).
It means, the industrial house invests in an emission reduction project outside of its sector. Carbon-offsets programs can include
- Renewable energy-sustainable development projects,
- Reforestation projects
- Methane capture/combustion projects

(iii) The Industrial house receives carbon credits for its investment in the form of a carbon-offset certificate.

**One carbon credit = one tonne of greenhouse gas emission reduction**

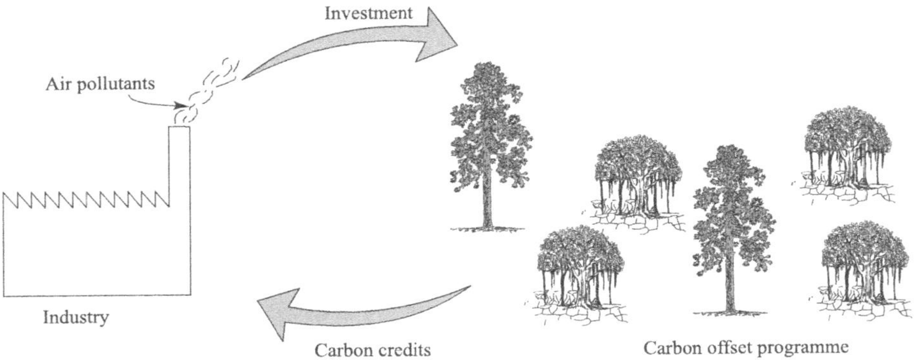


**Fig. 6.33** *Carbon credit concept*

### *Benefits of Carbon-Credits Concept*

(i) Global warming reduction
(ii) Desertification reduction
(iii) Environmental awareness
(iv) Biodiversity protection
(v) Reforestation

## 6.10 INDUSTRIAL SYMBIOSIS

***Industrial symbiosis** is a collaborative enterprise is which the byproducts or waste of one industry become valuable resources for one or several other industries.*

Industrial symbiosis is sharing of services, utility, and by-product resources among industrial units in order to add value, reduce costs and improve environment (Fig. 6.34).

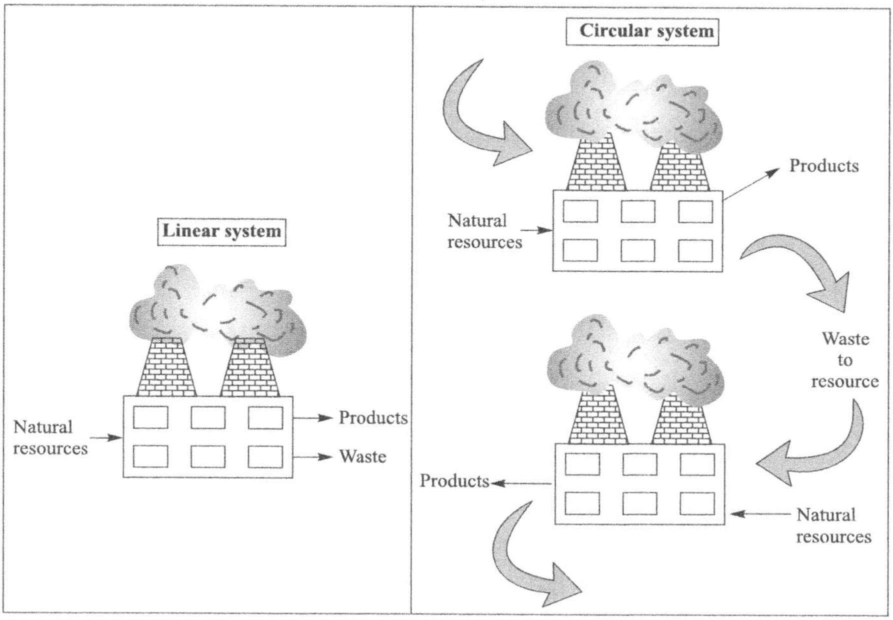


Fig. 6.34 *Industrial symbiosis*

## *Benefits*

### (i) Economic Benefits

(a) Industrial symbiosis turns a disposal cost into an income stream.
(b) It helps in reducing the cost of raw materials.
(c) It maximises use of under-utilised resources and facilities.
(d) It helps in spreading costs of new infrastructure (e.g. infrastructure cost of effluent treatment in shared).

### (ii) Environmental Benefits

(a) Industrial symbiosis reduces virgin resource use and net waste generation without compromising economic activity.
(b) Industrial symbiosis is a major step towards a more sustainable society.

### (iii) Other Benefits

(a) Industrial symbiosis helps in improving public relations.
(b) It facilitates integration of business.
(c) It helps in safeguarding and creating employment.
(d) It helps in sharing of laboratories, workshops, training facilities and other services.

Kalundborg is a medium-sized town in Denmark. Kalundborg's industrial symbiosis comprises eight core companies: (i) Power station, (ii) Refinery,

(iii) Pharmaceutical plant, (iv) Recycling company, (v) Sulphuric and plant, (vi) Municipality, (vii) Plasterboard factory, and (viii) Cement factory. Each company is bound to each other via an intricate network of flows; flows of steam, natural gas, water, gypsum, sludge and fly ash (Fig. 6.35).

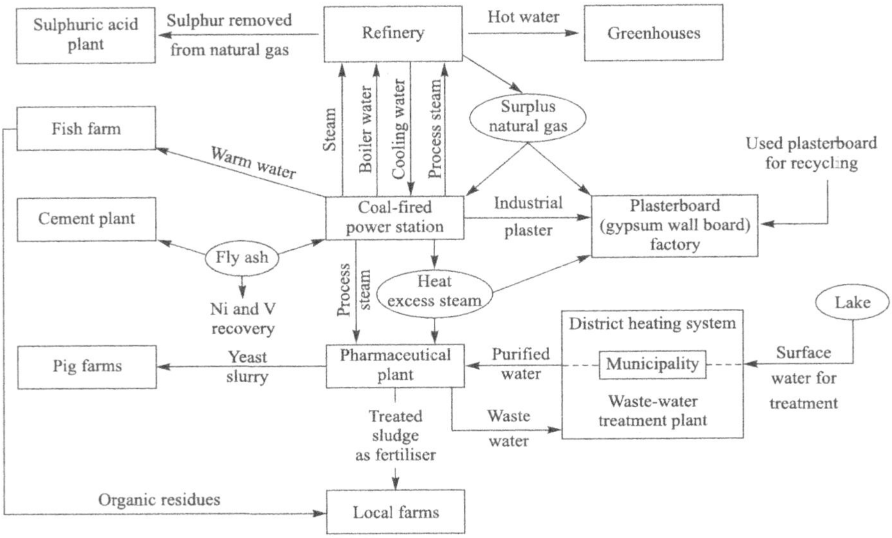


**Fig. 6.35** *Industrial symbiosis in Kalundborg*

Excess steam from power stations is exported to district heat and power supply, refinery, pharmaceutical plant and plasterboard factory. They use it as incoming heat source and export it back to the power station as condensed steam for cooling the plant.

The plasterboard factory receives surplus natural gas as an input energy source from refinery and industrial plaster as an input material source from power station. Calcium and recycled treated waste water are added to the sulphur extracted from the flue gas at power station to form industrial plaster.

$$CaCO_3 + SO_2 + 1/2\ O_2 + 2H_2O \rightarrow CaSO_4.2H_2O + CO_2$$

Insulin production at pharmaceutical plant releases yeast slurry material which is exported to surrounding farms as pig fodder. This replaces approximately 70% of the soy proteins in traditional food mixes. Pharmaceutical plant adds water, lactic acid bacteria and sugar to the yeast in order to make it a more effective and alternative food (Fig. 6.35).

## 6.11 INITIATIVES AND ROLES OF NONGOVERNMENTAL ORGANISATIONS (NGOs) IN ENVIRONMENTAL PROTECTION

*Privately owned organisations involved in providing financial and technical assistance to less developed countries are known as NonGovernmental Organisations (NGOs).*

They have no participation or representation of any government. An NGO is any nonprofit, voluntary citizen's group which performs a variety of services and humanitarian functions.

### 6.11.1 Functions and Advantages of NGOs

(i) They are good at reaching and mobilising the poor and remote communities.
(ii) They work with and strengthen local institutions.
(iii) They help empower poor people to gain control of their lives by counselling, support service, training, micro-credit, etc.
(iv) They carry out projects at lower costs and more efficiently than government agencies.
(v) They promote sustainable development, through economical development, social development and environmental protection.
(vi) They do funding of projects.
(vii) They help in critical analysis of social environments.

### 6.11.2 Roles of NGOs in Environmental Protection

(i) NGO's help in increasing local economic diversity.
(ii) They help in development of local markets, local production, local processing of previously imported goods, and greater cooperation among local economic entities. Thus, they help on attaining self-reliance.
(iii) They educate people on reduction in the use of energy and careful management with recycling of waste products.
(iv) They educate and motivate local people to protect and enhance biological diversity. They make them understand careful stewardship of natural resources.
(v) They give commitment of the sustainable communities to social justice.

The relationship between NGO functions, empowerment and sustainable community development is illustrated in Fig 6.36.

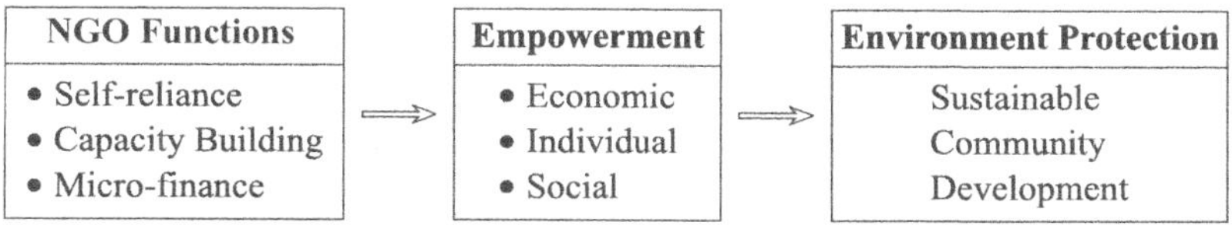


**Fig. 6.36** *Theoretical framework of the functions of NGOs in promoting environmental protection and sustainable community development*

NGOs, through some programmes and functions such as microfinance, capacity building and self-reliance, help communities to be empowered and finally contribute towards environmental protection.

## 6.12 ISSUES INVOLVED IN ENFORCEMENT OF ENVIRONMENTAL LEGISLATION

Regulatory measures in the form of legislation check environmental degradation. They also lead to the enacting of laws at the national or international levels to prevent pollution. To protect the environment, the role of the judiciary lies in formulation and enforcement of effective laws. The judiciary alone cannot improve the environment unless the states and citizens do their duties and obligations.

***Some important issues in the enforcement of environmental legislations are given here.***

**(i) Public Apathy** In contrast with conventional crimes such as rape, murder, dacoity, etc., pollution is treated as a white-collared crime. While conventional crimes are always taken seriously, the crime of pollution is generally taken for granted.

### (ii) Limitations of Regulating Agencies

***Poaching*** It is a big national problem despite the existence of the Wildlife Act. The Wildlife Department has no provision to punish poachers and unauthorized hunters unless they are caught red handed. In case they are caught, the standard excuse given by armed poachers is that they carried guns for self-defence from dangerous wild animals. Penalties by the Wildlife Department are nominal cash fines. Rich offenders continue poaching by paying such nominal cash fines to the Wildlife Department.

***Water Pollution*** The chairman of the state pollution control board is the key person for the enforcement of the Water Act, 1974. He/she should be professionally qualified and appointed on a full-time basis. However, several State Pollution Control Boards are headed by part-time chairpersons without requisite qualifications and experience. Often, the member secretaries of the Pollution Control Boards are drawn either from the administrative service or even from the forest service. They do not have the required technical background in pollution control. Thus, it becomes difficult for them to provide proper leadership and guidance to their subordinates. The enforcement action in such situations is obviously weak.

### (iii) Legal Loopholes

(a) The legal provision for penal action against the polluters requires that the State Pollution Control Board has to file a case before the lower court for action against a polluting unit. However, the lower courts are too busy to devote enough time for environment-related litigations. As a consequence, thousands of cases filed by the State Pollution Control Boards are still pending for years together. In some cases where decisions have been taken, the polluters were given the benefit of doubt due to absence of sufficient proof.

(b) For prevention of pollution, a provision is available for a citizen to approach a court. For this, the citizen is required to give a notice of not less than 60 days to the government, of his/her intention to make a complaint. If the government does not act on the notice, then only the citizen can go to the court. For meaningful result, the court must give directives to the government or the enforcing authority to collect a sample and submit a report.

Thus, this exercise would give the culprit a period of at least 60 days to clean up all traces of its offence and prepare itself for sample collection.

(c) The Factories Act, 1948, is an important legislation which provides for certain measures with respect to the industrial safety, health of the workers and welfare measures.

However, safety standards and maintenance procedures at various industries are not maintained. India has seen and suffered **Bhopal tragedy** on the night of December 23, 1984 which is considered the *worst industrial disaster in history*. The Bhopal tragedy was the result of a combination of legal, organisational, technological, and human errors. The accident occurred when toxic Methyl IsoCyanate (MIC) gas was accidentally released into the atmosphere. About 1 lakh people died and more than 2 lakh people were severely disabled. Even after 26 years of the accident, thousands of the poorest members of the population of shanty towns are still suffering.

**(iv) Lack of Knowledge** While urban citizens cry for stopping pollution and consumerism from one side, they watch television, see advertisements, purchase new vehicles, gadgets, luxury items and cosmetics whose ecological footprints are large. These urban citizens unknowingly became creators of pollution.

**(v) Lacuna in Implementation** Pollution Control Boards cannot take action against municipal corporations or other civic bodies because they are not empowered to do so. These Pollution Control Boards are empowered to stop industrial water pollution but they cannot fight with rich industrialists. Thus, the public is forced to suffer from use of polluted water.

Despite the various legislations, there is a depletion of forest resources because of commercial exploitation. The present and future generations are bound to suffer from deforestation-linked flood, soil erosion, siltation of water bodies, loss of biological and genetic diversity, etc.

**(vi) Intellectual Indifference** In the Environment Protection Act, all power and authority is reserved in the hands of the Central Government. For the efficient execution of the provisions of the Act, this excessive centralisation is a major burdle.

**(vii) Limitations of Environmental Risk Assessment** Scientific knowledge regarding the potential impacts of Persistent Organic Pollutants (POPs) and endocrine disrupting chemicals on human health and environment is not completely understood because of the complexity of natural ecosystems and limitations in experimental design.

For environmental protection, *The Precautionary Principle* is applied in such circumstances where there are reasonable grounds for concern that an activity is, or could, cause harm but risk is uncertain.

*The Precautionary principle directs that action should be taken to correct a problem the moment there is evidence that harm may occur, not after the harm has already occurred.*

For example, the German Government in 70's ordered to reduce power plant emissions when they realised that forests were suddenly dying because of the acid rain.

Thus, lack of full scientific certainty shall not be used as a reason for postponing cost-effective measures to prevent environmental degradation from substances or activities which cause threats of serious or irreversible damage.

**(viii) Poverty** Affordable food, shelter, clothing, medicine and minimum education for their children are the high-priority necessities of the poor. They do not have enough money to practice environmental conservation strategies used by the rich. They exploit the environment to generate income. For growing crops, they encroach watershed areas. They live in slums as they cannot afford proper sanitation and waste treatment. They can only purchase cheaper products made in factories where proper pollution control devices are lacking. Thus, survival activities of the poor indirectly contribute to environmental degradation.

*According to Polluter-Pays Principle, those who pollute the environment must pay for the damages they have caused.* This principle will burden the poor in order to create a better environment for the rich to enjoy. If enforced, this principle will certainly widen the real income gap and can even be interpreted as socially undesirable or morally unjust.

(a) Suppose forest encroachment is conducted by the poor landless farmers.

According to the Polluter-Pays Principle, they will be liable for causing flooding, soil sedimentation in irrigation and/or hydropower dams.

However, the wealthy people or factory owners will benefit from keeping the forests intact, namely in terms of water supply or hydropower electricity, or as a recreation site. Thus, according to the ***Beneficiary-Pays Principle***, the rich should pay for the cost of forest conservation.

(b) We know that for reducing air pollution, commuters should use public transport instead of personal vehicles. The poor commuters use public buses not for reducing pollution but because they cannot afford personal vehicles.

In order to finance better quality buses which cause minimum air pollution, the Polluter-Pays Principle requires the poor commuters to pay a higher charge as they use public buses.

The alternative Beneficiary-Pays Principle argues that cleaner and less smoky buses should be financed by all city residents and not just by bus commuters. This is because everyone will enjoy the clean air.

## 6.13 ANIMAL HUSBANDRY

*Animal husbandry is the science of taking care of domestic animals that are used primarily as food or product sources.* Anyone who takes care of domesticated animals especially in large groups, is practicing animal husbandry. Animal husbandry provides an understanding of how to care for and manage domestic animals so that the animal's requirements for good health and welfare, and humans' requirements for the use of these animals are met.

Modern animal husbandry is concerned with (a) increasing efficiency by increasing the number of animals raised per unit area; (b) using antibiotics to increase animal growth rates; and (c) decreasing labor costs by automated animal feeding, watering, housing, etc.

***Note:*** *Livestock* (also cattle) refers to a domesticated animal raised in an agricultural setting to produce food, fibre, or labour.

## Case Study

**Environmental and Social Impact of Livestock Revolution**

As per the Stanford University (March 2010) report on "Environmental and Social Impact of Livestock Revolution" *Science* daily: Worldwide more than $1.7 \times 10^9$ animals are used in livestock production. They occupy more than 25% of the earth's land. Production of animal feed consumes about 33% of total arable land. Livestock production accounts for approximately 40% of the global agricultural gross domestic product. The livestock sector, including feed production and transport, is responsible for about 18% of all greenhouse gas emissions worldwide.

### 6.13.1 Impacts of Livestock Revolution on Humanity

Although about $1 \times 10^9$ people worldwide derive at least some part of their livelihood from domesticated animals, the fast growth of commercialised industrial livestock has reduced employment opportunities for many. Many small, rural producers from India and China are under additional pressure from health authorities to meet the global food-safety standards.

Poultry, pork, beef, etc., provide about 33% of humanity's protein intake, but the impact on nutrition across the globe is highly variable. Too much animal-based protein is not good for human health, while too little is a problem for those on a protein-starved diet (applicable mainly for developing countries) (Fig. 6.37).

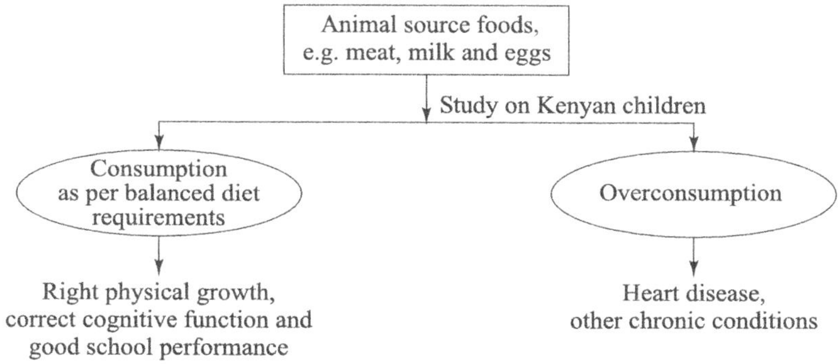


**Fig. 6.37** *Impacts of animal source foods on children*

Human health is also affected by pathogens and harmful substances transmitted by livestock.

### 6.13.2 Environmental Concerns of Animal Husbandry

The livestock sector is a major environmental polluter due to the following considerations:

(i) Much of the world's pastureland has been degraded by grazing or feed production.

(ii) Many forests have been clear-cut to make additional farmland.

(iii) Feed production for animals requires intensive use of water, fertilisers, fossil fuels, pesticides, etc.

(iv) Only 33% of the nutrients fed to animals are absorbed, and so animal waste is a leading factor in the pollution of land and water resources. For example, the total phosphorous excretions of animals are estimated to be 7–9 times greater than that of humans, with detrimental effects on environment.

(v) The poultry, pork and beef industries emit large amounts (about 51%) of greenhouse gases ($CO_2$, $CH_4$, etc.) and are responsible for climate change.

(vi) The quality of surface water and groundwater gets degraded by disposal of organic wastes and liquid effluents of livestock.

(vii) Huge quantity of fossil fuels are used to transport feed for livestock industry, applicable mainly for developed countries.

***Notes:***

- A person in America switching from a typical meat diet to a vegetarian diet with the same number of calories would prevent the emission of 1485 kg of carbon dioxide.
- A beef animal consumes about 100 kg of hay and 4 kg of grain per 1 kg of beef produced. It takes about 1000 litres of water to produce 1 kg of hay and grain, thus about 100, 4000 litres of water is required to produce 1 kg of beef meat.

### 6.13.3 Solutions

(i) Governments should implement policies that provide incentives for better management practices with focus on efficient water and fertiliser use, and land conservation.

(ii) Reduce or refuse the consumption of meat. Celebrate meat-free days. Promote vegetarian eating options.

(iii) Keep manure and urine away from household areas and water bodies. Collect and store manure for composting. Install a biogas plant.

(iv) Restrict animal access to fragile areas, allow rotational grazing so that plant re-growth is facilitated.

(v) Use integrated pest-management techniques. Apply fertilisers and pesticides at the correct time, in correct amounts to promote land use optimisation.

## *Important Definitions*

- *Sustainable development* means meeting the needs of the present without compromising the ability of future generations to meet their own needs.
- When the sum total of nature's resources (natural capital) is used up faster than it can be replenished, *degradation* of the environment occurs. However, if human activity only uses nature's resources at a rate at which they can be replenished naturally, *sustainability* occurs.

- A lifestyle that attempts to reduce an individual's or society's use of the earth's natural resource and his/her own resources is known as a *sustainable lifestyle*.
- *Equitable use of resources for sustainable lifestyle* requires that the rate of use of renewable resources do not exceed regeneration rates and rates of use of nonrenewable resources do not exceed rates of development of renewal substitutes.
- *Urbanisation* is defined as movement of people from rural to urban areas with population growth equating to urban migration or it can also be defined as the physical growth of urban areas as a result of global change.
- *Rainwater harvesting* means collecting rainwater and storing/conserving it for a later use.
- *Watershed* is a geographic area of land that collects, stores, and releases water.
- *Watershed management* refers to the conservation, protection and restoration of a watershed to secure water—both in quantity and quality for drinking, sanitation and agriculture in a sustained manner.
- *Weather* is the reflection of atmospheric humidity, temperature and rainfall.
- *Climate* is the average weather pattern over longer duration in a place.
- *Greenhouse effect* is a process by which infrared radiation leaving the earth's surface is trapped by some greenhouse gases; so the temperature is higher than it would be if direct heating by solar radiation were the only warming mechanism.
- *Greenhouse gases* are gases in an atmosphere that absorb and emit radiation within the thermal infrared range. Examples: Carbon dioxide, nitrous oxide, methane, water vapour, ozone.
- *Global warming* means a rise in temperature over the earth's surface. It is primarily caused by $CO_2$ emission by man-made activities.
- *Acid rain* is rain which is unusually acidic (pH of less than the natural range of 5 to 6); caused mainly by atmospheric pollution with sulphur dioxide and nitrogen compounds.
- The gradual thinning of the ozone layer and *ozone-hole* formation occurs by the destruction of ozone due to its reactions with nitric oxide, chlorine, hydroxyl radicals, etc., in the stratosphere.
- *Holocaust* means large-scale destruction of human lives by intense heat and fire.
- *Wasteland reclamation* is the process of converting sterile, barren wasteland into something that is fertile and suitable for habitation and cultivation.
- *Consumerism* is a process and habit of the chronic purchasing of new goods and services, with less attention to their true need, durability, origin of the product or the environmental impacts during manufacture and disposal.
- *Water pollution* means such contamination of water or such alteration of the physical, chemical or biological properties of water or such discharge of any sewage or trade effluent or of any other liquid, gaseous or solid substance into water (whether directly or indirectly) as may, or is likely to, create a nuisance or render such water harmful or injurious to public health or safety, or to domestic, commercial, industrial, agricultural or other uses or to the life and health of animals or plants or of aquatic organism.
- *Industrial symbiosis* is a collaborative enterprise is which the by-products or waste of one industry become valuable resources for one or several other industries.
- Privately owned organisations involved in providing financial and technical assistance to less developed countries are known as *Non-Governmental Organisations (NGOs)*.
- *Animal husbandry* is the science of taking care of domestic animals that are used primarily as food or product sources.
- *Livestock* (also cattle) refers to a domesticated animal raised in an agricultural setting to produce food, fibre, or labour.

# EXERCISES

**Based on role of government and legal aspects**

1. Describe the power of the state government to declare pollution control areas and restrictions or use of certain industrial plants as given in the Air Prevention and Control of Pollution Acts.
2. Briefly discuss the salient feature of the Environment (Protection) Act, 1986.
3. Discuss, in brief the salient features of the Air (Prevention and Control of Pollution) Act, 1981.
4. Briefly discuss the salient features of the Forest (Conservation) Act, 1980.
5. Write down the major highlights of the Wildlife Protection Act of India.
6. Discuss the role of government and legal aspects in environmental protection.
7. Which are the government organisations/departments responsible for the protection of the environment? Write brief details of these.
8. Write the aims and objectives of 'Family Welfare Programmes'.
9. Discuss briefly the salient features of the Water (Prevention and Control of Pollution) Act, 1984.
10. Mention two important environmental laws.
11. Discuss briefly the provision of the following acts:
    (a) Water (Prevention Control of Pollution) Act, 1974
    (b) Air (Prevention and Control of Pollution) Act, 1981
    (c) Wildlife Protection Act 1971
    (d) Forest Conservation Act of 1980

**Based on sustainable development and urbanisation**

1. What is sustainable environment? Define the carrying capacity of the earth.
2. Comment on the urban energy problems and discuss the effect of overpopulation over energy problems in India.

**Based on water conservation**

1. Discuss various water conservation techniques that can be practiced by individuals.
2. What are the major approaches to conserve water resources? Comment on water harvesting methods in India.
3. Define watershed management and explain its objectives.
4. Describe the parameters of water quality standards for drinking water in India and state their significances.
5. What is 'rainwater Harvesting'? Name and discuss in brief the types of rainwater harvesting.
6. What are the programs to uplift the status and lifestyle of people living in remote areas?
7. How is a habitation pattern developed? List all environmental factors governing human settlement.

**Based on global environmental pollution problems**

1. There is an impact of excessive use of fossil fuels on the environment. Comment and justify the statement.
2. How is the 'acid rain forming'? What are its effects on the atmosphere?
3. Explain, greenhouse effect by drawing sketch.
4. Write short notes on (a) acid rain (b) environmental degradation.
4. Enlist all global environmental pollution problems. Describe global warming and greenhouse effects in detail.
6. Explain the term "environmental degradation" and discuss the role of advanced technology in the degradation of the environment.

7. Compare the effect of different greenhouse gases in global warming.
8. What is ozone? How is it formed? Describe the main causes of depletion of ozone layer and briefly comment over control measures taken to prevent further depletion at the international level.
9. (a) What is acid rain? Write a brief note on effects of acid rain on the environment.
(b) What is global warming? Enlist its consequences.
10. Write various mechanisms involved in the formation and depletion of ozone in the atmosphere. What are the consequences of depletion of ozone layer? How can the ozone layer be protected?

**Based on Initiatives by Non-Governmental Organisations**

1. Discuss of role of NGOs in environmental protection.

# OBJECTIVE TYPE QUESTIONS

**I. Fill in the Blanks**

1. The Environment (Protection) Act was passed in the year _________.
2. By using a bicycle instead of a car, we _________ energy.
3. The Air (Prevention and Control of Pollution) Act, _________ .
4. The Water (Prevention and Control of Pollution) Act, _________ .
5. Wildlife (Protection) Act, _________.
6. The Forest (Conservation) Act, _________.
7. International Literacy Day is recalled on _________.
8. International Women's Day is recalled on _________.
9. Chlorofluorocarbon releases a chemical harmful to ozone is _________.
10. Greenhouse effect is related to _________ .
11. Increasing industrialisation is causing much danger to human life by _________ .
12. _____ is related to global warming.
13. The best known substance responsible for ozone-layer depletion is _______ .
14. Automobile pollution also causes _________ .
15. Global warming contributes to rise in sea level due to _________ of ocean and melting of _________ .
16. The thinning of stratospheric ozone layer during *springtime* is called ______ .
17. World Ozone Layer Preservation Day is celebrated on _________ .
18. The greenhouse gases are $CO_2$, $CH_4$, $O_3$, CFC and _________ .
19. World Population Day is recalled on _________ .
20. World Peace Day is recalled on _________ .

**II. Match the following terms.**

Match the terms of column I with appropriate terms of column II.

A.

| Column I | Column II |
|---|---|
| 1. 1972 | (a) Wildlife (Protection) Act |
| 2. 1974 | (b) The Water (Prevention and Control of Pollution) Act |
| 3. 1980 | (c) The Forest (Conservation) Act |
| 4. 1981 | (d) The Air (Prevention and Control of Pollution) Act |
| 5. 1986 | (e) Environment (Protection) Act |

B.

| Column I | Column II |
|---|---|
| 1. Acid rain | (a) $SO_2$ and $NO_2$ |
| 2. Global warming | (b) Deforestation |
| 3. Urbanisation | (c) Greenhouse gases |
| 4. Population explosion | (d) Food crisis |

C.

| Column I | Column II |
|---|---|
| 1. The gas having the highest global-warming potential | (a) CFC-11 |
| 2. The gas having the highest ozone-depleting potential | (b) HCFC |
| 3. The gas responsible for acid rain | (c) $CO_2$ |
| 4. The gas responsible for global warming | (d) $SO_2$ and $NO_2$ |

## III. Multiple Choice Questions

**1.** The entire National Capital Territory of Delhi has been declared as water pollution prevention control area under ________ of the Water Act.

(a) Section 19 (b) Section 21
(c) Section 23 (d) Section 24

**2.** The Forest (Conservation) Act extends to the whole of India except

(a) Haryana
(b) Jammu and Kashmir
(c) Delhi (d) UP

**3.** The provisions for environmental protection in the constitution were made in

(a) 1947 (b) 1950
(c) 1976 (d) 1982

**4.** The provisions for environmental protection in the constitution were made under

(a) Article 21B (b) Article 5A
(c) Article 27B (h)
(d) Article 48A and Article 51A (g)

**5.** Greenhouse effect is related to

(a) global warming
(b) green trees around house
(c) grasslands
(d) greenery in city

**6.** Which of the following is not a green-house gas?

(a) Carbon dioxide
(b) Oxygen
(c) Methane
(d) Chlorofluoro carbons

**7.** Ozone layer is present in the

(a) thermosphere
(b) mesosphere
(c) stratosphere
(d) troposphere

**8.** Formation of hole in the ozone layer is maximum over

(a) India (b) Pakistan
(c) Europe (d) Antarctica

**9.** The primary cause of acid rain around the world is

(a) $SO_2$ (b) $CO_2$
(c) CO (d) $O_3$

**10.** The average life expectancy around the world is presently

(a) stabilizing (b) increasing
(c) not changing (d) decreasing

**11.** Atmosphere of metropolitan cities is polluted by

(a) pesticides (b) radiations
(c) automobile exhaust
(d) domestic waste

**12.** Production, transformation and use of energy are the major problems of

(a) industrial activity
(b) acid rain
(c) global warming
(d) sustainable development

**13.** Agricultural activities such as harvesting, heating, etc., are direct consumers of

(a) water (b) energy
(c) air (d) heat

**14.** Benefits of EIA are
(a) cost-saving
(b) improved project performance
(c) healthier local environment
(d) all

**15.** Objectives of sustainable development are
(a) social development
(b) economic development
(c) environment protection
(d) all

**IV. Indicate True or False for the following statements:**

**1.** Only legislation is not sufficient for the control of environmental pollution. True/False

**2.** Increased government intervention is a must for solving environmental problems. True/False

**3.** Privately owned organisations involved in providing financial and technical assistance to less developed countries are known as nongovernmental organisations. True/False

**4.** Community-based environmental education is capable of protecting health and habitat from the various problems existing in the world. True/False

**5.** The act or process of imparting or acquiring knowledge, skill or judgement by women is known as women's education. True/False

**6.** Ozone is a major constituent of photochemical smog. True/False

**7.** Ozone protects us from harmful UV radiation of the Sun. True/False

## Answers to Objective Type Questions

**I. Fill in the Blanks**

1. 1986 2. Conserve
3. 1981 4. 1974
5. 1972 6. 1980
7. September, 8th 8. March, 8th
9. Chlorine
10. Global warming
11. Air and water pollution
12. Greenhouse effect
13. CFCs 14. Air pollution
15. Thermal expansion, glaciers
16. Ozone hole
17. September, 16th
18. Nitrous oxide 19. July 11th
20. January 1st

**II. Matching the terms**

**A.** 1. (a) 2. (b) 3. (c) 4. (d) 5. (e)
**B.** 1. (a) 2. (c) 3. (b) 4. (d)
**C.** 1. (b) 2. (a) 3. (d) 4. (c)

**III. Multiple Choice Questions**

1. (a) 2. (b) 3. (c) 4. (d)
5. (a) 6. (b) 7. (c) 8. (d)
9. (a) 10. (b) 11. (c) 12. (a)
13. (b) 14. (d) 15. (d)

**IV. True or False**

1. True 2. True 3. True 4. True
5. True 6. True 7. True

# 7

# HUMAN POPULATION AND THE ENVIRONMENT

## Learning Objectives

After studying this chapter, you should be able to
- explain the meaning of population growth and population stabilisation
- describe population explosion in the Indian context
- enumerate and describe the major periods of growth of human population
- write a short note on methods of projection of population
- explain the meaning of value and value education
- describe the role of information technology in environment and human health
- describe the various schemes launched for women's education in India
- state the aims and objectives of 'Family Welfare Programmes'
- describe the role of information technology in environment and human health

## 7.1 POPULATION GROWTH

*Population* is a group of organisms of a particular species, sharing a particular characteristic of interest, most often that of living in a given area at a specific time.

*Population growth* is the change in a population per unit time. Population growth can be positive, static or negative.

### 7.1.1 Major Periods of Growth of Human Population

As of November 2011, the human population of the world is 7 billion (estimated by the United States Census Bureau).

**World Population Milestones**

*World Population reached:*

| | | | |
|---|---|---|---|
| 1 | billion in | 1804 | |
| 2 | billion in | 1927 | (123 years later) |
| 3 | billion in | 1960 | (33 years later) |
| 4 | billion in | 1974 | (14 years later) |
| 5 | billion in | 1987 | (13 years later) |
| 6 | billion in | 1999 | (12 years later) |
| 7 | billion in | 2011 | (12 years later) |

The growth of human population is summarised in the following four periods:

**(i) Hunter-gatherer Era** It refers to the earliest period of human history. Humans were mainly hunters and used to live in forests.

**(ii) Agricultural Era** Humans learnt agricultural practices and cultures were evolved.

**(iii) Era of Industrial Revolution** This era was of scientific developments and industrial advances.

**(iv) Modern Industrial Era** Scientific and industrial revolution touched every sphere of human life: health, education, living style, etc.

Human populations in these periods are summarised in Table 7.1.

**Table 7.1** *Major periods of growth of human population*

| *Era* | *Period* | *Total population* |
|---|---|---|
| (i) Hunter-gatherer Era | (From evolution—9000 BC)# | Less than 1 million |
| (ii) Agricultural Era | 9000 BC–1600 AD | 500 million (by 1600 AD) |
| (iii) Era of Industrial Revolution | 1600 AD–1950 AD | 1800 million (by 1900 AD) |
| (iv) Modern Industrial Era | 1950 AD–till date | 7 billion (by 2011 AD) |

BC = Before Christ, AD = Anno Domini, 1 million = 10 lakhs; 1 billion = 1000 million = 100 crores

# 100,000–10,000 years ago

## 7.1.2 Population Explosion

*Population explosion* means extremely fast rise in the number of people.

**Table 7.2** *Population of India during 1951–2001*

| *Year* | 1951 | 1961 | 1971 | 1981 | 1991 | 2001 |
|---|---|---|---|---|---|---|
| *Population (Millions)* | 361 | 439 | 548 | 683 | 846 | 1027 |

***Population Explosion in Indian Context*** India alone has about 16% of the world's population. India has a population growth rate of about 2.15%. Population growth is the reason for every environmental problem faced by Indian citizens:

(i) About one-third of the total population is poor and is subject to live below the poverty line.
(ii) About 53% of India's total land area is prone to soil erosion.
(iii) Forests have been declining.
(iv) Water and other natural resources are diminishing.
(v) Major population lacks basic amenities of living such as water, food, health care, etc.
(vi) Ecosystems and biodiversity is in danger.
(vii) India is facing energy crisis.
(viii) Due to upcoming shelter needs for the growing population, agricultural land is shrinking and leading to food crisis.

(ix) Population explosion has resulted in overcrowding, creation of slums, etc.

(x) Because of unemployment, rural people are migrating to urban cities; so the government is not able to provide jobs to all.

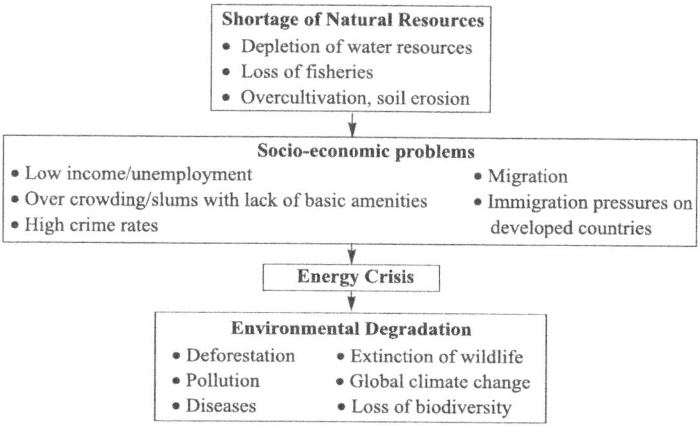


**Fig. 7.1** *Consequences of population explosion*

### 7.1.3 Population Policy

*Population policy means measures instituted by a government to influence size, growth, distribution or composition of population.*

***The objectives of a good population policy are the following:***

(i) Proper child care.

(ii) Provide universal *access* to family planning and reproductive health programmes and to information and education regarding these programmes.

(iii) Ensure that *men* fulfill their *responsibility* to ensure healthy pregnancies, *proper child care*, promotion of women's worth and dignity, etc.

(iv) Make women equal participants in all aspects of society—by increasing women's *education*, health and employment.

(v) *Recognise* that economic development is essential for environmental protection.

(vi) Provide information for *adolescents* (by increasing their access to education) to prevent unwanted pregnancies, unsafe abortion, and the spread of AIDS and sexually transmitted diseases.

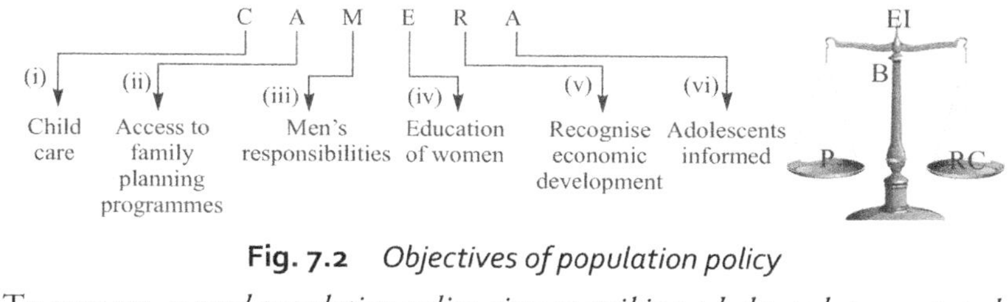


**Fig. 7.2** *Objectives of population policy*

To sum up, *a good population policy aims at striking a balance between population* (P) and resource consumption (RC) so that biodiversity (B) and ecological integrity (EI) is not lost.

### 7.1.4 Population Stabilisation

*Population stabilisation means the attainment of zero growth, in which the number of births in a population equals the number of deaths.*

Population stabilisation occurs when parents have enough children to replace them in population. In industrialised countries, a total fertility rate of 2.1 is considered to be a replacement-level fertility needed for population stabilization.

### 7.1.5 Population Structure

India alone has about 16% of the world's population which needs support from only 2.4% of the world's area, available in India. Population structure of a country is given below.

**Table 7.3** *Population structure of a country*

| S.No. | Population characteristics | = Description |
|---|---|---|
| 1. | Population size | = *No. of individuals* |
| 2. | Natality (birth rate) | = *No. of offsprings produced per female per unit time* |
| 3. | Mortality (death rate) | = *No. of deaths of individuals per unit time* |
| 4. | Population density | = *No. of Individuals per unit area or volume* |
| 5. | Population growth rate | = *Net result of births, deaths, and dispersals* |
| 6. | Total fertility rate | = *The average number of children each woman has over her lifetime, expressed as a yearly rate* |
| 7. | Population profile (age structure) | = *A bar graph plotting numbers of males and females for successive ages in the population, ending with the oldest at the top* |
| 8. | Crude Birth Rate (CBR) | = *The number of births per thousand of the population per year, when consideration is not given to what proportion of the population is young or old, female or male.* |
| 9. | Crude Death Rate (CDR) | = *The number of deaths per thousand of the population per year, when consideration is not given to what proportion of the population is young or old, female or male.* |
| 10. | Doubling time | = *The time it takes for a population to double its size when population is growing at a given growth rate* |

### 7.1.6 Population Pyramids

Age distribution influences both birth and death rates. In any ecological population, there are mainly three age groups: Pre-reproductive (0–14 years), reproductive (15–44 years) and post-reproductive (45 years and above). *The proportion of different age groups in any population is generally expressed graphically in the form of* ***population (or age) pyramids****.* There are three types of population pyramids.

***(A) Broad-based Pyramid or Expanding-age Pyramid*** In a rapidly growing population, birth rate is high, and population growth is exponential. So each successive generation will be more numerous than the preceding one, and the shape of the age structure is like a pyramid.

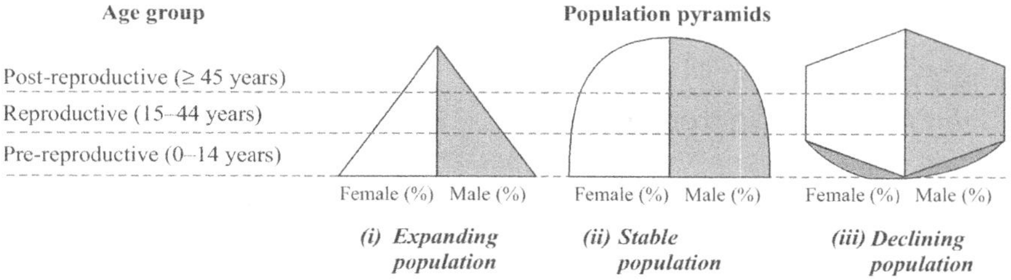


Fig. 7.3 *Age distribution and population pyramids*

***(B) Bell-shaped Polygon*** As the rate of growth of a population slows and stabilises, the reproductive and pre-reproductive age groups become almost equal in size while the post-reproductive group is the smallest and thus a stable age pyramid or bell-shaped polygon is formed.

***(C) Urn-shaped Pyramid*** If the birth rate is drastically reduced, the pre-reproductive group decreases in proportion to the reproductive and post-reproductive groups and thereby, an urn-shaped pyramid is formed. This type of age pyramid is also known as a *diminishing-age pyramid* and it is the representation of a population that is dying off.

## 7.1.7 Population Forecasting Methods

To plan services like designing a new water supply scheme and/or wastewater treatment plant, the population needs to be calculated after a few decades so that the scheme remains useful for the expected period. For population forecasting, data available from a municipality-based census is used.

Important methods of population forecasting are described below:

***(A) Arithmetical Increase Method*** This method is useful for metro cities which have reached their saturation. This method is based on the principle that the rate of change of population with time is constant. Future population is calculated by using the following formula:

$$\boxed{P_x = P_0 + x\ I}$$

where $P_x$ is forecasted population in the '$x$' decade
$P_0$ is present population
$x$ is number of decades between $P_0$ and $P_x$
$I$ is average increase of population in a decade

***(B) Geometric Increase Method*** This method is useful for developing countries which have high population growth. This method is based on the principle that the percentage growth rate of a population with time is constant. Future population as per this method is calculated by using the following formula:

$$P_x = P_0\left(1+\frac{r}{100}\right)^x$$

where $P_x$ is forecasted population in the '$x$' decade
$P_0$ is present population
$x$ is number of decades between $P_0$ and $P_x$
$r$ is the percentage growth rate of population

***(C) Incremental Increase Method*** This method is useful for cities of medium population growth. To have more accurate predictions, both arithmetic and geometrical increase of population is taken into consideration. As per this method, future population is calculated by using the following formula:

$$P_x = P_0 + x\ I + x I + \frac{x(x+1)r'}{2}$$

where $P_x$ is forecasted population in the '$x$' decade
$P_0$ is present population
$x$ is number of decades between $P_0$ and $P_x$
$I$ is average increment of a decade
$r'$ is the average incremental rate of a decade

***(D) Graphical Extension Method*** This method consists of the following steps:

Step (i): Draw a growth curve between population and time using the past data.
Step (ii): Obtain shape of the curve till the present population.
Step (iii): Extend the curve till the decade of population forecasting.

**Example 1** *Calculate the population of the world after two decades, i.e. forecasted population in 2031 by (i) arithmetical increase method, (ii) geometrical increase method, and (iii) incremental increase method.*

*Use following data for your calculations.*

| ***Year*** | 1981 | 1991 | 2001 | 2011 |
|---|---|---|---|---|
| ***World population (in billions)*** | 4.5 | 5.3 | 6.2 | 7.3 |

**Solution (i)** *Arithmetical increase method:*

| ***Year*** | ***Population (in Billions)*** | ***Increase in population*** |
|---|---|---|
| 1981 | 4.5 | – |
| 1991 | 5.3 | $5.3 - 4.5 = 0.8$ |
| 2001 | 6.2 | $6.2 - 5.3 = 0.9$ |
| 2011 | 7.3 | $7.3 - 6.2 = 1.1$ |
| Average increase of population in a decade ($I$) | | $\frac{0.8+0.9+1.1}{3} = 0.93$ |

Given present population (in 2011) = 7.3 billion = $P_0$

$\therefore$ number of decades between 2031 and 2011 = $\dfrac{2031-2011}{10} = 2 = x$

Thus, population in 2031 ($P_x$) = $P_0 + xI$

$\Rightarrow$ $P_x = 7.3 + 2 \times 0.93 =$ **9.16 billions**

**(ii)** ***Geometrical increase method:***

| *Year* | *Population (in billions)* | *Increase in population* | *Percentage increase in population (r)* |
|---|---|---|---|
| 1981 | 4.5 | – | |
| 1991 | 5.3 | 5.3 –4.5 = 0.8 | $\frac{0.8}{4.5} \times 100 = 17.78$ |
| 2001 | 6.2 | 6.2 – 5.3 = 0.9 | $\frac{0.9}{5.3} \times 100 = 16.98$ |
| 2011 | 7.3 | 7.3 – 6.2 = 1.1 | $\frac{1.1}{6.2} \times 100 = 17.74$ |
| Percentage growth rate of population ($r$) = $\frac{17.78+16.98+17.74}{3} = 17.5$ | | | |

Given, present population (in 2011) = 7.3 billion = $P_0$

Number of decades between 2031 and 2011 = $\dfrac{2031-2011}{10} = 2 = x$

Thus, population in 2031 ($P_x$) = $P_0\left(1+\dfrac{r}{100}\right)^x = 7.3\left(1+\dfrac{17.5}{100}\right)^2 =$ **10.08 billions**

**(iii)** ***Incremental increase method:***

| *Year* | *Population (in billions)* | *Increase in population* | *Incremental increase in population* |
|---|---|---|---|
| 1981 | 4.5 | – | – |
| 1991 | 5.3 | 5.3 – 4.5 = 0.8 | – |
| 2001 | 6.2 | 6.2 – 5.3 = 0.9 | 0.9 – 0.8 = 0.1 |
| 2011 | 7.3 | 7.3 – 6.2 = 1.1 | 1.1 – 0.9 = 0.2 |
| Average increase of population in a decade ($I$) = $\frac{0.8+1.9+1.1}{3} = 0.93$ | | | |
| Average incremental increase in population ($r'$) = $\frac{0.1+0.2}{2} = 0.15$ | | | |

Given present population (in 2011) = 7.3 billion = $P_0$

Number of decades between 2031 and 2011 = $\dfrac{2031-2011}{10} = 2 = x$

Thus, population in 2031 ($P_x$) = $P_0 + xI + \dfrac{x(x+1)r'}{2}$

$\Rightarrow$ $P_x = 7.3 + 2 \times 0.93 + \dfrac{2(2+1) \times 0.15}{2} =$ **9.61 billions**

## 7.1.8 Demographic Projections

***Demography*** is the field of collecting, compiling, and presenting information about populations.

***Epidemiologic transition*** is the shift from high death rates to low death rates in a population as a result of sanitary developments and medical advances.

***Epidemiology*** is the study of diseases in human societies.

***Fertility transition*** is the decline of birth rates from high levels to low levels in a population.

***Demographic transition*** is the tendency of a population to shift from high birth and death rates to low birth and death rates as a result of the epidemiologic and fertility transitions.

***Demographic projections*** are used to represent the increase in the population. The four stages of demographic projections are illustrated in Fig. 7.4 and discussed below:

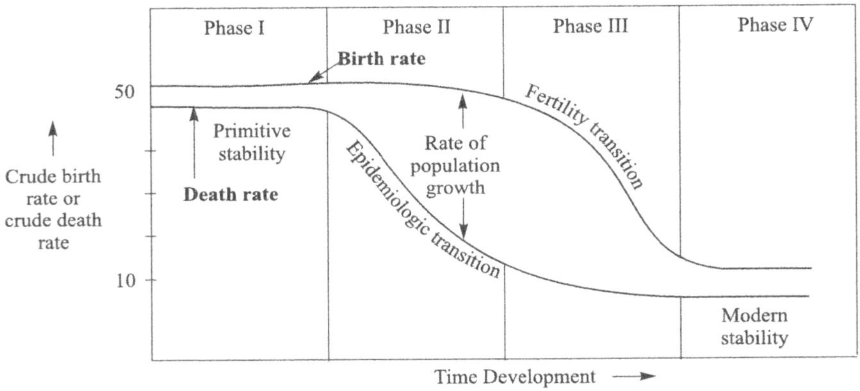


**Fig. 7.4** *Stages of demographic transition*

**Phase-I** Population remains more or less stable as both death and birth rates are high. The main features of this stage are backward economy where agriculture is the main occupation, low income, poor standard of living, inadequate and unbalanced diet, absence of educational opportunities and early marriages, etc.

**Phase-II** There is rapid growth of population. With the start of the developmental process, living standards of people improve, education expands, medical and health facilities increase; so death rates come down. But as education remains confined to a small section of the society, and attitude of the people towards the size of the family does not change radically, birth rates remain high. Population explosion occurs in this phase.

**Phase-III** In this stage, high reduction in birth rates are observed due to urbanisation, industrialisation, reduction in natural resources, women's education and use of family planning methods. Population growth is still significant.

**Phase-IV** High living standards, more jobs by women lead to reduction in birth rates. Medical facilities and balanced diet lead to reduction in death rates. Thus, Phase IV is reached, in which modern stability is achieved by a continuing low death rate and an equally low birth rate.

The demographic transition shows that there is a correlation between development and changing birth and death rates. It does not prove that development is necessary for the demographic transition to occur. Present stresses on the biosphere are largely a consequence of the consumption-oriented lifestyles of the high-income nations. If developing countries must modernise before population growth comes under control, their population growth, economic growth and a demand for resources and services, will result in profound consequences. A country is said to have an *optimum population* so long as the number of people is in balance with the available resources of the country.

# Case Studies

### *(A) India's Population*

India's population in 1901 was about 238.4 million, which has increased by more than four times in 110 years to reach a population of 1.2 billion in 2011. The growth rate of population for India in the last decade was 17.60%. The growth rate of population in rural and urban areas was 12.18% and 31.80% respectively.

(Source: http: //censusindia.gov.in/)

India has one of the highest population growth rates in the world. In Kerala, the population growth rate is slowing. This is because of the following initiatives:

**(i) Educated Women** About 50% of all Indian women are illiterate. However, in Kerala, 85% of women are literate. Better educated women are more likely to keep their children healthy. As children are surviving, families no longer have to have extra children to replace those that die.

**(ii) Better Status of Women** Women are no longer seen as a burden. They are regarded as an asset. When a women gets married, her family (traditionally in India) has to pay monetary dowry to the bridegroom's family. However, in Kerala it is the bridegroom's family who pays a dowry to the bride's family.

Family welfare means are available in Kerala.

[http: //www.geography.learnontheinternet.co.uk]

### *(B) The Fertility Rate*

The fertility rate means average number of children born per woman during her lifetime. India had a lower estimated fertility rate than Pakistan and Bangladesh in 2009. However, India had a higher fertility rate than China, Iran, Burma and Sri Lanka. The fertility rate in India has been in long-term decline, and had more than halved in the 1960–2009 period. From 5.7 in 1966, it declined to 3.3 by 1997 and 2.7 in 2009.

### (C) The Replacement Rate

The replacement rate means the total fertility rate at which newborn girls would have an average of exactly one daughter over their lifetimes. Andhra Pradesh, Goa, Tamil Nadu, Himanchal Pradesh, Kerala, Punjab and Sikkim are seven Indian states that have dipped below the 2.1 replacement rate level and no longer contributing to the Indian population growth.

(Source: http// en.wikipedia.org/wiki/family.planning.in.India)

## 7.2 FAMILY WELFARE PROGRAMMES

*Family* includes children, women, men, the aged, handicapped and less privileged.

*Welfare* means protection from hunger, poverty, undernourishment, underdevelopment, etc.

### 7.2.1 Aims of Family Welfare Programmes

Family welfare programmes aim at improving the quality of life by providing food, shelter, education, medical and developmental assistance.

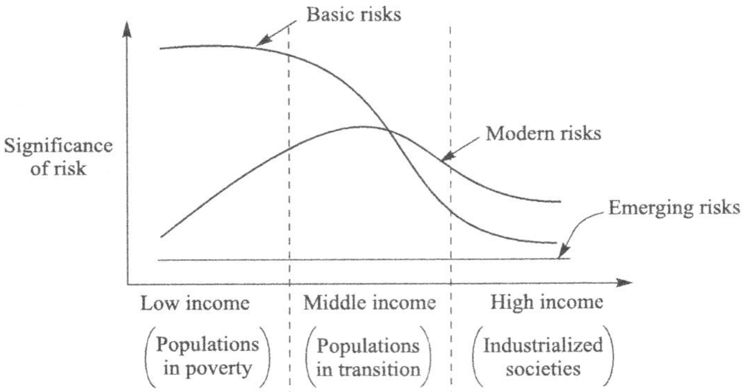


**Fig. 7.5** *Environmental health-risk transitions*

### 7.2.2 Objectives of Family Welfare Programmes

The objective of the National Family Welfare Programme, launched in 1951 in India has been *to stabilise the population at a level consistent with the requirement of the national economy by reducing the birth rate to the extent necessary.*

### 7.2.3 Problems of Family Welfare Programmes

There is no AIM with respect to the welfare of family.

| | | |
|---|---|---|
| A : Awareness (poor) : | (i) | Inadequate *awareness* about family welfare programmes, like Pulse-Polio Movement. |
| I : Infanticide (girl child) : | (ii) | Due to female *infanticide*, male–female sex ratio has reached an alarming stage at the national as well as state levels. |
| M : Male dominance : | (iii) | Due to *male dominance* in society, women are generally forced to adopt means of family planning. |

***Family Welfare Programme*** Jobs, safe water supply, environmental sanitation, healthy work conditions, and smart investments in education and healthcare are extremely effective in improving welfare of families. A focused approach in this direction will improve productivity and economic growth.

(Partha Das Sharma, http: //saferenvironment.wordpress.com/2009)

## 7.3 ENVIRONMENT AND HUMAN HEALTH

Resource depletion, waste generation, disturbance of ecosystems, consumerism, discharge of air or water pollutants, etc., are some of the human activities which have continuously been changing our environment. As a result of this, human health has been adversely affected.

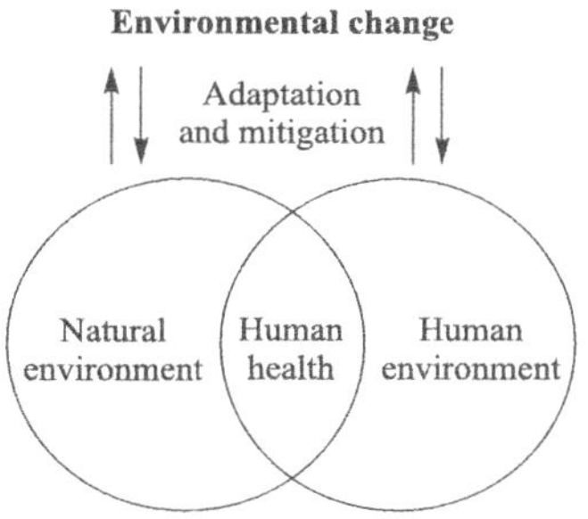


**Fig. 7.6** *Environment and human health*

The following facts are indicators which support that **health is an outcome of the interactions of humans with their environment**:

(i) *Due to exposure to the air pollutants released by industries, motor vehicles, smoking, etc.,* humans suffer from serious respiratory diseases such as tuberculosis and lung cancer.

(ii) *Due to consumption of impure water*, cholera, typhoid, diarrhoea, dysentery, etc., are caused.

(iii) *Due to contamination of water* through harmful pesticides, cancer, infertility and neurological diseases are caused.

(iv) *Due to scarcity of water and consequent unhygienic conditions*, tuberculosis, tetanus and leprosy are caused.

(v) *Due to stagnant water*, mosquitoes breed and spread malaria.

(vi) *Due to high-rise buildings*, visual pollution and mental strain is caused.

(vii) *Due to untreated human excreta*, several kinds of virus and bacteria grow which give rise to diseases like cholera, typhoid, jaundice, diarrhoea, etc.

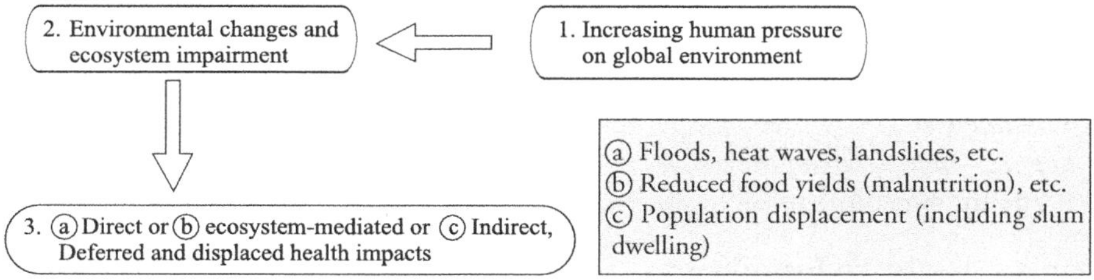


**Fig. 7.7** *Harmful effects of environmental change and ecosystem impairment on human health*

(viii) *By direct contact with blood of infected persons or by exchange of body fluids during sexual contact*, Acquired Immune Deficiency Syndrome (AIDS) is caused.

(ix) *By consuming arsenic-contaminated water for more than 5 years,* humans develop colour change on the skin, cancer of skin, bladder, kidney, lungs and legs.

(x) *Deforestation* has resulted in biodiversity loss and depleted flora and fauna. Thus, the source of large number of medicines which are essential for maintaining human health are badly affected.

We must work for a sustainable environment, which ultimately will result in good health for all.

*A sustainable environment and good health is achieved through*

| | | |
|---|---|---|
| **C** | (i) | avoiding **c**onsumerism, |
| **A** | (ii) | ***a**nti-smoking* movements and campaigns, |
| **P** | (iii) | ***p**opulation* control, |
| **T** | (iv) | using public ***t**ransport,* controlling transport emissions, |
| **A** | (v) | ***a**fforestation* and *reforestation,* |
| **I** | (vi) | preventing ***i**ndustrial* pollution, preferred use of *renewable* nonconventional energy instead of thermal power generation, and |
| **N** | (vii) | reduction in the consumption of ***n**atural resources,* protection and conservation of natural resources, |

# Case Study

### Environment and Human Health

Human health is facing new challenges due to problems resulting from environmental change. Air pollution and polluted water are shortening our lives. In the poorest regions of the world an estimated one in five children will not live to see their fifth birthday due to malaria, diarrhoea, etc., that are preventable. Environmental degradation is an important factor contributing to the burden of disease.

Environmental degradation exaggregates the imbalance between population and resources. It worsens the severity of poverty and population health.

(Source: http://www.sercc.com/health)

## 7.4 FUNDAMENTAL RIGHTS

***Fundamental rights*** *are those rights which are very essential for the development, happiness and welfare of the people.*

The Indian constitution has given the following six rights to its citizens:

### 7.4.1 Right to Equality

Articles 14 to 18 of the Indian constitution refer to this right:

*Article 14* establishes equality before law. It states *the state shall not deny to any person equality before the law.*

*Article 15* prohibits any sort of discrimination among the citizens on grounds of caste, race, religion, place of birth, sex, etc.

*Article 16* guarantees equality of opportunity in all fields of public employment.

*Article 17* prohibits practising of untouchability in any form. This practice of untouchability is now an offence punishable by law.

*Article 18* puts an end to all titles like Khan Bahadur, Rai Sahib, etc.

### *Exceptions to the Right to Equality*

- Exception for Article 15 is stated below:
  "Nothing in this article shall prevent the state from making any special provision for children and women; socially and educationally backward classes or for the Scheduled Castes and the Scheduled Tribes."
  The above classes need special protection because most of the times they have been victims of unequal treatment.
- Exception for Article 16 is stated below:
  "Nothing in this article shall prevent the state from making any provision for the reservation of appointment or posts in favour of any backward class of citizens, which, in the opinion of the state, is not adequately represented in the services under the state."

## 7.4.2 Right to Freedom

Our constitution has guaranteed various kinds of individual and collective freedoms to the citizens because freedom is the very essence of democracy.

The right to freedom is a cluster of the following six freedoms:

(i) Freedom of speech and expression
(ii) Freedom to assemble peacefully and without arms
(iii) Freedom to form associations or unions
(iv) Freedom to move freely throughout the territory of India
(v) Freedom to reside and settle in any part of the country of India
(vi) Freedom to practice any profession, or to carry on any occupation, trade or business

With respect to freedom (iv) above, a restriction is laid down which states that *Indian citizens need permission to visit some border areas of the country for reasons of security.*

With respect to freedom (v) above, a restriction is laid down which states that *Outsiders are not allowed to buy property in some areas to protect the interest of the local population.*

Especially in hilly and tribal areas, the people are very poor and innocent. If rich people are allowed to settle there, they would buy each and every piece of their land and render them homeless in their own land. Thus, this restriction on outsiders to buy land in certain areas is justified.

## 7.4.3 Cultural and Educational Rights

Indian society is very diverse. To preserve its culture, every community has been given educational and cultural rights to preserve its culture, language and script.

Any group, sect or minority in the country can open its own educational institutions and can teach its children accordingly. However, it has been made obligatory on such institutions to give admission to any student desirous to join them, irrespective of his/her religion, race, caste, sex or language. Such students cannot be forced to learn what does not fit in their own cultural framework.

## 7.4.4 Right to Religious Freedom

India is a secular state. The constitution confers on the people of India the freedom to follow any religion of their choice, to preach and practise it according to their specific ways.

The only restriction on this right is that nothing should be done to malign or ignite reaction among the followers of other religions.

## 7.4.5 Right Against Exploitation

This right saves people from any type of exploitation, It includes the following:

**(i) Right Against Employment of Children Below the Age of 14** This has been done to save children from greedy employers who exploit then in factories or mines, etc.

**(ii) Right Against Forced Labour or 'Begar'** Earlier certain influential people used to adopt such a practice in backward areas. However, our constitution now has declared begar as a crime punishable by law.

## 7.4.6 Right to Constitutional Remedies

This is a right to secure other rights. Had this right not been there, the other rights would have been quite meaningless.

According to this right to constitutional remedies, if any of the fundamental rights is encroached upon by the government, the citizens can move any court. The courts have been vested with the power to issue orders, directions and writs in order to protect the rights of the complainants.

### *Chief Characteristics of the Fundamental Rights*

(i) They are for each and every citizen of the republic without any consideration of religion, caste, colour or sex.

(ii) Certain restrictions are imposed on each of them to safeguard encroachment by others upon the similar rights of one or more citizens.

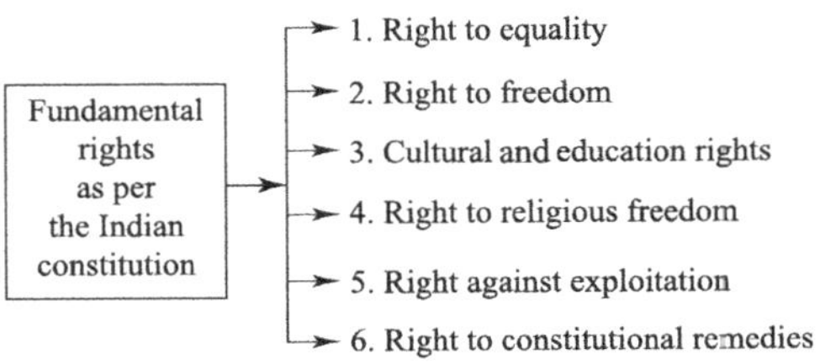


**Fig. 7.8** *Fundamental rights*

(iii) In case of encroachment of our rights by a person or a group of persons, or by the state, citizens have been given the right to knock at the doors of the judiciary and get them redressed.

(iv) Fundamental rights have a comprehensive approach. They tend to safeguard our social, economic, cultural and religious interests very carefully.

(v) In the interests of the safety and integrity of India, the fundamental rights of the citizens can be suspended in the event of a national emergency.

**Table 7.4** *Differences between fundamental rights and human rights*

| *Fundamental Rights* | *Human Rights* |
|---|---|
| 1. Country-specific | 1. Universally applicable |
| 2. Guaranteed by constitution of a country | 2. Relatively new |
| 3. Specific and have legal sanction, and are enforceable in courts | 3. No consensus, not enforceable in courts |

## 7.5 HUMAN RIGHTS

*Human rights are the rights a person has which he or she must enjoy on this earth because he or she is a human being.*

### Rights Guaranteed by the Indian Constitution (Fundamental Rights)

The rights guaranteed by the Indian constitution are called *fundamental rights* because

(i) these rights are quite essential for the all-round development of the citizens,

(ii) no government can abridge or abolish them, and

(iii) democratic government is not possible to run without these rights being given to the citizens.

The United Nations General Assembly adopted the *Universal Declaration of Human Rights* in1948. The declaration states that *the inherent dignity of all members of the human family is the foundation of freedom, justice and peace in the world.*

With the main objective of teaching the common language of humanity to people and to build a universal culture of human rights, the United Nations has initiated efforts to promote human rights education. It is believed that

(i) **knowledge** (Information about human rights and the mechanisms that exists to protect rights) helps in development of (ii) **values, beliefs and attitudes** (towards the establishment of a peaceful and harmonious society) which promotes an (iii) **action** (encouraging people to defend human rights and prevent human rights abuses).

**Fig. 7.9** *Advantages of human rights*

**The National Human Rights Commission (NHRC)** conducts carefully designed orientation and training programmes for the officers of police, armed forces, etc., to sensitise them to human rights.

***Some of the important articles of the Universal Declaration of Human Rights are summarised below:***

**Article 1** All human beings are born free and are equal in dignity and rights.

**Article 2** Everyone is entitled to all the rights and freedoms, without distinction of any kind, such as race, colour, sex, language, religion, political or other opinion, national or social origin, property, birth or other status.

**Article 3** Everyone has the right to life, liberty and security of person.

**Article 4** No one shall be held in slavery or servitude; slavery and the slave trade shall be prohibited in all their forms.

**Article 5** No one shall be subjected to torture or to cruel, inhuman or degrading treatment or punishment.

**Article 6** Everyone has the right to recognition everywhere as a person before the law.

**Article 7** All are equal before the law and are entitled without any discrimination to equal protection of the law.

**Article 8** Everyone has the right to an effective remedy by the competent national tribunals for acts violating the fundamental rights granted to him/her by the constitution or by law.

**Article 9** No one shall be subjected to arbitrary arrest, detention or exile.

**Article 10** Everyone is entitled in full equality to a fair and public hearing by an independent and impartial tribunal, in the determination of his/her rights and obligations and of any criminal charge against him/her.

**Article 11**

- Everyone charged with a penal offence has the right to be presumed innocent until proved guilty according to law in a public trial at which he/she has had all the guarantees necessary for his/her defence.
- No one shall be held guilty of any panel offence on account of any act or omission which did not constitute a penal offence, under national or international law, at the time when it was committed. Nor shall a heavier penalty be imposed than the one that was applicable at the time the penal offence was committed.

**Article 12** No one shall be subjected to arbitrary interference with his/her privacy, family, home or correspondence, nor to attacks upon his/her honour and reputation. Everyone has the right to the protection of the law against such interference or attacks.

**Article 13**

- Everyone has the right to freedom of movement and residence within the borders of each state.
- Everyone has the right to leave any country, including his/her own, and to return to his/her country.

**Article 14**

- Everyone has the right to seek and to enjoy in other countries asylum from persecution.
- This right may not be invoked in the case of prosecutions genuinely arising from nonpolitical crimes or from acts contrary to the purposes and principles of the UN.

**Article 15**

- Everyone has the right to a nationality.
- No one shall be arbitrarily deprived of his/her nationality nor denied the right to change his/her nationality.

**Article 16**

- Men and women of full age, without any limitation due to race, nationality or religion, have the right to marry and to found a family. They are entitled to equal rights as to marriage, during marriage and at its dissolution.
- Marriage shall be entered into only with the free and full consent of the intending spouses.
- The family is the natural and fundamental group unit of society and is entitled to protection by society and the state.

**Article 17**

- Everyone has the right to own property alone as well as in association with others.
- No one shall be arbitrarily deprived of his/her property.

**Article 18** Everyone has the right to freedom of thought, conscience and religion; this right includes freedom to change his/her religion or belief, and freedom, either alone or in community with others and in public or private, to manifest his/her religion or belief in teaching, practice, worship and observance.

**Article 19** Everyone has the right to freedom of opinion and expression; this right includes freedom to hold opinions without interference and to seek, receive and impart information and ideas through any media and regardless of frontiers.

**Article 20**

- Everyone has the right to freedom of peaceful assembly and association.
- No one may be compelled to belong to an association.

**Article 21**

- Everyone has the right to take part in the government of his/her country, directly or through freely chosen representatives.
- Everyone has the right of equal access to public service in his/her country.
- The will of the people shall be the basis of the authority of government; this will shall be expressed in periodic and genuine elections which shall be by universal and equal suffrage and shall be held by a secret vote or by equivalent free voting procedures.

**Article 22** Everyone, as a member of society, has the right to social security and is entitled to realisation, through national effort and international co-operation and in accordance with the organisation and resources of each state, of the economic, social and cultural rights indispensable for his dignity and the free development of his personality.

**Article 23**

- Everyone has the right to work, to free choice of employment, to just and favourable conditions of work and to protection against unemployment.
- Everyone, without any discrimination, has the right to equal pay for equal work.
- Everyone who works has the right to just and favourable remuneration ensuring for himself/herself and his/her family an existence worthy of human dignity, and supplemented, if necessary, by other means of social protection.
- Everyone has the right to form and to join trade unions for the protection of his/her interests.

**Article 24** Everyone has the right to rest and leisure, including reasonable limitation of working hours and periodic holidays with pay.

**Article 25**

- Everyone has the right to a standard of living adequate for the health and well being of himself/herself and of his/her family, including food, clothing, housing and medical care and necessary social services, and the right to security in the event of unemployment, sickness, disability, widowhood, old age or other lack of livelihood in circumstances beyond his/her control.
- Motherhood and childhood are entitled to special care and assistance. All children, whether born in or out of wedlock, shall enjoy the same social protection.

**Article 26**

- Everyone has the right to education. Education shall be free, at least in the elementary and fundamental stages. Elementary education shall be compulsory. Technical and professional education shall be made generally available and higher education shall be equally accessible to all on the basis of merit.
- Education shall be directed to the full development of the human personality and to the strengthening of respect for human rights and fundamental freedoms. It shall promote understanding, tolerance and friendship among all nations, racial or religions groups, and shall further the activities of the UN for the maintenance of peace.
- Parents have a prior right to choose the kind of education that shall be given to their children.

**Article 27**

- Everyone has the right to freely participate in the cultural life of the community, to enjoy the arts and to share in scientific advancement and its benefits.

- Everyone has the right to the protection of the moral and material interests resulting from any scientific, literary or artistic production of which he/she is the author.

**Article 28** Everyone is entitled to a social and international order in which the rights and freedoms set forth in this declaration can be fully realised.

**Article 29** Everyone has duties to the community in which alone the free and full development of his/her personality is possible.

- In the exercise of his/her rights and freedoms, everyone shall be subject only to such limitations as are determined by law slowly for the purpose of securing due recognition and respect for the rights and freedoms of others and of meeting the just requirements of morality, public order and the general welfare in a democratic society.
- These rights and freedoms may in no case be exercised contrary to the purposes and the principles of UN.

**Article 30** Nothing in this declaration may be interpreted as implying for any state, group or person any right to engage in any activity or to perform any act aimed at the destruction of any of the rights and freedoms set forth herein.

## Case Study

**Human Rights**

Human Rights watch in its report on human rights in India during 2010 stated India had "significant human rights problems". They identified lack of accountability for security forces and harm for abusive policing including "police brutality, extra judicial killings, and torture" as major problems. In 2006, the Supreme Court ordered police reforms in response to the poor human rights record of Indian police. The Supreme Court ordered extensive orders to implement the Right to Food in 2001.

In India, it is possible to solve various socio-economic issues by giving human rights to Indian citizens.

Source: http: //www.rightsined.org.nz/
http://en.wikipedia.org/wik:/human.rights.in.India

## 7.6 VALUE EDUCATION

### 7.6.1 Value

*Value* means 'the ultimate worth' of an action or a thing. For example, the nonviolence movements of Mahatma Gandhi were of high value.

However, the misuse of power, killing of millions of persons and other such actions of Hitler, leading to World War II have no value or have a negative value.

Thus, *values are one's own beliefs, feelings, perceptions, principles and behaviour to judge what is right or wrong.*

### 7.6.2 Value Education

Value education is defined as *the education that develops moral, spiritual and cultural sense; and makes one able to take right judgements in one's own life.*

In the context of the environment, value education teaches us values for nature, culture, social justice, human heritage, equitable use of resources, and sharing common natural resources.

It also teaches us to avoid consumerism, wastefulness and overexploitation of nonrenewable natural resources.

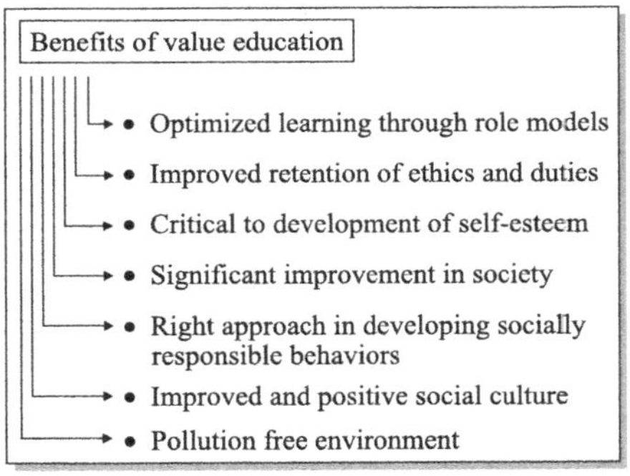


**Fig. 7.10** *Benefits of value education*

### 7.6.3 Illustrations and Examples of High and Low Values

The following examples and illustrations help us to understand high and low values.

**Table 7.5** *Examples and Illustrations of high values and low values*

| *High Values* | *Low Values* |
|---|---|
| (i) Helping others | (i) To be selfish |
| (ii) Serving older and needy people | (ii) Not caring for older and needy people |
| (iii) Invention and practice of life-saving drugs | (iii) Invention and use of bombs and explosives for killing innocent people |
| (iv) Development of technology for curing AIDS or cancer | (iv) Development of biological weapons |
| (v) Use of dynamite to pave way for constructing railways and roadways in hilly regions | (v) Use of dynamite to kill innocent humans and animals |
| (vi) To work for the benefit of society and environment | (vi) To exploit nature |
| (vii) Generating hard-earned money and doing some charity | (vii) Money making by unfair means |

### 7.6.4 Goals and Functions of the National Resource Centre on Value Education (NRCVE)

The goals and functions of NRCVE are

(i) To develop educational materials and other *teaching aids*, to document and disseminate information

(ii) To design strategies for *effective* implementation

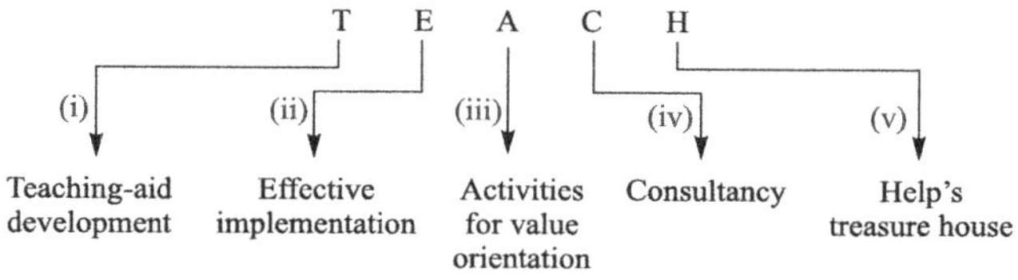


**Fig. 7.11** *Goals and functions of NRCVE*

(iii) To develop plans, *activities* and programmes for value orientation of school education
(iv) To provide extension and *consultancy* services
(v) To serve as a treasure *house* for any help

## 7.6.5 Objective of Value Education

The major objective of value education is to inculcate good values in individuals and to help them lead a life as responsible future citizens of India with feelings of universal brotherhood.

## Case Study

### Value Education

Seeing the acts of environmental degradation committed by humans all around us, we are left to wonder if most humans have forgotten their responsibility towards the environment.

Children learn best through the good manner exhibited by their parents/teachers and through their ethically correct conduct. Most people are unknowingly harming the environment. A lot of difference can be made by talking to these young people who are looking for guidance and love.

Environmentally friendly behaviour will automatically develop in the society by teaching core values like honesty, trust, respect, integrity, commitment, open minded, individuality and equality to the youth.

[Ayesha Parveen, Shaijavalikatri.blogspot.com/2011]

## 7.7 HIV/AIDS

*HIV stands for Human Immunodeficiency Virus.* Under normal circumstances, $CD_4$ cells (or $CD_4$ helper lymphocyte cells: a type of defence cells in the body) help the immune system to function normally and fight off certain kinds of infections by acting as messengers to other immune-system cells telling them to become active and fight against an invading germ.

A person infected with HIV is referred to as an HIV positive person. In them, the HIV attaches to these $CD_4$ defence cells, infects them and uses them to multiply resulting in loss of ability of $CD_4$ cells to do their job of fighting infections. As the immune system becomes weak, such people are unable to fight off many infections, particularly cancers, pneumonia, tuberculosis, meningitis, etc. The name for this condition is *Acquired Immuno Deficiency Syndrome (AIDS)*. In the absence of an immune system, a minor disease may be fatal.

AIDS is one of the most destructive epidemics in recorded history. The Joint United Nations Programme on HIV/AIDS and the World Health Organization estimated that AIDS has killed more than 25 million people as of January 2006 since it was first recognised on December 1, 1981.

***(A) Transmission of HIV*** The most common ways for the transmission of HIV from one person to another are described below:

(i) From an infected *mother* to her baby before birth, during birth and after birth. Breast milk can also transmit HIV infection to the infant.
(ii) HIV is mostly transmitted through semen and vaginal fluids during *unprotected* sex.
(iii) Sharing of *syringes* and needles among intravenous drug users can transmit HIV from an infected person to a normal person. Some nurses/doctors have become infected after being stuck with needles containing HIV infected blood or through splashes inside his or her nose or into their eyes.
(iv) *By transfusion of blood having HIV*, the virus can be transmitted to healthy persons.

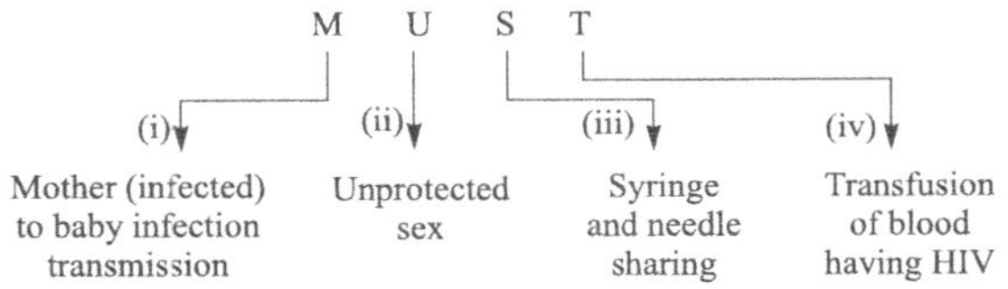


**Fig. 7.12** *Transmission of HIV*

It is very difficult to stop the spread of HIV in India because of poverty, illiteracy and poor health.

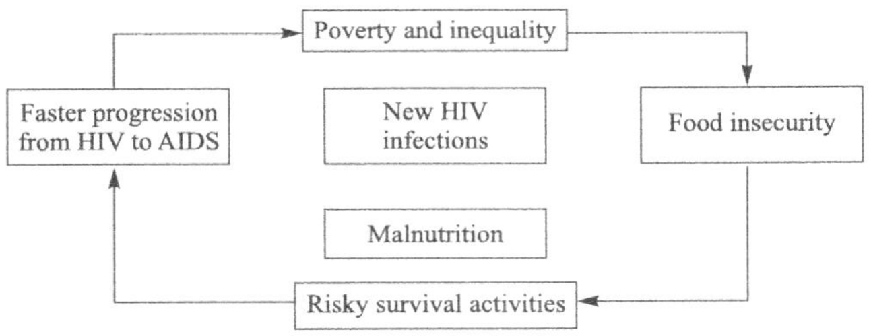


**Fig. 7.13** *Spread of HIV in India*

*www.icrisat.org/what.we.do/satrends/apr 2007.htm*

***(B) Symptoms of AIDS*** In a person infected with AIDS, symptoms can include

(i) Sweating at night
(ii) Swollen lymph glands
(iii) White spots in the mouth or throat
(iv) Loss of memory

(v) Consistent cough
(vi) Rapid weight loss
(vii) Extreme weakness or fatigue
(viii) Frequent long fevers
(ix) Chronic diarrhoea that lasts for more than a week
(x) Minor infections that cause skin rashes and sores in the mouth, annus or genitals
(xi) Pneumonia
(xii) Depression and other neurological disorders

It should be remembered that each or any of the above symptoms can be related to other illnesses. A test for HIV infection is the only way to certainly find out whether a person has AIDS or not.

***(C) Prevention of AIDS*** To ensure an HIV/AIDS free society, the following awareness and proactive actions need to be implemented:

(i) Having a *faithful monogamous sexual relationship* with an uninfected partner
(ii) Spreading *awareness*, proper medical care for HIV positive pregnant women can prevent HIV infection to the newborn.
(iii) Use of *condoms* (safe sex)
(iv) Transfusion of unaffected blood ensured by proper *test* for HIV freeness.
(v) Use of *sterilised dispensable syringes*

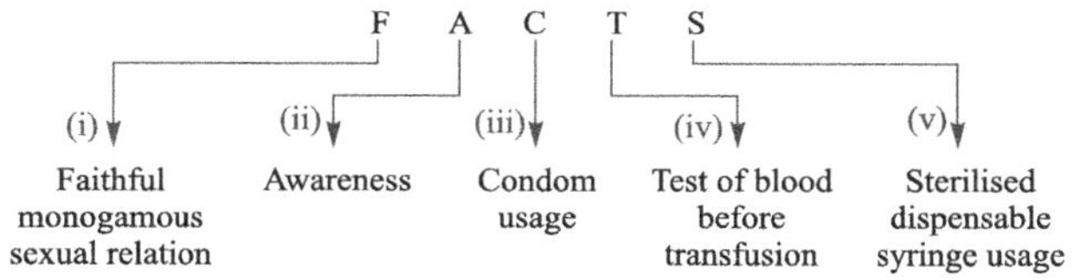


**Fig. 7.14** *HIV/AIDS prevention*

***(D) Social and Economic Impacts of AIDS*** Impacts of AIDS are briefly summarised below:

(i) Millions of young people are *dying* every year due to AIDS. Increased mortality of earning members results in loss of family income.
(ii) *Expenditures* on treating the sick, caring for AIDS orphans, training to replace sick workers keeps on growing.
(iii) Victims of AIDS, who are still alive are unable to work. They require special medical care. Newly trained workers have little knowledge and work experience so the productivity reduces, increasing pressure for the state's *finances*.
(iv) Many *orphans* are left behind.
(v) The *resources* available for public expenditures (such as education) reduces.
(vi) Slower growth of *economy* is the result.
(vii) *Social unrest* in the society is the outcome.
(viii) *Taxable population* reduces as a result of mortalities, due to AIDS.

| *Impacts of AIDS* | | | | | | | |
|---|---|---|---|---|---|---|---|
| "D | E | F | O | R | E | S | T" |
| Death | Expenditures | Finance Problems | Orphans | Resource reduction | Economical loss | Social unrest | Tax collection reduction |

## Case Study

### HIV/AIDS

India is one of the largest and most populated countries in the world, with over one billion inhabitants. It is estimated that around 2.4 million people are living with HIV/AIDS, in 2009.

HIV emerged in 1986 in India. Infection rates kept rising during the 1990s, and today HIV affects all sectors of Indian society, not just some groups—such as sex workers and truck drivers—with which it was originally associated.

(Source: http: //www.avert.org/acidsindia.htm)

## 7.8 ENVIRONMENTAL EDUCATION

Community-based environmental education helps in building knowledge and skills. It also helps in building an infrastructure for change that is sustainable, equitable and empowering. The simplified framework is illustrated in Fig. 7.15. It shows that *community based environmental education is capable of protecting health and habitat from the various problems existing in the world.*

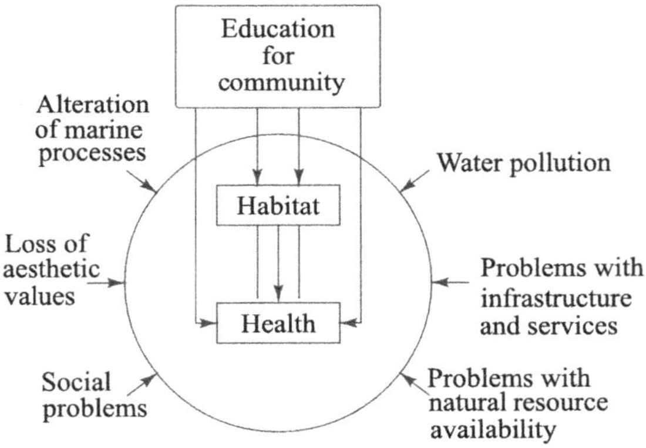


**Fig. 7.15** *Significance of environmental education*

### 7.8.1 Challenges

In India, the development and environmental protection challenges are enormous due to the following reasons:

**(i) Poverty** It is a big challenge in reaching out to large population cost-effectively because financial sources are very limited.

**(ii) Increasing Population** India's annual population increase is equal to the population of Australia.

**(iii) Less Land** With about 16% of the world population and a little over 2% of its land, there is already enormous pressure on our resources.

**(iv) Low Literacy Levels** The environmental educators face many challenges to spread awareness regarding conservation and environmental management.

**(v) Low Awareness** Poor Indian citizens have low or no awareness about importance of environment.

**(vi) Less Resources and Corruption** Putting environmental education on the agenda of educational decision makers and policy makers is also a big challenge primarily because of less resources and more corruption.

**(vii) No Applicability of Global Solutions** The environmental educators face difficulties in meeting the objectives of effective and local specific environmental education because environmental conditions and environmental concerns vary from one region of the state to another.

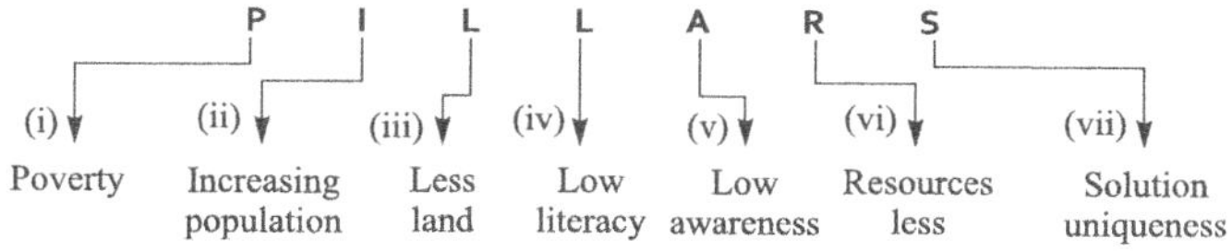


**Fig. 7.16** *Challenges for development and environment protection are weak pillars*

## 7.8.2 Environmental Education and Its Focuses

*Environmental education refers to organised efforts to teach how natural environments function and how people can manage their behaviour and ecosystems in order to live sustainably.*

Environmental education focuses on efforts to make the world a heaven like Kashmir is in India:

(i) Increasing people's awareness and knowledge about the environment and environmental challenges
(ii) Developing necessary skills and expertise to address the challenges
(iii) Fostering attitudes, motivations, and commitments to make informed decisions and take responsible action for solving environment-related problems

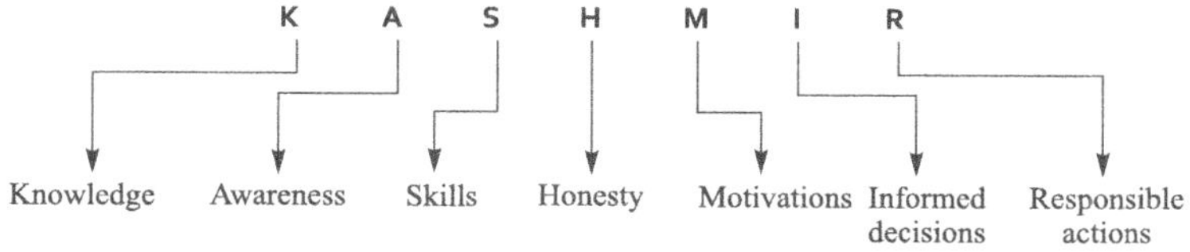


**Fig. 7.17** *Focuses of environmental education*

## 7.8.3 Role of Environmental Education for Environment Protection

Environmental education is a process of recognising values and clarifying concepts in order to develop skills and added tools necessary to understand and appreciate the inter-relationship among humans, their culture and their biophysical surrounding. It is through this process of education that people can be sensitised about environmental issues. Awareness and understanding of environmental issues help in practicing right actions needed for development that meets the needs of the present without compromising the ability of future generations to meet their own needs. These concepts are illustrated in Fig. 7.18.

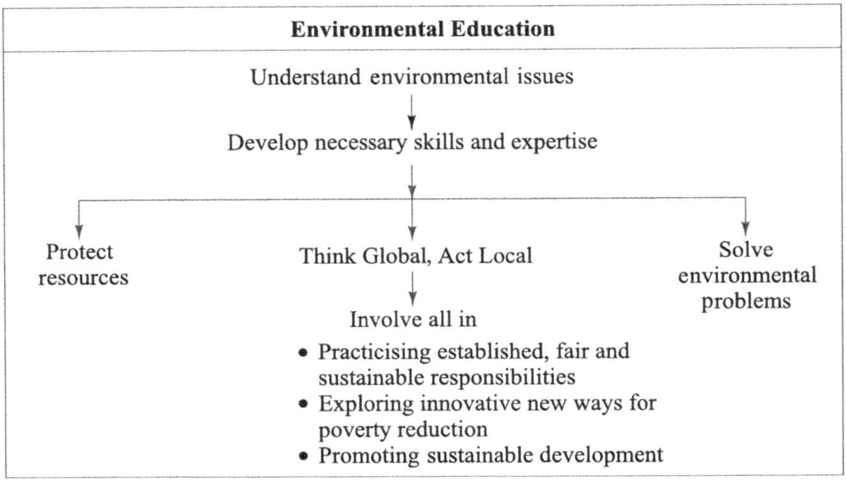


**Fig. 7.18** *Roles of environmental education for environmental protection*

# 7.9 WOMEN'S EDUCATION

*The act or process of imparting or acquiring knowledge, skill, or judgment to women is known as women's education.*

## 7.9.1 Various Schemes Launched for Women's Education in India

(i) The Sarva Shiksha Abhiyan (SSA) serves as an umbrella scheme for schemes directly and indirectly beneficial to the girl child:
   (a) The National Programme for the Education of Girls at an Elementary Level (NPEGEL) provides free uniforms and text books.
   (b) The Education Guarantee Scheme under SSA also aims to provide vocational and nonformal education to out-of-school children, of which, girls are the main beneficiaries.

(ii) The Kasturba Gandhi Balika Vidyalaya sets up residential schools at the upper primary region—primarily for girls from SC, ST, and OBC families as well as minority communities.

(iii) The Early Childhood Care and Education (ECCE) aims at setting up preschools to prepare children for schooling. It has an indirect bearing on education for girls as with her siblings in school, the girl child need not assume sibling care responsibilities during school hours and can therefore, attend school.

(iv) Mahila Samakhya (MS) Programme—It seeks to benefit women of all ages, especially those from economically and socially marginalised groups. It aims to integrate formal and nonformal education for girls, education schemes for adult women and vocational training for girls and women.

(v) The Mid-Day Meal (MDM) Scheme—The presence of midday meals in the schools
  (a) increases chances for girls attending schools
  (b) reduces caste biases as it forces children of all castes to eat together
(vi) The education schemes of the Ministry of Women and Child Development (MWCD).

The Balika Samriddhi Yojana, the Integrated Child Development Services and the Kishori Shakti Yojana are designed and funded by MWCD.

### 7.9.2 Obstacles in Educating Women

(i) The society dominated by males *fears* that its power will be taken away by women with education.
(ii) The societies are *afraid* to lose their cultural identity by women's education and globalisation.
(iii) *Poverty* and *scarcity of resources* force the poor to send their children for work and not to school.
(iv) Girls are trapped in a *vicious downward circle* of denied rights which generates a number of problems like social unrest, high maternal mortality rates, etc. (Fig. 7.6).

### 7.9.3 Advantages of Women's Education

(i) Education provides girls and women with an understanding of basic health, nutrition and family planning. It leads directly to better reproductive health, improved family health, economic growth, lower rates of child mortality and malnutrition. It is also a key in the fight against the spread of HIV and AIDS.
(ii) Educating girls and women is an important step in overcoming poverty and raising incomes. Educated girls have better skills needed for most of the new job categories.
(iii) Educated women are more aware of the problems in society and have ideas on how to solve them. They contribute better to a society.
(iv) A woman with knowledge is more respectable. She has the power to make people listen to her and the charisma that makes people follow her.
(v) An educated mother can educate the whole family.

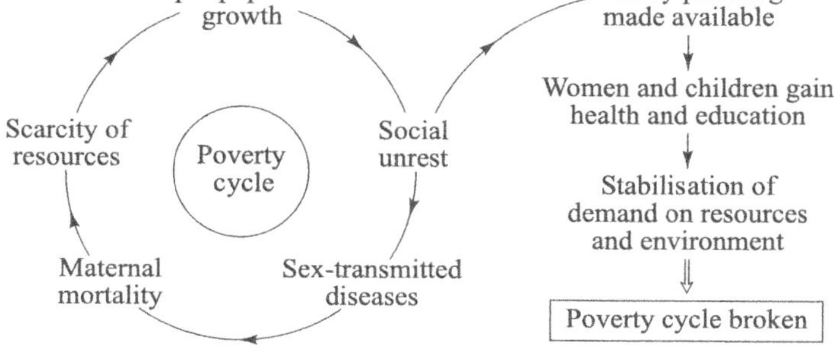


**Fig. 7.19** *Obstacles in educating women and advantages of women's education*

### 7.9.4 Roles of Women in Environment Protection

(i) An educated woman can easily *motivate* other women (who are generally shy).

(ii) She can conduct different *campaigns* (health care, environment protection, etc.) for local people located in the rural and urban areas. Education will enhance *awareness* for the preservation of natural resources.

(iii) She can raise the *interest* of her family members towards education.

(iv) Only she can *discuss* sensitive issues like family planning and the relevant precautionary and preventive measures needed.

(v) She is capable of attracting the attention of media, government, NGO's, etc., regarding initiation of developmental activities for *sustainable development* (e.g. proper waste disposal, cleanliness, tree plantation, etc.). She can *mobilise funds* through voluntary donations for social activities.

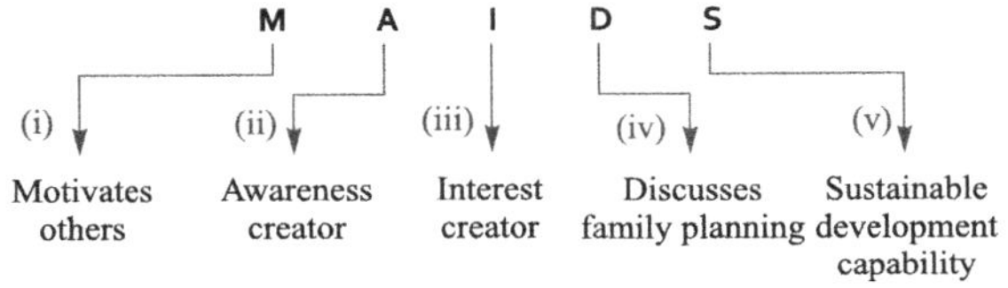


Fig. 7.20 *Roles of women in environment protection*

## Case Study

### Women and Child Welfare

The World Health Organisation estimates that 1.6 billion early deaths occur annually from cooking-stove pollution. Between 4 to 5.5 lakh children less than 5 years old, and women die each year in India due to indoor smoke.

*Chula* smoke is the third highest cause of disease and death after dirty water and lack of sanitation.

Over half the diseases and premature death could be avoided in India by providing access to clean water, sanitation, food and well-ventilated homes.

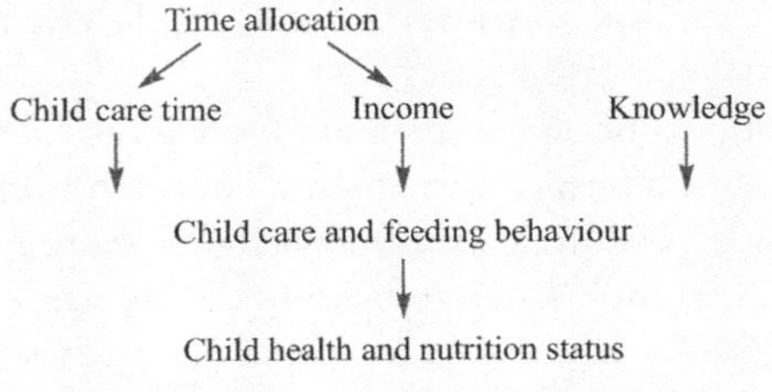


(*Source:* http://llwww.seribd.com/doc)

Fig. 7.21 *Role of mother in the welfare of child*

## 7.10 ROLE OF INFORMATION TECHNOLOGY IN ENVIRONMENT AND HUMAN HEALTH

### 7.10.1 Applications of IT in the Environment

Some of the important applications of IT in the field of environment and ecology are listed below:

(i) *Weather forecasting* through Geographical Information System (GIS) for agricultural production, water resource management, etc.

(ii) *Exploring* the possible availability of crude oils, gold mines, metal ores, geothermal power sources, etc., using Remote Sensing Information System (RSIS). Optimum selection of sites for railways or industry, etc. Biodiversity conservation by mapping and monitoring various natural resources—*flora and fauna*.

(iii) *Disaster management* in calamity-hit areas by extracting information. Monitoring of environmental pollution through remote sensing.

(iv) *Simulation* of environmental scenarios for analysis, prediction, decision making and development activities. Collaboration, communication and coordination among environmental scientists for decision-making.

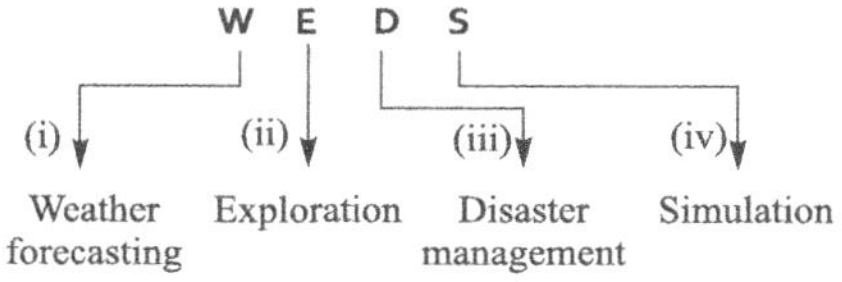


**Fig. 7.22** *IT for environment*

## 7.10.2 Applications of IT in Human Health

Some of the applications in which IT is playing an important role for better human health are listed below:

(i) Information on *health*, epidemics and their prevention is maintained on *web sites* of the World Health Organization.

(ii) Through electronic media; dengue fever, bird flu and other epidemics are brought to the *attention* of people.

(iii) *Dates of immunisation and sanitation programmes* are transmitted to public using television, computers, satellite communication, etc. Bioinformatics is used in the Human Genome Project (HGP) to create a map of the entire set of genes (genome) in the human cell by decoding the three billion units of human DNA.

(iv) Help and expert opinion can be obtained from expert *doctors* of any part of the world through telemedicine.

(v) *Health training* is imparted using satellite communication system.

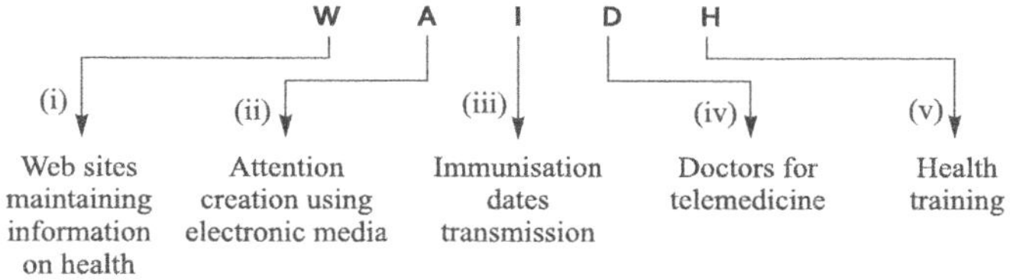


**Fig. 7.23** *IT for human health*

# Case Study

**Role of Information Technology in Environment and Human Health**

The world is facing many environmental problems. Humans are exploring and discovering innovative solutions for these problems using information technology. They are working in exploration and discovery teams and making observations, asking questions, exploring the literature, finding inspiration, sharing data and ideas.

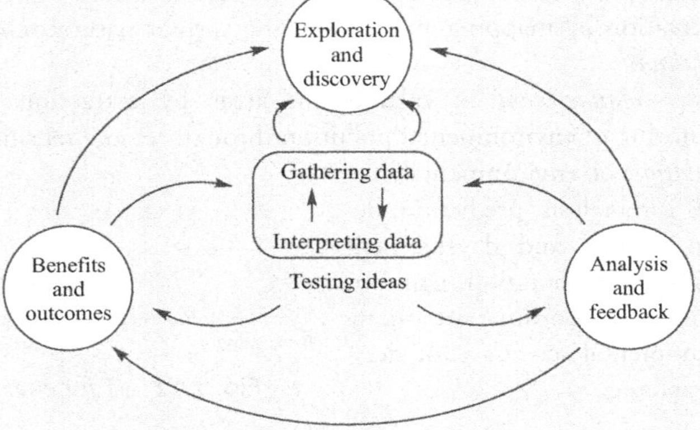


**Fig. 7.24** *IT is helping in every domain of research in finding solutions of various environment-related problems*

*For testing ideas*, they are *gathering data* and are working on various hypotheses and comparing observations and results of expected and actual findings:

(1) Supportive, (2) Contradictory, (3) Surprising, or (4) inconclusive. This is helping them in interpreting the data to (1) support, or (2) oppose, or (3) inspire revised/new hypothesis, or (4) inspire revised assumptions, respectively.

**Testing of hypotheses and ideas** is done through discussions with colleagues, publication, replication, peer review analysis and feedback.

The above testing of ideas helps experts in coming up with new questions and/ or ideas. It also helps in theory building.

When curiosity is satisfied, knowledge is built and technology is developed, social issues are addressed, policies are made and informed, and then everyday problems gets solved. These are the real benefits and outcomes of the exploration and discovery process.

(Source: http: //evolvingcomplexityii.files.wordpress.com/2009)

## Important Definitions

- *Population* is a group of organisms of a particular species, sharing a particular characteristic of interest, most often that of living in a given area at a specific time.
- *Population growth* is the change in a population per unit time. Population growth can be positive, static or negative.
- *Population explosion* means extremely fast rise in the number of people.
- *Population policy* means measures instituted by a government to influence size, growth, distribution or composition of population.
- *Population stabilisation* means the attainment of zero growth, in which the number of births in a population equals the number of deaths.
- The proportion of different age groups in any population is generally expressed graphically in the form of *population (or age) pyramids*.

- *Demography* is the field of collecting, compiling, and presenting information about populations.
- *Epidemiologic transition* is the shift from high death rates to low death rates in a population as a result of sanitary developments and medical advances.
- *Epidemiology* is the study of diseases in human societies.
- *Fertility transition* is the decline of birth rates from high levels to low levels in a population.
- *Demographic transition* is the tendency of a population to shift from high birth and death rates to low birth and death rates as a result of the epidemiologic and fertility transitions.
- *Demographic projections* are used to represent the increase in the population.
- *Family includes children*, women, men, the aged, handicapped and less privileged.
- *Welfare* means protection from hunger, poverty, undernourishment, underdevelopment, etc.
- *Fundamental rights* are those rights which are very essential for the development, happiness and welfare of the people.
- *Human rights* are the rights a person has which he or she must enjoy on this earth because he or she is a human being.
- *Values* are one's own beliefs, feelings, perceptions, principles and behaviour to judge what is right or wrong.
- *Value education* is defined as the education that develops moral, spiritual and cultural sense; and makes one able to take right judgements in one's own life.
- The act or process of imparting or acquiring knowledge, skill, or judgment to women is known as *women's education*.

# EXERCISES

**Based on Population**

1. Enlist the different adverse effects of population explosion.
2. What is population growth? How is it calculated? Describe the factors affecting population growth.
3. Give reasons for overpopulation and mention problems created by overpopulation.
4. Comment on the urban problems related to energy and discuss the effect of overpopulation on energy problems in India.
5. (a) Explain the equation $N_t = N_o e^{rt}$
(b) Enlist the assumptions of Malthusian theory of maximum population.
6. Write and explain the IPAT equation.
7. Describe in detail the environmental factors governing human settlement. Give effects of overpopulation.
8. Write a short note on (a) methods of projection of population, and (b) population and food production.
9. Discuss the impact of population, affluence and technology on environment with suitable examples.
10. Define the term *population explosion* and enumerate the causes of rapid population growth. Enlist the different methods of population forecasting.

**Based on Environmental Education**

1. "Environment education can play an important role in environmental protection". Explain it.

**Based on Women's Education**

1. Briefly describe the various schemes launched for women's education in India.

**Based on Value Education**

1. What is meant by 'value' and 'value education'? Discuss their concepts with the help of suitable illustrations.

**Based on the Roles of Information Technology**

1. Discuss the role of information technology on human health.

# OBJECTIVE TYPE QUESTIONS

## I. Fill in the Blanks

1. __________ is a group of organisms of a particular species, sharing a particular characteristic of interest, most often that of living in a given area at a specific time.
2. Population growth is the change in a population per unit __________.
3. Population explosion means __________ rise in the number of people.
4. __________ means measures instituted by a government to influence size, growth, distribution or composition of population.
5. Population stabilisation means the attainment of __________, in which the number of births in a population equals the number of deaths.
6. __________ is a field of collecting, compiling, and presenting information about populations.
7. Epidemiology is the study of __________ in human societies.
8. __________ are the rights a person has which he or she must enjoy on this earth because he or she is a human being.

## II. Match the following terms.

Match the terms of column I with appropriate terms of column II.

A.

| Column I | Column II |
|---|---|
| 1. Population growth | (h) Change in a population per unit time. |
| 2. Population explosion | (g) Extremely fast rise in the number of people. |
| 3. Population stabilisation | (f) Attainment of zero growth. |
| 4. Demography | (e) A field of collecting, compiling, and presenting information about populations. |
| 5. Epidemiology | (d) study of diseases in humans |
| 6. Fundamental rights | (c) Rights which are very essential for the development, happiness and welfare of the people. |
| 7. Human rights | (b) The rights a person has which he or she must enjoy on this earth because he or she is a human being. |
| 8. Value education | (a) The education that develops the moral, spiritual and cultural sense; and makes one able to take right judgements in own's own life. |

## III. Multiple Choice Questions

1. The World AIDS Day is recalled on
   (a) Ist December (b) Ist January
   (c) Ist February (d) Ist March
2. The highest number of people with HIV infection have been recorded from
   (a) India (c) China
   (b) Africa (d) Pakistan
3. The maximum number of individuals that can be supported by a given environment is called

(a) population size
(b) fertility rate
(c) carrying capacity
(d) mortality

4. The number of babies produced per thousand individuals is called
(a) immigration (c) mortality
(b) emigration (d) natality

5. One of the important mechanisms by which the environment controls population of a species is
(a) supply of food
(b) spread of disease
(c) check on death rate
(d) supply of oxygen

6. Population growth in developing countries as compared to developed countries is
(a) slower (c) same
(b) faster (d) negligible

7. Equity and social justice are essential components of
(a) fundamental rights
(b) ethics
(c) human rights
(d) human values

8. Prevention of AIDS is possible through
(a) condom usage
(b) use of sterilised dispensable syringes
(c) faithful monogamous sexual relation
(d) all of the above

9. AIDS stands for
(a) Acquired Immuno Deficiency Syndrome
(b) Acquired Immunity Development System
(c) Acquired Insuline Deficiency Syndrome
(d) Acquired Insuline Development System

10. The human rights include the following:
(a) Everyone has the right to life, liberty and security of person.
(b) Everyone has the right to own property alone as well as in association with others.
(c) Everyone has the right to education.
(d) All of the above.

11. One's own belief, principles, perceptions, feeling and behaviour to judge what is right and wrong is called
(a) primary education
(b) secondary education
(c) value education
(d) tertiary education

12. The threat to global environmental balance and fast depletion of natural resources are all outcomes of
(a) pollution (b) corruption
(c) government inaction
(d) population explosion

13. Telemedicine and weather forecasting are examples of
(a) application of IT in environment and healthcare
(b) application of IT in society
(c) application of IT in industry
(d) application of IT in medicine

14. HIV destroys a kind of defence cells in the body called
(a) white blood cells
(b) CD4 helper lymphocyte cells
(c) Red blood cells
(d) nerve cells

15. HIV does not spread through
(a) Contaminated blood transfusion
(b) Unprotected sex
(c) Sharing food vessels and eating food cooked by the infected person
(d) From infected mother to her baby

**IV. Indicate True or False for the following statements:**

1. The more developed countries have higher fertility rates. True/False

2. The human population has increased fivefold during the last 150 years. True/False

3. The zero population growth due to equal birth and death rates is called demographic transition. True/False

4. Population explosion is world's number one problem. True/False

5. Demography is the study of trends in human population growth and prediction of future growth. True/False

6. The world population touched the 6 billion mark on October 12, 1999. True/False

7. South Africa will be amongst the most populous countries in 2050. True/False

8. Carrying capacity is the maximum number that a habitat can sustain. True/False

9. Many environmental problems arise from the abuse, misuse and overuse of natural resources by humans. True/False

## Answers to Objective Type Questions

**I. Fill in the Blanks**

1. Population 2. time
3. extremely fast 4. Population policy
5. zero growth 6. Demography
7. diseases 8. Human rights

**II. Matching the terms.**

A. 1. (h) 2. (g) 3. (f) 4. (e)
5. (d) 6. (c) 7. (b) 8. (a)

**III. Multiple Choice Questions**

1. (a) 2. (b) 3. (c) 4. (d)
5. (a) 6. (b) 7. (c) 8. (d)
9. (a) 10. (d) 11. (c) 12.(d)
13. (a) 14. (b) 15. (c)

**IV. True or False**

1. False 2. True 3. True 4. True
5. True 6. True 7. False 8. True
9. True

# 8

# FIELD WORK

## Learning Objectives

After studying this chapter, you should be able to

- describe the features of water resource ecosystem you have studied during the field visit
- explain the aspects to be studied and the procedure for collecting information when you want to document the environmental features and resource assets of an ecosystem
- explain the methodology to be followed for study of cause and effects of a polluted site
- write about any polluted site you have visited and describe your findings in detail

## 8.1 INTRODUCTION

By supplementing classroom knowledge, field work can empower students with

(i) In-depth knowledge of our nature and its richness by observing real things
(ii) Leadership and cooperation capabilities
(iii) Future planning capabilities
(iv) Innovative solutions to existing problems
(v) Self-confidence

### *Guidelines for the Field Work*

The study of environment can be done precisely by making timely accurate observations. An effective field study trip should consist of planning, preparation, execution, follow-up, and evaluation. These simple steps are briefly discussed below:

| GUIDELINES | |
|---|---|
| 1. | Planning |
| 2. | Preparation |
| 3. | Execution |
| 4. | Follow-up |
| 5. | Evaluation |

### Step (i) Planning a Field Study Trip

Planning should be done by the student under the guidance of a teacher.

The following guidelines can be of some use:

| | | | |
|---|---|---|---|
| **S:** | Students | (i) | How many *students* need to be allowed? |
| **P:** | Permission | (ii) | Have students taken *permission* from their parents? |
| **E:** | Expense | (iii) | What will be the total *expense*? |
| **N:** | Necessary things | (iv) | Are *necessary things* like luggages ready as per the climate conditions of the place where field work has to be carried out? |
| **D:** | Days | (v) | How many *days* are needed to complete the field work? |

### Step (ii) Getting Prepared for a Field Study Trip

Only a good teacher can clarify the purpose and importance of the field work. When the importance or significance is known, initiative from the student side will automatically come for the field work. This will be the right time for the teacher to direct the students for field work.

### Step (iii) Tips for Successful Execution

| | | | |
|---|---|---|---|
| **S:** | Smaller groups | (i) | Teacher should divide the students in *smaller groups with one leader each.* |
| **E:** | Execution plans | (ii) | The respective leaders can clarify their doubts from the teacher, if any, and make out their respective *execution plans.* |
| **A:** | Accomplishable assignments | (iii) | Teams should cooperate among themselves, and divide total work into easily *accomplishable assignments.* |

### Step (iv) Necessity of a Well-planned Follow-up:

*The collected information is converted into useful knowledge* by a follow-up exercise, which includes

| | | | |
|---|---|---|---|
| **M:** | Models | (i) | Preparation of charts and *models.* |
| **A:** | Article | (ii) | *Article* writing for the magazine or for the bulletin board. |
| **D:** | Discussions | (iii) | Class *discussions.* |
| **E:** | Essay | (iv) | *Essay* writing. |

### Step (v) Getting Evaluated

| | | | |
|---|---|---|---|
| **F:** | Future | (i) | Evaluation helps in equipping students for a better *future.* |
| **A:** | Appreciated | (ii) | During class discussions, students should be *appreciated* for their right achievements during a field trip. Views should also be exchanged for avoiding any future discrepancies. |
| **T:** | Teacher | (iii) | Field work should be evaluated by the *teacher.* |

## 8.2 VISIT TO A LOCAL AREA TO DOCUMENT ENVIRONMENTAL ASSETS: RIVER/FOREST/ GRASSLAND/MOUNTAIN

Students can visit a local area to document environmental assets. They should observe the nature in its true form. They should also get feedback from local people of that area.

## Possible Local Areas for Field Work

| | | |
|---|---|---|
| ➢ River | ➢ Grassland | ➢ Mountain |
| ➢ Forest | ➢ Hill | ➢ Pond |

## Guidelines for Reporting Important Information of Field Work on Environmental Assets

(1) Area:........................................................................................................

| | | | |
|---|---|---|---|
| **F:** | Flora/Fauna | (i) | ........................................................................ ........................................................................ |
| **A:** | Physical details of *area* | (ii) | ........................................................................ ........................................................................ |
| **V:** | Vegetation | (iii) | ........................................................................ ........................................................................ |
| **O:** | Ownership | (iv) | ........................................................................ ........................................................................ |
| **U:** | Usage | (v) | ........................................................................ ........................................................................ |
| **R:** | Rainfall pattern | (vi) | ........................................................................ ........................................................................ |
| **E:** | Importance in ecosystem | (vii) | ........................................................................ ........................................................................ |
| **R:** | Role of human activities | (viii) | ........................................................................ ........................................................................ |

(2) Natural assets: .........................................................................................

(3) Usage of natural assets: ..............................................................................

(a) Used by.........................................................................................................

(b) Used during....................................................................................................

(4) Contamination of natural assets: ...................................................................

| | | | |
|---|---|---|---|
| **P:** | Pollutants | (i) | ........................................................................ ........................................................................ |
| **E:** | Extent of Pollution | (ii) | ........................................................................ ........................................................................ |
| **P:** | Problems | (iii) | ........................................................................ ........................................................................ |
| **S:** | Source | (iv) | ........................................................................ ........................................................................ |

(5) Mitigation possibilities ..................................................................................

- Best solution ..............................................................................................

(6) Lifestyle of indigenous people: sustainable or unsustainable .................. ..........................

(7) Best methodology for convincing indigenous people for the necessity of replacement of their unsustainable lifestyle ............................................. .................................

(8) Time frame for completion of field work..................................................
(9) Total cost..................................................................................
(10) Achievements..............................................................................

## 8.3 VISIT TO A LOCAL POLLUTED SITE–URBAN/ RURAL/INDUSTRIAL/AGRICULTURAL

Students can visit a local polluted site and report important observations.

### Guidelines for Reporting Important Information of Industrially Polluted Site

1. Name of the industry:..................................................................
2. Capacity:...............................................................................
3. Year of establishment:..................................................................
4. Products manufactured:..................................................................
5. Number of workers employed:.............................................................
6. Type of wastes/emissions produced:......................................................
7. Names of pollutants emitted:............................................................
8. Dumping sites:..........................................................................
9. Effluent Treatment Plant (ETP) for the treatment of industrial waste before discharge:...............................................................................
10. Chimneys or stacks installed:...........................................................
11. Green belt: ............................................................................
12. Distance of industry from city:.........................................................
13. Health of workers:......................................................................
14. Health impacts on people living near the industry:.....................................
15. Causes of illness

| | | | |
|---|---|---|---|
| **P:** | Thermal *pollution* | (i) | ................................................ |
| **U:** | Untreated discharge of *wastes* | (ii) | ................................................ |
| **N:** | Excessive *noise* | (iii) | ................................................ |
| **C:** | *Contamination* of drinking water | (iv) | ................................................ |
| **H:** | Hazardous pollution | (v) | ................................................ |
| **E:** | Emission of toxic *gases* | (vi) | ................................................ |
| **D:** | Deforestation | (vii) | ................................................ |

16. Impacts on environment: .................................................................
17. Tragedies or disaster happened in past: .................................................
18. Best alternative technology available in world: .........................................
19. Time consumed for this project: .........................................................
20. Achievements: ...........................................................................

## 8.4 STUDY OF COMMON PLANTS, INSECTS, BIRDS

Plants, insects and birds have tremendous potential in terms of productive, social, ethical, ecological and consumptive value. It is worthwhile to study these species present in our surroundings.

### Guidelines for Reporting Important Information of Common Plants in an Area

(1) Climate of area:..........................................................................................
(2) Soil type of area:.........................................................................................
(3) Type of plant (herbs/shrubs/trees):.........................................................
(4) Medicinal value of plants as per the indigenous knowledge of local people: ............
(5) Important trees which produce timber wood:...........................................
(6)

| Use | Names of plants |
|---|---|
| (a) Gum | |
| (b) Fibre | |
| (c) Rubber | |
| (d) Dye | |
| (e) Resin | |
| (f) Fuel | |

(7) Any threat from possible pollutants:........................................................
(8) Best solution:...............................................................................................
(9) Time needed for this study:.......................................................................
(10) Cost:.............................................................................................................
(11) Achievement:...............................................................................................

## 8.5 STUDY OF SIMPLE ECOSYSTEMS–POND, RIVER, HILL SLOPES

*"An ecosystem is defined as a natural unit that consists of living and nonliving parts which interact to form a stable system. It is the minimal grouping of diverse organisms that interact and function together so as to sustain life."*

### Guidelines for Reporting Important Information of Ecosystems

(1) Background history:....................................................................................
(2) Abiotic components:...................................................................................
  (a) Topography:...........................................................................................
  (b) Climatology:...........................................................................................
  (c) Latitude:.................................................................................................
  (d) Physiography:.........................................................................................

(3) Biotic components:.......................................................................
  (a) Producers:.......................................................................
  (b) Consumers:.......................................................................
  (c) Decomposers:.......................................................................
(4) (a) Food chains:.......................................................................
  (b) Food webs:.......................................................................
  (c) Ecological pyramids:.......................................................................
(5) Biogeochemical cycle:.......................................................................
(6) Utilisation of biota by humans:.......................................................................
(7) Conservation potential:.......................................................................
(8) Mitigation possibilities:.......................................................................
(9) Time requirement for completion of study:.......................................................................
(10) Cost:.......................................................................
(11) Achievements:.......................................................................

## Important Definitions

- The study of environment can be done precisely by making timely accurate observations. An effective *field study trip* should consist of planning, preparation, execution, follow-up, and evaluation.
- Students can *visit* a local area to document environmental assets. They should observe the nature in its true form. They should also get feedback from local people of that area.
- Plants, insects and birds have tremendous potential in terms of productive, social, ethical, ecological and consumptive value. It is worthwhile to study these species present in our surroundings.
- An *ecosystem* is defined as a natural unit that consists of living and nonliving parts which interact to form a stable system. It is the minimal grouping of diverse organisms that interact and function together so as to sustain life.

# EXERCISES

1. **(a)** Explain the aspects to be studied and the procedure for collecting information when you want to document the environmental features and resource assets of a water resource ecosystems during a field visit.
**(b)** Describe, in the format given above, the features of any such ecosystem you have studied during a field visit.

2. **(a)** What is the methodology to be followed for study of cause and effects of a polluted site? Write the observations for various aspects and data to be collected.
**(b)** Write about any polluted site you have visited and describe your findings in detail.

# OBJECTIVE TYPE QUESTIONS

## I. Fill in the Blanks

1. __________ can empower students with in-depth knowledge of our nature and its richness by observing real things.
2. An effective field study trip should consist of planning, preparation, execution, follow-up and __________.
3. Only a __________ can clarify the purpose and importance of the field work.
4. __________ helps in equipping students for better future.
5. The plants, insects and __________ have tremendous potential in terms of productive, social, ethical, ecological and consumptive value.
6. An __________ is defined as a natural unit that consists of living and nonliving parts which interact to form a stable system.
7. __________ are small-sized plants with soft stems.
8. __________ are medium-sized plants with profuse branches.
9. __________ are tall with woody trunks.
10. __________ are nongreen plants, which grow on other plants and obtain food from them.

## II. Match the following terms.

Match the terms of column I with appropriate terms of column II.

**(A)**

| Column I | Column II |
|---|---|
| 1. Gully | (a) Refers to the surface behaviour of the earth such as steepness of slope, altitude, direction of mountain chains |
| 2. Meadows | (b) A well-developed grass cover interspersed with scattered shrubs or small trees |
| 3. Pasture | (c) A land for grazing cattle |
| 4. Savanna | (d) A Piece of flat grass land |
| 5. Topography | (e) Refer to a deep channel worn by water. |

## III. Multiple Choice Questions

1. The aquatic ecosystem polluted by industrial waste may
   (a) smell bed
   (b) has floating matter
   (c) has oil on surface
   (d) look fresh
2. The area where solid waste is dumped may look
   (a) clayey and yellowish
   (b) dark and odorous
   (c) reddish (d) brown
3. An ecosystem that comprises of trees, shrubs and herbs is called
   (a) desert (c) forest
   (b) grassland (d) pond
4. An aquatic ecosystem whose water is flowing is called.
   (a) lake (c) estuary
   (b) ocean (d) river
5. The bird which is an indicator of human habitation is
   (a) crow (c) koel
   (b) peacock (d) parrot
6. In the grass → dear → tiger Food chain, there are
   (a) 2 trophic levels
   (b) 3 trophic levels
   (c) 4 trophic levels
   (d) 5 trophic levels

7. The common pollutants present in ponds nearby agricultural fields are
(a) dust
(b) oil
(c) pesticides and chemical fertilisers
(d) dyes

8. The soil on a hill mountain site is characterised by the presence of
(a) silt
(b) sand
(c) clay
(d) pebbles and silt

9. Birds are known for exhibiting a variety of behaviours like
(a) singing
(b) nest building
(c) parental care and migration
(d) all of the above

10. The correct food chain(s) is
(a) flower → butterflies → spiders
(b) seeds → rodents → birds
(c) grass → deer → tiger
(d) all of the above

**IV. Indicate True or False for the following statements:**

1. For a field study trip, students are not required to take permission from their parents. (True/False)
2. Knowledge of climate conditions of the place where field work has to be carried out is not necessary. (True/False)
3. The collected information from field work is converted into useful knowledge by a follow-up exercise (True/False)
4. Evaluation of field work by their teacher helps students equipping themselves for a better future. (True/False)
5. It is not worthwhile to study common plants, insects and birds present in our surroundings. (True/False)
6. Ecosystem is the minimal grouping of diverse organisms that interact and function together so as to sustain life. (True/False)

## Answers to Objective Type Questions

**I. Fill in the Blanks**

1. Field work
2. evaluation
3. good teacher
4. Evaluation
5. birds
6. ecosystem
7. Herbs
8. Shrubs
9. Trees
10. Parasites

**II. Matching the terms**

A. 1. (e) 2. (d) 3. (c) 4. (b) 5. (a)

**III. Multiple Choice Questions**

1. (a) 2. (b) 3. (c) 4. (d)
5. (a) 6. (b) 7. (c) 8. (d)
9. (d) 10. (d)

**IV. True or False**

1. False 2. False 3. True 4. True
5. False 6. True

# INDEX

## D



## E

## F

## G



## H



## I

## S



## T

Made in the USA
Monee, IL
04 November 2026

67384357R00197